Computational Complexity of Counting and Sampling

Computational Complexity of Counting and Sampling

István Miklós

Rényi Institute, Budapest, Hungary

CRC Press
Taylor & Francis Group
Boca Raton London New York

CRC Press is an imprint of the
Taylor & Francis Group, an **informa** business

CRC Press
Taylor & Francis Group
6000 Broken Sound Parkway NW, Suite 300
Boca Raton, FL 33487-2742

© 2019 by Taylor & Francis Group, LLC
CRC Press is an imprint of Taylor & Francis Group, an Informa business

Printed on acid-free paper
Version Date: 20190201

International Standard Book Number-13: 978-1-138-03557-7 (Paperback)
International Standard Book Number-13: 978-1-138-07083-7 (Hardback)

Library of Congress Cataloging-in-Publication Data

Names: Miklos, Istvan (Mathematician), author.
Title: Computational complexity of counting and sampling / Istvan Miklos.
Description: Boca Raton : Taylor & Francis, 2018. | Includes bibliographical references.
Identifiers: LCCN 2018033716 | ISBN 9781138035577 (pbk.)
Subjects: LCSH: Computational complexity. | Sampling (Statistics)
Classification: LCC QA267.7 .M55 2018 | DDC 511.3/52--dc23
LC record available at https://lccn.loc.gov/2018033716

Visit the Taylor & Francis Web site at
http://www.taylorandfrancis.com

and the CRC Press Web site at
http://www.crcpress.com

To the memory of my beloved wife, Ágnes Nyúl
(1972–2018)

Contents

Preface

The idea to write a book on the computational complexity of counting and sampling came to our mind in 2016 February, when Miklós Bóna co-organized a Dagstuhl seminar with Michael Albert, Einar Steingrímsson, and me. We realized that many of the enumerative combinatorists know little about computer science, and clearly, there is a demand for a book that introduces the computational aspects of enumerative combinatorics. Similarly, there are physicists, bioinformaticians, engineers, statisticians, and other applied mathematicians, who develop and use Markov chain Monte Carlo methods, but are not aware of the theoretical computer scientific background of sampling.

The aim of this book is to give a broad overview of the computational complexity of counting and sampling, from very simple things like linear recurrences, to high level topics like holographic reductions and mixing of Markov chains. Since the book starts with the basics, eager MSc, PhD students, and young postdoctoral researchers devoted to computer science, combinatorics, and/or statistics might start studying this book. The book is also unique in the way that it focuses equally on computationally easy and hard problems, and highlights those easy problems that have hard variants. For example, it is easy to count the generations of a regular grammar that produce sequences of length n. On the other hand, it is hard to count the number of sequences of length n that a regular grammar can generate.

There is a special emphasis on bioinformatics-related problems in the hope of bringing theory and applications closer. A bunch of open problems are drawn to the attention of theorists, who might find them interesting and challenging enough to work on them. We also believe that there will be applied mathematicians who want to deepen their understanding of the theory of sampling, and will be happy to see that the theory is explained via examples they already know.

Many of the topics are introduced via worked-out examples, and a long list of exercises can be found at the end of each chapter. Exercises marked with * have a detailed solution, while hints can be found on exercises marked with ∘. Unsolved exercises vary from very simple to challenging. Therefore, instructors will find appropriate exercises for students at all levels.

Although the book starts with the basics, it still needs prerequisites. Background in basic combinatorics, graph theory, linear and abstract algebra, and probability theory is expected. A discussion on computational complexity is very briefly presented at the beginning of the book. However, Turing Machines and/or other models of computations are not explained in this book.

We wanted to give a thorough overview of the field. Still, several topics are omitted or not discussed in detail in this book. As the book focuses on classifying easy and hard computational problems, very little is presented on improved running times and asymptotic optimality of algorithms. For example, divide and conquer algorithms, like the celebrated "four Russians speed-up", cannot be found in this book. Similarly, the logarithmic Sobolev inequalities are not discussed in detail in the chapter on the mixing of Markov chains. Many beautiful topics, like stochastic computing of the volume of convex bodies, monotone circuit complexity, #BIS-complete counting problems, Fibonacci gates, path coupling, and coupling from the past are mentioned only very briefly due to limited space.

Writing this book was great fun. This work could not have been accomplished without the help of my colleagues. I would like to thank Miklós Bóna for suggesting to write this book. Also, the whole team at CRC Press is thanked for their support. Special thanks should go to Jin-Yi Cai, Catherine Greenhill, and Zoltán Király for drawing my attention to several papers I had not been aware of. I would like to thank Kálmán Cziszter, Mátyás Domokos, Péter Erdős, Jotun Hein, Péter Pál Pálfy, Lajos Rónyai, and Miklós Simonovits for fruitful discussions. András Rácz was volunteered to read the first two chapters of the book and to comment, for which I would like to warmly thank him. Last but not least, I will always remember my beloved wife, Ágnes Nyúl, who supported the writing of this book till the end of her last days, and who, unfortunately, passed away before the publication of this book.

List of Figures

List of Tables

Chapter 1

Background on computational complexity

In computational complexity theory, we distinguish decision, optimization, counting and sampling problems. Although this book is about the computational complexity of counting and sampling, counting and sampling problems are related to decision and optimization problems. Counting problems are always at least as hard as their decision counterparts. Indeed, if we can tell, say, the number of perfect matchings in a graph G, then naturally we can tell if there exists a perfect matching in G: G contains a perfect matching if and only if the number of perfect matchings in G is at least 1.

Optimization problems are also related to counting and sampling. As we are going to show in this chapter, it is hard to count the cycles in a directed graph as well as sampling them since it is hard to find the longest cycle in a graph. This might be surprising in the light that finding a cycle in a graph is an easy problem. There are numerous other cases when the counting version of an easy decision problem is hard since finding an optimal solution is hard in spite of the fact that finding one (arbitrary) solution is easy. Although we briefly review the main complexity classes of decision and optimization problems in Sections 1.2 and 1.5, we assume readers have prior knowledge on them. Possible references on computational complexity are [8, 140, 160].

When we are talking about easy and hard problems, we use the convention of computational complexity that a problem is defined as an easy computational problem if there is a polynomial running time algorithm to solve it. Very rarely we can unconditionally prove that a polynomial running time algorithm

1

does not exist for a computational problem. However, we can prove that no polynomial running time algorithm exists for certain counting problems if no polynomial running time algorithm exists for certain hard decision problems. This fact also underlines why discussing decision problems is inevitable in a book about computational complexity of counting and sampling.

When exact counting is hard, approximate counting might be easy or hard. Surprisingly, hard counting problems might be easy to approximate stochastically, however, there are counting problems that we cannot approximate well. We conjecture that they are hard to approximate, and this is a point where stochastic approximations are also related to random approaches to decision problems. Particularly, if no random algorithm exists for certain hard decision problems that run in polynomial time and is any better than random guessing, then there is no efficient good approximation for certain counting problems.

In this chapter, we give a brief introduction to computational complexity and show how computational complexity of counting and sampling is related to computational complexity of decision and optimization problems.

1.1   General overview of computational problems

A *computational problem* is a mathematical object representing a collection of questions that computers might be able to solve. The questions belonging to a computational problem are also called *problem instances*. An example of a *decision problem* is the triangle problem which asks if there is triangle in a finite graph. The problem instances are the finite graphs and the answer for any problem instance is "yes" or "no" depending on whether or not there is a triangle in the graph. In this computational problem, a triangle in a graph is called a *witness* or *solution*. In general, the witnesses of a problem instance are the mathematical objects that certify that the answer for the decision problem is "yes". An example for an *optimization problem* is the clique problem which asks what the largest clique (complete subgraph) is in a finite graph. The problem instances are again the finite graphs and the solutions are the largest cliques in the graphs.

Any decision or optimization problem has its natural counting counterpart problem asking the number of witnesses or solutions. For example, we can ask how many triangles a graph has, as well as how many largest cliques a graph has.

Computational problems might be solved with algorithms. We can classify algorithms based on their properties. Algorithms might be exact or approximate, might be deterministic or random, and probably their most important feature is if they are feasible or infeasible. To define feasibility, we have to define how to measure it. Larger problem instances might need more computational steps, also called running time. Therefore, it is natural to measure the complexity of an algorithm with the necessary computational steps as a function of the input (problem instance) size. The size of the problem instance is defined as the number of bits necessary to describe it. A computational problem is defined as *tractable* if its running time grows with a polynomial function of the size of the input, and *intractable* if its running time grows exponentially or even more with the size of the input. This definition ignores constant factors, the order of the polynomial and typical input sizes. This means that theoretically tractable problems might be infeasible in practice, and *vice versa*, theoretically intractable problems might be feasible in practice if the typical input sizes are small. Interested readers can find a series of exercises exploring this phenomena at the end of the chapter (Exercises 1–3). In practice, most of the tractable algorithms run in at most cubic time, and their constant factor is less than 10. These algorithms are not only theoretically tractable but also feasible in practice. The given definitions of tractable and intractable problems do not cover all algorithms as there are functions that grow faster than any polynomial function but slower than any exponential function. Such functions are called *superpolynomial* and *subexponential*. Although there are remarkable computational problems, most notably the graph isomorphism problem [9, 10], which is conjectured to have superpolynomial and subexponential running time algorithms in the best case, such problems are relatively rare, and not discussed in detail in this book.

Observe that both the size of the problem instance and the number of computational steps are not precisely defined. Indeed, a graph, for example, might be encoded by its adjacency matrix or by the list of edges in it. These encodings might have different numbers of bits. Similarly, on many computers, different operations might have different running times: the time necessary to multiply two numbers might be much more than the time needed to add two numbers. To get rigorous mathematical definitions, theoretical computer science introduced mathematical models of computations; the best known are the Turing machines. In this book, we avoid these formal descriptions of computations. The reason for this is that we are interested in only the *order* of the running time, and constant factors are hidden in the O (big O, ordo) notation. Even if sizes are defined in different ways, different definitions almost never have exponential (or more precisely, superpolynomial) gaps. For example, if a graph has n vertices, then it might have $O(n^2)$ edges. However, it does not make a theoretical difference if an algorithm on graphs runs in $O(n^3)$ time or $O(m^{1.5})$ time, where n is the number of vertices and m is the number of edges: both functions are polynomials. The only difference when there is an exponential gap between two concepts of input sizes is when we distinguish the value of the number and the number of bits necessary to describe a number. When we would like to emphasize that the input size is the value of the number, we will say that the input numbers are given *in unary*. A typical example is the *subset sum* problem, where we ask if we can select a subset of integers whose sum is a prescribed value W. There is a dynamic programming algorithm to solve this problem whose running time is polynomial with the value of W. However, it is a hard decision problem if W is not given in unary [108].

1.2 Deterministic decision problems: P, NP, NP-complete

Definition 1. *In computational complexity theory, P is the class that contains the decision problems solvable in polynomial time.*

Examples for decision problems in P are the following:

- The perfect matching problem asks if a graph has a perfect matching. A perfect matching is a set of independent edges that covers all vertices [60].

- The substring problem asks if a sequence A is a substring of sequence B. A substring is a series of consecutive characters of a sequence, for example, $A = aba$ is a substring of $B = bbabaaab$ since the third, fourth and fifth characters of B is indeed sequence A.

- The primality testing problem asks if a positive integer number is a prime number. Surprisingly, this problem can be solved in polynomial time even if the input size is the number of digits necessary to write down the number [3].

One of the most important and unsolved questions in theoretical computer science is whether or not P is equal to NP. Formally, the complexity class NP contains the problems that can be solved in polynomial time with non-deterministic Turing machines. The name NP stands for "non-deterministic polynomial". Since we do not introduce Turing machines in this book, an alternative, equivalent definition is given here.

Definition 2. *The complexity class NP contains the decision problems for which solutions can be verified in polynomial time.*

This definition is more intuitive than the formal definition using Turing machines. Examples for problems in NP are the following.

- The k-clique problem asks if there is a clique of size k in a graph, that is a subgraph isomorphic to the complete graph K_k.

- The two partitioning problem asks if there is a partitioning of a finite set of integer numbers into two subsets such that the sum of the numbers in the two subsets is the same.

- The feasibility of an integer programming question asks if there is a list of integer numbers $x_1, x_2, \ldots x_n$ satisfying a set of linear inequalities having the form

$$\sum_{i=1}^{n} c_i x_i \leq b. \tag{1.1}$$

It is easy to see that these problems are indeed in NP. If somebody selects vertices $v_1, v_2, \ldots, v_k$, it is easy to verify that for all $i, j \in \{1, 2, \ldots, k\}$, there is an edge between v_i and v_j. If somebody provides a partitioning of a set of numbers, it is easy to calculate the sums of the subsets and check if the two sums are the same. Also, it is easy to verify that assignments to the variables $x_1, x_2, \ldots, x_n$ satisfy any inequality under (1.1).

In many cases, finding a solution seems to be harder than verifying a solution. There are problems in NP for which no polynomial running time algorithm is known. We cannot prove that such an algorithm does not exist, however, we can prove that these hard computational problems are as hard as any problems in NP. To precisely state this, we first need the following definitions.

Definition 3. *Let A and B be two computational problems. We say that A has a* polynomial reduction *to B, if a polynomial running time algorithm exists that solves any problem instance $x \in A$ by generating problem instances $y_1, y_2, \ldots, y_k$ all in B and solves x using the solutions for $y_1, y_2, \ldots y_k$. The computational time generating problem instances $y_1, y_2, \ldots y_k$ counts in the running time of the algorithm, but the computational time spent in solving these problem instances is not considered in the overall running time. We also say that A is* polynomially reducible *to B.*

Example 1. *An independent set in a graph is a subset of the vertices such that no two vertices in it are adjacent. The k-independent set problem asks if there is an independent set of size k in a graph.*

The k-independent set problem is polynomially reducible to the k-clique problem. Indeed, a graph contains an independent set of size k if and only if its complement contains a clique of size k. Taking the complement of a graph can be done in polynomial time.

Similarly, the k-clique problem is also polynomially reducible to the k-independent set problem.

Polynomial reduction is an important concept in computational complexity. If a computational problem A is polynomially reducible to B and B can be solved in polynomial time, then A also can be solved in polynomial time. Similarly, if B is polynomially reducible to A, and A can be solved in polynomial time, then B can be solved in polynomial time, as well. Therefore, if A and B are mutually polynomially reducible to each other, then either both of them or none of them can be solved in polynomial time. These thoughts lead to the following definitions.

Definition 4. *A computational problem is in the complexity class* NP-hard *if every problem in NP is polynomially reducible to it. The* NP-complete *problems are the intersection of NP and NP-hard.*

What follows from the definition is that P is equal to NP if and only if there is a polynomial running time algorithm that solves an NP-complete problem. It is widely believed that P is not equal to NP, and thus, there are no polynomial running time algorithms for NP-complete problems.

It is absolutely not trivial that NP-complete problems exist. Below we define a decision problem and state that it is NP-complete.

Definition 5. *In Boolean logic, a* literal *is a logical variable or its negation. A* disjunctive clause *is a logical expression of literals and OR operators* $(\vee)$.

A conjunctive normal form *or* CNF *is a conjunction of disjunctive clauses, that is, disjunctive clauses connected with the logical AND* $(\wedge)$ *operator. A conjunctive normal form* Φ *is* satisfiable *if there is an assignment of logical variables in* Φ *such that the value of* Φ *is TRUE. Such an assignment is called a* satisfying assignment. *The decision problem if there is a satisfying assignment of a conjunctive normal form is called the* satisfiability *problem and denoted by* SAT.

Theorem 1. *For any decision problem A in NP and any problem instance x in A, there exists a conjunctive normal form* Φ *such that* Φ *is satisfiable if and only if the answer for the problem instance x is "yes". Furthermore, for any x, such a conjunctive normal form can be constructed in polynomial time of the size of x. Since verifying that an assignment of the logical variables is a satisfying assignment can be clearly done in polynomial time, and thus, SAT is in NP, this also means that SAT is an NP-complete problem.*

We do not prove this theorem here; the proof can be found in any standard textbook on computational complexity, see for example [72]. The satisfiability is the only problem for which we can directly prove NP-completeness. For all other decision problems, NP-completeness is proved by polynomial reduction of the SAT problem or other NP-complete problems to those decision problems. Indeed, the following theorem holds.

Theorem 2. *Let A be an NP-complete problem and let B be a decision problem in NP. If A is polynomially reducible to B, then B is also NP-complete.*

Proof. The proof is based on the fact that the sum as well as the composition of two polynomials are also polynomials. □

Stephen Cook proved in 1971 that SAT is NP-complete [44], and Richard Karp demonstrated in 1972 that many natural computational problems are NP-complete by reducing SAT to them [108]. These famous Karp's 21 NP-complete problems drove attention to NP-completeness and initiated the study of the P versus NP problem. The question whether or not P equals NP has become the most famous unsolved problem in computational complexity theory. In 2000, the Clay Institute offered \$1 million for a proof or disproof that P equals NP [1].

Below we give a list of NP-complete problems that we are going to use in proofs of theorems about computational complexity of counting and sampling.

Definition 6. *Let* $\vec{G} = (V, E)$ *be a directed graph. A* Hamiltonian path *is a directed path that visits each vertex exactly once. A* Hamiltonian cycle *is a directed cycle that contains each vertex exactly once.*

Based on this definition, we can define the following two problems.

Problem 1.
Name: H-PATH.
Input: a directed graph, $\vec{G} = (V, E)$.
Output: "yes" if $\vec{G}$ has a Hamiltonian path, "no" otherwise.

Problem 2.
Name: H-CYCLE.
Input: a directed graph, $\vec{G} = (V, E)$.
Output: "yes" if $\vec{G}$ has a Hamiltonian cycle, "no" otherwise.

Theorem 3. *[108] Both* H-PATH *and* H-CYCLE *are in NP-complete.*

It is also hard to decide if a graph contains a large independent set.

Problem 3.
Name: LARGEIS.
Input: a positive integer m and a graph G in which every independent set has size at most m.
Output: "yes" if G has an independent set of size m, and "no" otherwise.

Theorem 4. *[73] The decision problem* LARGEIS *is in NP-complete.*

The subset sum (see below) is an infamous NP-complete problem. It is polynomially solvable if the weights are given in unary, however, it becomes hard for large weights.

Problem 4.
Name: SUBSETSUM.
Input: a set of numbers, $S = \{x_1, x_1, \ldots, x_n\}$ and a number m.
Output: "yes" if there is a subset $A \subseteq S$ such that $\sum_{x \in A} x = m$, otherwise "no".

Theorem 5. *[108] The decision problem* SUBSETSUM *is in NP-complete.*

1.3 Deterministic counting: FP, #P, #P-complete

Definition 7. *The complexity class #P contains the counting problems that ask for the number of witnesses of the decision problems in NP. If A denotes a problem in NP, then #A denotes its counting counterpart.*

Since the decision versions of #P problems are in NP, there is a witness that can be verified in polynomial time. This does not automatically imply that all witnesses can be verified in polynomial time, although it naturally holds in many cases. When it is questionable that all solutions can be verified in polynomial time, a polynomial upper bound must be given, and only those witnesses count that can be verified in that time.

For example, #SAT denotes the counting problem that asks for the number of satisfying assignments of conjunctive normal forms. Some counting problems are tractable. Formally, they belong to the class of tractable function problems.

Definition 8. *A function problem is a computational problem where the output is more complex than a simple "yes" or "no" answer. The complexity class FP (Function Polynomial-Time) is the class of function problems that can be solved in polynomial time with an algorithm.*

We can define the #P-hard and #P-complete classes analogously to the NP-hard and NP-complete classes.

Definition 9. *A computational problem is in #P-hard if any problem in #P is polynomially reducible to it. The #P-complete class is the intersection of #P and #P-hard.*

As one can naturally guess, #SAT is a #P-complete problem. Indeed, the following theorem holds.

Theorem 6. *For every problem #A in #P, and every problem instance x in #A, there exists a conjunctive normal form Φ such that the number of satisfying assignments of Φ is the answer for x. Furthermore, such a Φ can be constructed in polynomial time of the size of the problem instance x. Since #SAT is in #P, this means that #SAT is a #P-complete problem.*

It is clear that $\#P \subseteq FP$ if and only if there exists a polynomial running time algorithm for a #P-complete problem. It is also trivial to see that $\#P \subseteq FP$ implies P = NP. However, we do not know if the reverse is true, namely, whether or not P = NP implies that counting the witnesses of any #P-complete problem is easy. Still, we have the following non-trivial result.

Theorem 7. *[161] If P = NP, then for any problem #A in #P and any polynomial p, there is a polynomial time algorithm for #A such that it approximates any instance of #A within a multiplicative approximation factor $1 + \frac{1}{p}$.*

By assuming that P is not equal to NP, we cannot expect a polynomial running time algorithm counting the witnesses of an NP-complete problem. Naturally, the counting versions of many NP-complete problems are #P-complete. However, we do not know if it is true that for any NP-complete problem A, its counting version $\#A$ is #P-complete. On the other hand, there are easy decision problems whose counting version is #P-complete. Below we show two of them.

Definition 10. *The* permanent *of an $n \times n$ matrix $M = \{m_{i,j}\}$ is defined as*

$$per(M) := \sum_{\sigma \in S_n} \prod_{i=1}^{n} m_{i,\sigma(i)} \qquad (1.2)$$

where S_n is the set of permutations of numbers $1, 2, \ldots, n$.

Theorem 8. *Computing the permanent is #P-hard. Computing the permanent of a 0-1 matrix is still #P-hard.*

We are going to prove this theorem in Chapter 4. Calculating the permanent is related to counting the perfect matchings in a graph, as stated and proved below.

Theorem 9. *Computing the number of perfect matchings in a bipartite graph is a #P-complete counting problem.*

Proof. We reduce the permanent of a 0-1 matrix to computing the number of perfect matchings in a bipartite graph.

Let $A = \{a_{i,j}\}$ be an arbitrary $n \times n$ matrix containing 0s and 1s. Construct a bipartite graph $G = (U, V, E)$ such that there is an edge between u_i and v_j if and only if $a_{i,j} = 1$.

Matrix A contains only 0s and 1s, therefore for any permutation σ,

$$\prod_{i=1}^{n} a_{i,\sigma(i)} \qquad (1.3)$$

is 1 if each $a_{i,\sigma(i)}$ is 1 and 0 otherwise. Let S' denote the subset of permutations for which the product is 1. Let $\mathcal{M}$ denote the set of perfect matchings in G. Clearly, there is bijection between S' and $\mathcal{M}$: if $\sigma \in S'$, then assign the perfect matching to σ that contains the edges $(u_i, v_{\sigma(i)})$. This is indeed a perfect matching, since each $(u_i, v_{\sigma(i)}) \in E$ due to the definition of S' and A, and each vertex is covered by exactly one edge since σ is a permutation. It is also clear that this mapping is an injection, if $\sigma_1 \neq \sigma_2$, then their images are also different.

Similarly, if $M \in \mathcal{M}$ is a perfect matching, then for each i, it contains an edge (u_i, v_j). Then assign to M the permutation that maps i to j. It is indeed a permutation due to the definition of perfect matching, and the so obtained σ is indeed in S' due to the definition of S' and A.

Therefore, the number of perfect matchings in G is the permanent of A. Since constructing G can be clearly done in polynomial time, this is a polynomial reduction, and thus, computing the number of perfect matchings in a bipartite graph is #P-hard. Since this counting problem is also in #P, it is in #P-complete. □

Leslie Valiant defined the classes #P-hard and #P-complete, and proved that computing the permanent of a matrix is #P-hard, it is still #P-hard to compute the permanent of a matrix if the entries are restricted to the $\{0, 1\}$ set, and thus, counting the perfect matchings in a bipartite graph is #P-complete [175]. This is quite surprising, since deciding if there is a perfect matching in a bipartite graph is in P [94].

We introduce another #P-complete counting problem, which is even more surprising in the sense that its decision version is absolutely trivial. The problem is also related to finding the volume of a convex body. It is also the first example for the fact that hard counting problems might be easy to approximate, as we are going to discuss later on in this chapter.

Definition 11. *A partially ordered set or short: a poset is a pair* $(A, \leq)$ *where A is a set and $\leq$ is a reflexive, antisymmetric and transitive relation on A, that is, for any $a, b, c \in A$*

(a) $a \leq a$,

(b) *if $a \neq b$ and $a \leq b$, then the relation $b \leq a$ does not hold,*

(c) $a \leq b \wedge b \leq c \implies a \leq c$.

The meaning of the name is that there might be elements $a, b \in A$ such that neither $a \leq b$ nor $b \leq a$ hold. An example for partial ordered sets is when $A = 2^X$ for some set X, and the relation is $\subseteq$. Further examples are: A is the natural numbers and $a \leq b$ if $a|b$, A is the set of subgroups of a group and $a \leq b$ is a is a subgroup of b, etc.

It is easy to see that any poset can be extended to a total ordering, such that for any $a, b \in A$, $a \leq b$ implies that $a \leq_t b$, where $\leq_t$ is the defining relation of the total ordering. Such a total ordering is called a *linear extension* of the poset. We can ask how many linear extensions a poset has.

Problem 5.
Name: #LE.
Input: a partially ordered set $(P, \leq)$.
Output: the number of linear extensions of $(P, \leq)$.

Observe that the decision version of #LE is trivial: whatever the poset is, the answer is always "yes" to the question if the poset has a linear extension. Therefore, it is very surprising that the following theorem holds.

Theorem 10. *#LE is a #P-complete problem.*

We are going to prove this theorem in Chapter 4. #LE is related to computing the volume of a convex body. We can define a polytope for each finite poset in the following way.

Definition 12. *Let $(A, \leq)$ be a finite poset of n elements. The poset polytope is a convex body in the Euclidian space $\mathbb{R}^n$ in which each point $(x_1, x_2, \ldots, x_n)$ satisfies the inequalities*

1. *for all i, $0 \leq x_i \leq 1$,*

2. *for all $a_i \leq a_j$, $x_i \leq x_j$.*

Theorem 11. *The volume of the poset polytope of a poset $P = (A, \leq)$ is $\frac{1}{n!}$ times the number of linear extensions of P where $n = |A|$.*

Proof. Any total ordering is also a partial ordering, so we can define its polytope. The intersection of the polytopes of two total orderings has 0 measure (the possible common facets of the polytopes). Therefore, it is sufficient to prove that the poset polytope of any total ordering of a set of size n has volume $\frac{1}{n!}$. There is a natural bijection between the permutations of length n and the total orderings of a set of size n: $a_i \leq a_j$ in a total ordering $(A, \leq)$ if and only if $\sigma(i) < \sigma(j)$ in permutation σ. The union of the $n!$ poset polytopes is the n-dimensional unit cube, which has volume 1. Each poset polytope has the same volume since they can be transformed into each other with linear transformations preserving the volume. Indeed, the matrices of these linear transformations in the standard basis are permutation matrices. Therefore, the determinant of them in absolute value is 1. The intersections of these polytopes have 0 measure. Then indeed, the volume of the poset polytope of any total ordering of a set of size n is $\frac{1}{n!}$. The poset polytope of any partial ordering is the union of the poset polytopes of its linear extensions, therefore its volume is indeed $\frac{1}{n!}$ times the number of linear extensions of the poset. $\square$

Polytopes are related to other counting problems, too, see for example, Exercise 5.

1.4 Computing the volume of a convex body, deterministic versus stochastic case

The corollary of Theorem 10 is that it is #P-hard to find the volume of a convex body defined as the intersection of half spaces given by linear inequalities. Computing the volume of a convex body is intrinsically hard. However, the computation becomes easy if stochastic computations are allowed. In this section, we present two results highlighting that there might be exponential gaps between random and deterministic computations.

Consider a computational model with an oracle. The oracle can be asked if any point in $\mathbb{R}^d$ is in the convex set $K \subset \mathbb{R}^d$. The oracle answers "yes" if the point is in K. If the point is not in K, then it gives a hyperplane separating the point from K. Since $\mathbb{R}^d$ is infinite, we need further information about the convex set: it is promised that the convex set is in the hypersphere RB and contains the hypersphere rB where $0 < r < R$ are real numbers, and B is the Euclidian unit ball around the center, defined by the inequality

$$\sum_{i=1}^{d} x_i^2 \le 1. \tag{1.4}$$

The central question here is how many oracle calls are needed to approximate the volume of K. We are going to show a negative result. It says that there is no deterministic, polynomial running time algorithm that can reasonably approximate the volume of a convex body in this computational model. The extremely surprising fact is that in the same computational model, approximating the volume with a random algorithm is possible in polynomial time.

Before we state and prove the theorem, we need the following lemma.

Lemma 1. *Let $v_1, v_2, \ldots, v_n$ be points on the surface of the d $(d \le n + 1)$ dimensional Euclidian unit ball B around the center. Let K be the convex hull of the given points. Then*

$$vol(K) \le n \left(\frac{1}{2}\right)^d vol(B) \tag{1.5}$$

where $vol(K)$ and $vol(B)$ are the volumes of the convex hull and the unit ball, respectively.

Proof. Consider the balls $B_1, B_2, \ldots, B_n$ whose radii are $\frac{1}{2}$ and centers are $\frac{v_1}{2}, \frac{v_2}{2}, \ldots, \frac{v_n}{2}$. Here each point v_i is considered as a d-dimensional vector. We claim that

$$K \subseteq \cup_{i=1}^{n} B_i. \tag{1.6}$$

Indeed, assume that there is a vertex $v \in K$, but $v \notin \cup_{i=1}^{n} B_i$. Then for each i, the angle Oxv_i is strictly smaller than $\pi/2$. Now consider the hyperplane P whose normal vector is Ox, and contains x, and let H be the open halfspace determined by P, containing O. Since each Oxv_i angle is strictly smaller than $\pi/2$, each v_i is in H. But then K, the convex hull of v_is, cannot contain x, a contradiction.

$\square$

Now we are ready to prove the following theorem.

Theorem 12. *Let $\frac{1}{d}B \subseteq K \subseteq B$ be a d-dimensional convex body. Assume that an oracle is available which tells for any point in $\mathbb{R}^d$ whether the point* *is in K. Then any deterministic algorithm that has only poly(d) oracle calls cannot give an estimation f of the volume of K satisfying*

$$\frac{vol(K)}{1.999^{\frac{d}{2}}} \leq f \leq 1.999^{\frac{d}{2}} vol(K) \tag{1.7}$$

where $vol(K)$ denotes the volume of K.

Proof. Since $K \subseteq B$, we can assume that all points submitted to the oracle are in B. Let $z_1, z_2, \ldots z_n$ denote the submitted vertices. Let $v_1, v_2, \ldots, v_n$ be the normalizations of these points, normalized to length 1. Then $v_1, v_2, \ldots, v_n$ are on the surface of B. Let $s_1, s_2, \ldots s_{d+1}$ be the vertices of a regular d-simplex on the surface of B. Consider two convex bodies, $K_1 = B$, and K_2, which is the convex hull of the vertices $v_1, \ldots, v_n, s_1, \ldots, s_{d+1}$. Both convex bodies contain $\frac{1}{d}B$, K_1 trivially, and K_2 because the inscribed hypersphere of the regular d-simplex is $\frac{1}{d}B$ (see also Exercise 4). From Lemma 1, we know that

$$\frac{vol(K_1)}{vol(K_2)} \geq \frac{1}{n+d+1} 2^d. \tag{1.8}$$

In both cases, the oracle tells us that all points $z_1, z_2, \ldots, z_n$ are inside the convex body. Let f be the value that the algorithm generates, based on the answers of the oracle. Then either

$$f > 1.999^{\frac{d}{2}} vol(K_2) \tag{1.9}$$

or

$$f < \frac{vol(K_1)}{1.999^{\frac{d}{2}}}. \tag{1.10}$$

If both inequalities failed, than it would be true that

$$1.999^d \geq \frac{\frac{vol(K_1)}{f}}{\frac{vol(K_2)}{f}} = \frac{vol(K_1)}{vol(K_2)} \geq \frac{1}{n+d+1} 2^d, \tag{1.11}$$

which is a contradiction since n is only a polynomial function of d. Therefore no algorithm can generate an approximation f based on a polynomial number of oracle calls that satisfy the inequality in Equation (1.7) for every input K. □

The above introduced proof is based on the work of György Elekes [61]. Later, Imre Bárány and Zoltán Füredi extended this result for a stronger one proving that there is no polynomial algorithm that gives lower and upper bounds, $\underline{vol}(K)$ and $\overline{vol}(K)$, for the volume of K satisfying

$$\frac{\overline{vol}(K)}{\underline{vol}(K)} \leq \left(c\frac{d}{\log(d)} \right)^d \tag{1.12}$$

where c does not depend on the dimension d [12].

The situation completely changes if random computation is allowed. Random algorithms sample random points from B and estimate the volume of K based on the answers of the oracle to the randomly generated points. Although it cannot be guaranteed that the resulting random estimation will be always in between two multiplicative approximations to the exact answers, still this happens with high probability. Ravi Kannan, László Lovász and Miklós Simonovits proved the following theorem [105].

Theorem 13. *There is an algorithm with parameters $\epsilon, \delta > 0$ which returns a real number f satisfying the inequality*

$$P\left(\frac{vol(K)}{1+\epsilon} \le f \le (1+\epsilon)vol(K)\right) \ge 1 - \delta. \qquad (1.13)$$

The algorithm uses

$$O\left(\frac{d^5}{\epsilon^2} \ln^3\left(\frac{1}{\epsilon}\right) \ln\left(\frac{1}{\delta}\right) \ln^5(d)\right) \qquad (1.14)$$

oracle calls.

Later on, László Lovász and Santosh Vempala gave an $O^*(d^4)$ running time algorithm to estimate the volume of a convex body in the same computational model [122]. Here O^* is for hiding logarithmic terms. Their algorithm still runs in polynomial time with $\frac{1}{\epsilon}$ and $-\log(\delta)$. We can set, say, ϵ to $\frac{1}{d}$ and δ to $\frac{1}{e^d}$, and still, the running time will be a polynomial function of d. Let us emphasize again that this is a very striking result: there is no polynomial running time deterministic algorithm that could approximate the volume of a convex body within anything better than an approximation factor growing exponentially with the dimension. On the other hand, if random computation is allowed, polynomial running time is sufficient to have an approximation algorithm whose approximation factor actually can tend to 1 polynomially fast with the dimension. The price we have to pay is that the algorithm might be wrong with a small probability, however, this probability might even tend to 0 exponentially fast with the dimension.

This highlights the importance of random algorithms in counting and sampling problems.

1.5 Random decision algorithms: RP, BPP. Papadimitriou's theorem

In this section, we introduce the two basic complexity classes for random decision algorithms: the BPP and RP classes, and prove a theorem that is an exercise in Papadimitriou's book on computational complexity [140]. The theorem says the following: if we can stochastically solve any NP-complete problem with anything better than random guessing, then we could solve any problem in the NP class with probability almost 1. We will use this theorem to prove that certain counting problems cannot be well approximated stochastically unless RP = NP.

Definition 13. *A decision problem is in the* BPP (Bounded-error Probabilistic Polynomial) *class if a random algorithm exists such that*

(a) *it runs in polynomial time on all inputs,*

(b) *if the correct answer is "yes" it answers "yes" with probability at least 2/3,*

(c) *if the correct answer is "no" it answers "no" with probability at least 2/3.*

Any such algorithm is also called a BPP algorithm.

The 2/3 in the definition of BPP is just a convention. The number 2/3 is the rational number p/q between 1/2 and 1 such that $p+q$ is minimal. Indeed, any fixed constant number between 1/2 and 1 would suffice or even it would be enough if one of the probabilities was strictly 1/2 and the probability for the other answer would converge to 1/2 only polynomially fast, see Exercise 6. Similarly, if a BPP algorithm exists for a decision problem, then also a random algorithm exists that runs in polynomial time and gives the wrong answer with very small probability (say, with probability $\frac{1}{2^{100}}$), see Exercise 7.

Definition 14. *A decision problem is in* RP (Randomized Polynomial time) *if a random algorithm exists such that*

(a) *it runs in polynomial time on all inputs*

(b) *if the correct answer is "yes" it answers "yes" with probability at least 1/2*

(c) *if the correct answer is "no" it answers "no" with probability 1.*

Any such algorithm is also called RP algorithm.

Again, the 1/2 in the definition of RP is only a convention; it is the rational number p/q between 0 and 1 such that $p+q$ is minimal. Just like in the case of the BPP class, any fixed constant probability or a probability that converges only polynomially fast to 0 would suffice, see Exercise 10. Also, the existence of an RP algorithm means that the correct answer can be calculated with very high probability (say, with probability $1 - \frac{1}{2^{100}}$) in polynomial time, see Exercise 11.

We know that

$$P \subseteq RP \subseteq NP \tag{1.15}$$

however, we do not know if any containment is proper. Surprisingly, we do not know if BPP $\subseteq$ NP or NP $\subseteq$ BPP. Throughout this book, we will assume the following conjecture.

Conjecture 2. *For the decision classes* P, RP *and* NP, *the relation*

$$P = RP \subset NP \tag{1.16}$$

holds. The P $\neq$ NP *assumption also means that*

$$\text{NP-complete} \cap P = \emptyset \tag{1.17}$$

and

$$\#\text{P-complete} \cap FP = \emptyset. \tag{1.18}$$

The conjecture that RP = NP also implies that RP $\cap$ NP-complete = $\emptyset$. This comes from the following theorem and from the easy observation that RP $\subseteq$ BPP.

Theorem 14. *(Papadimitriou's theorem) If the intersection of* NP-complete *and* BPP *is not empty, then* RP = NP.

Proof. We prove this theorem in three steps.

1. First we prove that a BPP algorithm for any NP-complete problem would prove that BPP = NP.

2. In particular, SAT would be in BPP. We show that a BPP algorithm for SAT would provide an RP algorithm for SAT.

3. Finally, we show that an RP algorithm for SAT would mean that RP = NP.

Concerning the first point, assume that there is an NP-complete problem A for which a BPP algorithm exists. Let B be an arbitrary problem in NP. Since A is NP-complete, for any problem instance x in B, there is a series of problem instances $y_1, y_2, \ldots y_k \in A$ such that these problem instances can be constructed from x in polynomial time with the size of x. Particularly, $k = O(poly(|x|))$ and for each $i = 1, 2, \ldots, k$, $|y_i| = O(poly(|x|))$. Furthermore, from the solutions of these problems, the solution to x can be achieved in polynomial time. If only random answers are available for $y_1, y_2, \ldots, y_k$, then only a random answer for x can be generated. One wrong solution for any y_i might result in a wrong answer for x. Therefore, to get a BPP algorithm for solving x, we need that the probability that all y_i are answered correctly must be at least 2/3. For this, each y_i must be answered correctly with probability at least $\left(\frac{2}{3}\right)^{\frac{1}{k}}$. We can approximate it with

$$1 - \frac{1}{\frac{2k}{\log\left(\frac{3}{2}\right)}} \tag{1.19}$$

since

$$\left(1 - \frac{1}{\frac{2k}{\log\left(\frac{3}{2}\right)}}\right)^k = \left(\left(1 - \frac{1}{\frac{2k}{\log\left(\frac{3}{2}\right)}}\right)^{\frac{2k}{\log\left(\frac{3}{2}\right)}}\right)^{\frac{\log\left(\frac{3}{2}\right)}{2}} >$$

$$\left(\frac{1}{e}\right)^{\frac{\log\left(\frac{3}{2}\right)}{2}} = \sqrt{\frac{2}{3}} > \frac{2}{3}. \tag{1.20}$$

To achieve this probability, we repeat the BPP algorithm for each y_i m times, and take the majority answer. The number of correct answers follows a binomial distribution with parameter $p \geq \frac{2}{3}$. We can use Chernoff's inequality to give an upper bound on the probability that less than half the times the BPP algorithm generates the correct answer. Recall that Chernoff's inequality is

$$P(Y_m \leq mp(1 - \epsilon)) \leq \exp\left(-\frac{1}{2p}\frac{(mp - mp(1 - \epsilon))^2}{m}\right) \tag{1.21}$$

where Y_m is a binomial variable, p is the parameter of the binomial distribution and ϵ is an arbitrary number between 0 and 1. In our case, $p = 2/3$, therefore ϵ should be set to $1/4$ to get that $p(1 - \epsilon) = 1/2$. We want to find m satisfying

$$\exp\left(-\frac{1}{2p}\frac{(mp - mp(1 - \epsilon))^2}{m}\right) \leq 1 - \left(1 - \frac{1}{\frac{2k}{\log\left(\frac{3}{2}\right)}}\right) = \frac{1}{\frac{2k}{\log\left(\frac{3}{2}\right)}} \tag{1.22}$$

namely,

$$\exp\left(-\frac{m}{48}\right) \leq \frac{1}{\frac{2k}{\log\left(\frac{3}{2}\right)}}. \tag{1.23}$$

We get that

$$m \geq 24\log\left(\frac{2k}{\log\left(\frac{3}{2}\right)}\right) \tag{1.24}$$

which is clearly satisfied if

$$m \geq \frac{96}{\log\left(\frac{3}{2}\right)}k. \tag{1.25}$$

This means that the following is a BPP algorithm for problem instance x in problem B:

1. Construct problem instances $y_1, y_2, \ldots y_k$ in problem A.

2. Solve each y_i $\frac{96}{\log\left(\frac{3}{2}\right)}k$ times with the BPP algorithm available for A, and take the majority answer as the answer to problem instance y_i.

3. Solve x using the answers to each y_i.

The first and the last step can be done in polynomial time due to the definition of NP-completeness. Since $k = O(poly(|x|))$ and for each i, $|y_i| = O(poly(|x|))$, the second step also runs in polynomial time. Therefore, the overall running time is polynomial with the size of x. It is also a BPP algorithm since the probability that all y_i are answered correctly is at least $2/3$.

Next, we prove that a BPP algorithm for SAT provides an RP algorithm for SAT. Let Φ be a conjunctive normal form. Consider the conjunctive normal form Φ_1 that is obtained from Φ by removing all clauses that contain the literal x_1 and removing all occurrences of the literal $\overline{x_1}$. Clearly, Φ_1 is satisfiable if and only if Φ has a satisfying assignment in which x_1 is TRUE. Consider also the conjunctive normal form Φ_1' that is obtained from Φ by removing all clauses that contain the literal $\overline{x_1}$ and removing all occurrences of the literal x_1. Φ_1' is satisfiable if and only if Φ has a satisfying assignment in which x_1 is FALSE. We can solve the decision problem if Φ_1 is satisfiable with a very high probability by repeating the BPP algorithm an appropriate number of times, and we can also do the same with Φ_1'. If one of them is satisfiable, then we can continue with the variable x_2, and decide if there is a satisfying assignment of Φ_1 or Φ_1' in which x_2 is TRUE or FALSE. Iterating this procedure, we can actually build up a candidate for satisfying assignment. By repeating the BPP algorithm sufficiently many times, we can achieve that the probability that all calculations are correct is at least $1/2$. After building up the candidate assignment, we can deterministically check if this candidate is a satisfying assignment, and answer "yes" to the decision question if Φ is satisfiable only if we verified that the candidate is indeed a satisfying assignment.

We claim that this procedure is an RP algorithm for the SAT problem. If there are n logical variables, then the candidate assignment is built up in n iterations. We can use again Chernoff's inequality to show that $O(n)$ repeats of the BPP algorithm in each iterative step is sufficient for having at least $1/2$ probability that all computations are correct. (Detailed computations are skipped here; the computation is very similar to the previous one.) Therefore, if Φ is satisfiable, then we can construct a candidate assignment which is indeed a satisfying assignment with probability at least $1/2$. We can verify that the candidate assignment is indeed a satisfying assignment in polynomial time, and thus, we answer "yes" with probability at least $1/2$. If Φ is not satisfiable, then either we conclude with this somewhere during the iteration or we construct a candidate assignment which is actually not a satisfying one. However, we can deterministically verify it, and therefore, if Φ is not satisfiable, then with probability 1 we answer "no".

Finally, we claim that an RP algorithm for SAT provides an RP algorithm for any problem in NP. This is the direct consequence of Theorem 1. Indeed, let A be in NP, and let x be a problem instance in A. Then construct the conjunctive normal form Φ which is satisfiable if and only if the answer for x is "yes". By solving the satisfiability problem for Φ with an RP algorithm, we also solve the decision problem for x. $\qquad\square$

1.6 Stochastic counting and sampling: FPRAS and FPAUS

In a random computation, we cannot expect that the answer be correct with probability 1, and we even cannot expect that the answer have a given approximation ratio with probability 1. However, we might require that the computation have small relative error with high probability. This leads to the definition of the following complexity class.

Definition 15. *A counting problem is in* FPRAS (Fully Polynomial Randomized Approximation Scheme) *if for any problem instance x and parameters $\epsilon, \delta > 0$ it has a randomized algorithm generating an approximation $\hat{f}$ for the true value f satisfying the inequality*

$$P\left(\frac{f}{1+\epsilon} \le \hat{f} \le f(1+\epsilon)\right) \ge 1 - \delta \qquad (1.26)$$

and the running time of the algorithm is polynomial in $|x|$, $\frac{1}{\epsilon}$ and $-\log(\delta)$.

An algorithm itself with these prescribed properties is also called FPRAS.

An example for an FPRAS is the algorithm approximating the volume of a convex body in Theorem 13.

The following theorem shows that we cannot expect a counting problem to be in FPRAS if its decision version is NP-complete.

Theorem 15. *If there exists an NP-complete decision problem A such that #A is in FPRAS, then RP = NP.*

Proof. By Theorem 14, it is sufficient to show that an FPRAS algorithm for the number of solutions provides a BPP algorithm for the decision problem. Clearly, let x be a problem instance in A, then the answer for x is "no" if the number of solutions is 0 and the answer is "yes" if the number of solutions is a positive integer. Consider an FPRAS with input x, $\epsilon = 1/2$ and $\delta = 1/3$. Such an FPRAS runs in polynomial time with the size of x, and provides an answer larger than $1/2$ with probability at least $2/3$ if there is at least one solution for x, namely, if the correct answer for the decision problem is "yes". Furthermore, if the correct answer for the decision problem is "no", then the FPRAS returns 0 with probability at least $2/3$. Therefore, the algorithm that sends the input x, $\epsilon = 1/2$ and $\delta = 1/3$ to an FPRAS and answers "no" if FPRAS returns a value smaller that $1/2$ and answers "yes" if the FPRAS returns a value larger than or equal to $1/2$ is a BPP algorithm even if the running time of the FPRAS is included in the running time. $\square$

There are also counting problems whose decision versions are easy (they are in P), however, they cannot be approximated unless RP = NP. Jerrum,

Valiant and Vazirani already observed in 1986 that it is hard to count the number of cycles in a directed graph [103]. Below we introduce this hardness result.

Problem 6.
Name: CYCLE.
Input: a directed graph $\vec{G}$.
Output: "yes" if $\vec{G}$ contains a cycle, "no" otherwise.

Theorem 16. *The* CYCLE *problem is in P. On the other hand, the following also holds. If there is a deterministic algorithm such that*

(a) *its input is a directed graph $\vec{G} = (V, E)$,*

(b) *it runs in polynomial time of the size of $\vec{G}$,*

(c) *it gives an estimation $\hat{f}$ of the number of directed cycles in $\vec{G}$ such that*

$$\frac{f}{poly(n)} \leq \hat{f} \leq f \times poly(n)$$

where f is the number of directed cycles in $\vec{G}$ and n is the number of vertices in G,

then $P = NP$. With other words, it is NP-hard to approximate the number of cycles in a directed graph even with a polynomial approximation ratio.

Also, the following holds: if there is an FPRAS algorithm for #CYCLE then $RP = NP$.

Proof. In Chapter 2, we are going to prove that there is a polynomial running time algorithm that finds the shortest path from any vertex from v_i to v_j in a directed graph. We can exclude the 0 length path from the paths, and then the shortest path from v_i to itself is a cycle. For each v_i, we can ask if there is a cycle starting and ending in v_i. Since there are only polynomially many vertices in a graph, this proves that the CYCLE problem is in P.

To prove the hardness part of the theorem, we reduce the Hamiltonian cycle problem to #CYCLE. Consider any directed graph $\vec{G} = (V, E)$, and let $n = |V|$. We are going to "blow up" this graph, using the following gadget, see also Figure 1.1. Define the directed diamond motif as a graph containing 4 vertices and 4 edges. There are 2 edges going from vertex v_s to w and u, furthermore, there are 2 edges going from w and u to v_t. The gadget graph H contains n^2 diamond motifs, a source vertex s and a sink vertex t. There is an edge from vertex s to the v_s vertex of the first diamond motif, and for all $i = 1, 2, \ldots, n^2 - 1$, there is an edge from the v_t vertex of the ith diamond motif to the v_s vertex of the $i + 1^{\text{st}}$ diamond motif. Finally, there is an edge from the v_t vertex of the last diamond motif to vertex t. It is easy to see that there are 2^{n^2} number of paths from s to t: there are two paths for getting through

FIGURE 1.1: The gadget graph replacing a directed edge in the proof of Theorem 16. See text for details.

on each diamond motif, and any combination of them provides a path from s to t.

Construct the directed graph $\vec{G}'$ by replacing each edge (u, v) with the gadget graph H. For any cycle in $\vec{G}$ with k edges, there are $\left(2^{n^2}\right)^k$ corresponding cycles in $\vec{G}'$. Therefore, if there is no Hamiltonian cycle in G, then there are at most

$$\sum_{k=2}^{n-1} \binom{n}{k} k! \left(2^{n^2}\right)^k < n(n-1)! \left(2^{n^2}\right)^{(n-1)} = n! \left(2^{n^2}\right)^{(n-1)} \tag{1.27}$$

cycles in $\vec{G}'$. Indeed, there are at most $\binom{n}{k}k!$ cycles of length k in $\vec{G}$. On the other hand, if $\vec{G}$ has a Hamiltonian cycle, then there are at least

$$\left(2^{n^2}\right)^n \tag{1.28}$$

cycles in $\vec{G}'$. The ratio of this latter lower bound and the upper bound in case of no Hamiltonian cycles is

$$\frac{\left(2^{n^2}\right)^n}{n! \left(2^{n^2}\right)^{(n-1)}} = \frac{2^{n^2}}{n!} > 2^{\frac{n^2}{2}} \tag{1.29}$$

which grows faster than any exponential function, and thus, any polynomial function. Therefore, any polynomial approximation for the number of cycles in $\vec{G}'$ would provide an answer for the question whether there is a Hamiltonian cycle in $\vec{G}$. Furthermore, the size of $\vec{G}'$ is only a polynomial function of $\vec{G}$, thus H-CYCLE has a polynomial reduction to the polynomial approximation for #CYCLE. That is, it is NP-hard to have a polynomial approximation for the number of cycles in a directed graph.

An FPRAS approximation for the number of cycles would provide a BPP algorithm for the H-CYCLE problem. Indeed, an FPRAS approximating the number of cycles in $\vec{G}'$ with parameters $\epsilon = 1$ and $\delta = 1/3$ would separate the cases when $\vec{G}$ has and does not have a Hamiltonian cycle with probability at least 2/3. Since H-CYCLE is NP-complete, the intersection of NP-complete and BPP could not be empty which would imply that RP = NP, according to Theorem 14. $\square$

The situation will remain the same if we could generate roughly uniformly distributed cycles from a directed graph. Below we state this precisely after defining the necessary ingredients of the statement.

Definition 16. *The* total variation distance *of two discrete distributions p and π over the same (countable) domain X is defined as*

$$d_{TV}(p, \pi) := \frac{1}{2} \sum_{x \in X} |p(x) - \pi(x)|. \qquad (1.30)$$

It is easy to see that the total variation distance of two distributions is between 0 and 1, it is 0 if and only if the two distributions are pointwise the same (for all x, $p(x) = \pi(x)$), and it is 1 if and only if the two distributions have disjoint support (that is, $p(x) \neq 0$ implies that $\pi(x) = 0$ and vice versa). It is also easy to see that the total variation distance is indeed a metric, see also Exercise 12.

Definition 17. *A sampling problem $\#X$ is in* FPAUS *(Fully Polynomial Almost Uniform Sampler) if $\#X$ is in $\#P$, and there is a sampling algorithm that for any problem instance in $\#X$ and $\epsilon > 0$, it generates a random witness following a distribution p satisfying the inequality*

$$d_{TV}(p, U) \leq \epsilon \qquad (1.31)$$

where U is the uniform distribution of the witnesses. The algorithm must run in polynomial time both with the size of the problem instance and $-\log(\epsilon)$. This algorithm itself is also called FPAUS.

There is a strong relationship between the complexity classes FPRAS and FPAUS. In Chapter 7 we will show that for a large class of counting problems, called self-reducible counting problems, there is an FPRAS algorithm for a particular counting problem if and only if there is an FPAUS algorithm for it. Here we state and prove that an FPAUS for sampling cycles from a directed graph would have the same consequence as the existence of an FPRAS algorithm for counting the cycles in a directed graph.

Theorem 17. *If there is an FPAUS algorithm for $\#CYCLE$, then $RP = NP$.*

Proof. According to Theorem 14, it is sufficient to show that an FPAUS would provide a BPP algorithm for H-CYCLE. Assume that there is an FPAUS for sampling cycles from a directed graph. For any directed graph $\vec{G}$, construct the same graph $\vec{G}'$ that we constructed in the proof of Theorem 16. Apply the FPAUS algorithm on $\vec{G}'$ using $\epsilon = 1/10$, and generate one cycle. Draw back this cycle to $\vec{G}$. If this is a Hamiltonian cycle, then answer "yes", otherwise answer "no".

We claim that this is a BPP algorithm (actually, an RP algorithm). If there is no Hamiltonian cycle in $\vec{G}$, then the algorithm answers "no" with probability 1. If there are $c \geq 1$ Hamiltonian cycles in $\vec{G}$, then the cycles in $\vec{G}'$ corresponding to Hamiltonian cycles in $\vec{G}$ have probability at least

$$\frac{c\left(2^{n^2}\right)^n}{n!\left(2^{n^2}\right)^{n-1} + c\left(2^{n^2}\right)^n} = \frac{c2^{n^2}}{n! + c2^{n^2}} \geq \frac{16}{18} \tag{1.32}$$

assuming that $\vec{G}$ has at least 2 vertices. Note that the distribution p has at least $2/3$ probability on the cycles in $\vec{G}'$ that correspond to Hamiltonian cycles in $\vec{G}$, otherwise we would have

$$\frac{1}{10} \geq d_{TV}(p, U) \geq \frac{1}{2}\sum_{x \in H}|U(x) - p(x)|$$

$$\geq \frac{1}{2}\left(\sum_{x \in H}U(x) - \sum_{x \in H}p(x)\right) \geq \frac{1}{2}\left(\frac{16}{18} - \frac{2}{3}\right) = \frac{1}{9} \tag{1.33}$$

where H is the set of cycles in $\vec{G}'$ that corresponds to Hamiltonian cycles in $\vec{G}$. The inequality $\frac{1}{10} \geq \frac{1}{9}$ is clearly a contradiction. Therefore if there is a Hamiltonian cycle in $\vec{G}$, then the FPAUS algorithm with $\epsilon = \frac{1}{10}$ will generate a cycle in $\vec{G}'$ that corresponds to a Hamiltonian cycle in $\vec{G}$ with probability at least $2/3$. Generating $\vec{G}'$, running the FPAUS algorithm, drawing back the sampled cycle to $\vec{G}$ and checking if the so-obtained cycle is a Hamiltonian one can all be done in polynomial time. Therefore, this procedure is indeed a BPP for the Hamiltonian cycle problem. This would imply that RP = NP, according to Theorem 14. □

Still, there are #P-complete problems that do have FPRAS algorithms. The careful reader might observe that we already introduced such a problem. Indeed, #LE is a #P-complete counting problem on the one hand; on the other hand, it is in FPRAS since approximating the volume of a convex body is in FPRAS with an oracle that tells if a point is in the convex body. However, in the case of the poset polytope, we do not need an oracle, since for any point we can deterministically decide if it is in the poset polytope: we have to check the inequalities in the Definition 12. This clearly can be done in polynomial time for any point. Thus, there is an FPRAS for approximating the volume of a poset polytope. Dividing with $n!$ keeps the relative error. Therefore, there is also an FPRAS for counting the linear extensions of a poset.

One might ask if the fact that the intersection of #P-complete and FPRAS is not empty implies that RP = NP. We could see in the proof of Papadimitriou's theorem that we can nicely handle probabilities in polynomial reductions. However, we might not be able to handle relative errors. Indeed, there are operations that do not keep the relative error, most notably, the subtraction and modulo prime number calculations. We will see that such operations appear in each #P-completeness proof of a counting problem that is also in FPRAS. Therefore the polynomial reductions used in such proofs cannot be used to propagate FPRAS approximations to other counting problems in #P: we will lose the small relative error. Thus, it does not follow that any counting problem in #P has an FPRAS. In particular, we cannot prove that an FPRAS algorithm exists for a counting problem whose decision version is NP-complete.

On the other hand, if a #P-completeness proof for a problem #A preserves the relative error, it also proves that there is no FPRAS for #A unless RP = NP. The relative error can be preserved with a one-to-one, a one-to-many or a many-to-one reduction or even in a way as seen in the proof of non-approximability of the number of cycles in a directed graph.

1.7   Conclusions and the overview of the book

The central question in the computational complexity theory of counting and sampling is this: Which counting problems are in FP, which are in #P-complete, and if a problem is in #P-complete, is it in FPRAS and/or FPAUS or not (assuming that RP is not equal to NP)? Although there is no strict trichotomy, most of the counting problems fall into one of the following three categories:

1. The counting problem is in FP. Typically, if the counting problem is in FP, then there is a polynomial running time algorithm that can sample witnesses from exactly the uniform distribution.

2. The counting problem is in #P-complete, however, it is also in FPRAS. Typically, such a counting problem is also in FPAUS. There is a large class of counting problems for which we can prove that all of them are in

$$(\text{FPRAS} \cap \text{FPAUS}) \cup (\overline{\text{FPRAS} \cup \text{FPAUS}}),$$

   namely, we can prove that any counting problem in this class is in FPRAS if and only if it is also in FPAUS.

3. The counting problem is in #P-complete, and there is no FPRAS algorithm for it, unless RP = NP. Typically, these problems also do not have an FPAUS, unless RP = NP.

If a decision problem is NP-complete, then there is no FPRAS algorithm for its counting version, unless RP = NP. We conjecture that the counting version of any NP-complete problem is also #P-complete.

Easy decision problems might also be hard to count, even approximately. This book is about the state-of-the-art of our knowledge about which decision problems have their counting version in the above-mentioned three categories. The book contains the following chapters.

(a) Chapter 2 describes the easiest counting problems. These are the problems whose decision, optimization and counting versions can be universally solved with dynamic programming algorithms. The computations in these dynamic programming algorithms use only additions and multiplications, which we call monotone computations. We are going to show that from an algebraic point of view, the logical OR and AND operations can be considered as additions and multiplications. Similarly, addition and multiplication can be replaced with minimization and addition without changing the algebraic properties of the computations. We are going to show that a large class of dynamic programming algorithms have some universal algebraic properties, therefore essentially the same algorithms can solve the decision, optimization and counting versions of a given problem. If the universal algorithm has polynomial running time, then the problem it solves has a decision version in P and a counting version in FP, and furthermore, optimal solutions can be found also in polynomial time.

(b) Chapter 3 introduces counting problems solvable in polynomial time using subtraction. Particularly, the number of spanning trees in a graph as well as the number of Eulerian circuits in a directed Eulerian graph are related to the determinant of certain matrices, and the number of perfect matchings in a planar graph is the Pfaffian of an appropriately oriented adjacency matrix of the graph. We are also going to show that both the determinant and the Pfaffian can be computed in polynomial time using only additions, subtractions and multiplications, and therefore computations can be generalized to arbitrary commutative rings. These algorithms can also be viewed as monotone computations on some certain combinatorial objects. The signed and weighted sums of these combinatorial objects coincide with the determinants and Pfaffians of matrices via cancellations.

(c) We give a comprehensive list of #P-complete problems in Chapter 4. We highlight those problems for which approximation preserving #P-completeness proofs exist, and therefore, there is no FPRAS approximations for these problems unless RP = NP.

(d) Chapter 5 is about a relatively new topic, holographic algorithms. A holographic reduction is a linear algebraic many-to-many mapping between two sets. Such a mapping can prove that the sizes of the two sets are the same without explicitly giving a one-to-one correspondence between the two sets. Therefore, if the cardinality of one of the sets can be obtained in polynomial time, it can be done for the other set. Usually the holographic reduction maps a set to the (weighted) perfect matchings of planar graphs. Computing the sum of the weighted perfect matchings provides a tractable way to obtain the cardinality of the set. Holographic reductions are also used to obtain equalities of the cardinalities of two sets where finding the cardinality of one of the sets is known to be #P-hard. This provides a proof that finding the cardinality of the other set is also #P-hard.

(e) We turn to sampling methods in Chapter 6. We show how to sample uniformly combinatorial objects that can be counted with algebraic dynamic programming. This is followed by the introduction of ways of random generations providing techniques of almost uniform generations. Here Markov chains are the most useful techniques.

(f) Chapter 7 is devoted to the theory of the mixing of Markov chains. It turns out that in many cases, rapidly mixing Markov chains provide almost uniform generations of certain combinatorial objects. Markov chains are used to build a dichotomy theory for self-reducible problems: a self-reducible counting problem either has an FPRAS or cannot be approximated within a polynomial approximation factor. We also show that any self-reducible counting problem has an FPRAS if and only if it has an FPAUS. The consequence is that in many cases, we can prove that a counting problem is in FPRAS by showing that there exists a Markov chain which converges rapidly to the uniform distribution of the witnesses.

(g) Finally, Chapter 8 provides a comprehensive list of counting problems for which FPRAS exists.

1.8   Exercises

1. Algorithm A has running time $100000n$, and algorithm B has running time $0.1n^3$, where n is the input size. Which algorithm is faster if the input size is typically between 30 and 80?

2. Algorithm A has running time n^3, and algorithm B has a running time 1.01^n, where n is the input size. Which algorithm is faster if the input size is 1000?

3. An algorithm has running time n^{81}, where n is the input size. What is the largest input size for which the algorithm finishes in a year running on a supercomputer achieving 100 exaflops? An exaflop means 10^{18} floating point operations per second. Assume that one computational step needs one floating point operation. How many years does it take to run this algorithm on the mentioned supercomputer if the input size is $n = 3$?

4. * Let B be the unit Euclidian ball in a d-dimensional space. Let S be

a regular d-simplex whose circumscribed hypersphere is B. Prove that the inscribed hypersphere of S has radius $\frac{1}{d}$.

5. * A degree sequence is a list of non-negative integers. A graph G is a realization of a degree sequence D if the degrees of the vertices in G are exactly the elements of D. Show that for any degree sequence D, the number of realizations is the number of integer points in a convex body.

6. ○ Assume that the decision problem A has a random algorithm with the following properties:

   (a) Its running time grows polynomially with the size of the input.

   (b) If the correct answer is "yes", it answers "yes" with probability $1/2$.

   (c) If the correct answer is "no", it answers "no" with probability $\frac{1}{2} + \frac{1}{n^3}$, where n is the size of the problem instance.

   Show that A is in BPP.

7. ○ Assume that problem A is in BPP. Show that there is a random algorithm, that for any problem instance x in A and $\epsilon > 0$,

   (a) its running time grows polynomially with the size of the input.

   (b) it runs in polynomial time with $-\log(\epsilon)$,

   (c) if the correct answer is "yes", it answers "yes" with probability at least $1 - \epsilon$, and

   (d) if the correct answer is "no", it answers "no" with probability at least $1 - \epsilon$.

8. * An RP algorithm answers "yes" for some input problem instance. What can we say about the probability that the answer is correct?

9. * An RP algorithm answers "no" for some input problem instance. What can we say about the probability that the answer is wrong?

10. ○ Assume that the decision problem A has a random algorithm with the following properties:

   (a) Its running time grows polynomially with the size of the input.

   (b) If the correct answer is "yes", it answers "yes" with probability $\frac{1}{n^3}$, where n is the size of the problem instance.

   (c) If the correct answer is "no", it answers "no" with probability 1.

   Show that A is in RP.

11. ○ Assume that problem A is in RP. Show that there is a random algorithm, that for any problem instance x in A and $\epsilon > 0$,

    (a) it runs in polynomial time with the size of x,

    (b) it runs in polynomial time with $-\log(\epsilon)$,

    (c) if the correct answer is "yes", it answers "yes" with probability at least $1 - \epsilon$, and

    (d) if the correct answer is "no", it answers "no" with probability 1.

12. Prove that the total variation distance is indeed a distance on the space of distributions over the same countable domain. For any such domain X, let $\mathcal{F}$ be the set of distributions over X. Let $p_1, p_2, p_3 \in \mathcal{F}$. Then the following equations hold.

$$\begin{aligned}
d_{TV}(p_1, p_1) &= 0 \\
d_{TV}(p_1, p_2) &\geq 0 \\
d_{TV}(p_1, p_2) &= d_{TV}(p_2, p_1) \\
d_{TV}(p_1, p_2) + d_{TV}(p_2, p_3) &\geq d_{TV}(p_1, p_3).
\end{aligned}$$

13. * Let π and p be two arbitrary distributions over the same countable domain. Prove that

$$d_{TV}(p, \pi) = \max_{A \subseteq X}(p(A) - \pi(A))$$

where

$$p(A) := \sum_{x \in A} p(x).$$

14. Prove that the inequality in Equation (1.32) holds for any $n \geq 2$ and $c \geq 1$.

15. ○ Assume that the correct answer f of a problem instance x in the counting problem $\#A$ is an integer, and naturally bounded by 1 and $c^{poly(|x|)}$ for some $c > 1$. Furthermore, assume that $\#A$ is in FPRAS. Show that a random estimation $\hat{f}$ to the solution can be given with the following properties:

    (a) For the expected value of $\hat{f}$, $\overline{\hat{f}}$, it holds that

$$\frac{|\overline{\hat{f}} - f|}{f} \leq \frac{1}{poly(|x|)}.$$

(b) For the variance of $\hat{f}$, $\sigma_{\hat{f}}^2$ it holds that

$$\frac{\sigma_{\hat{f}}^2}{f} \leq \frac{1}{poly(|x|)}.$$

(c) The estimation $\hat{f}$ can be given in polynomial time.

16. * Let $\vec{G}$ be a directed graph. A cycle cover is a disjoint union of directed cycles that covers all vertices. Let A be the adjacency matrix of $\vec{G}$. Show that the number of cycle covers in $\vec{G}$ is the permanent of A.

17. Show that for any directed graph $\vec{G}$, there exists a graph H such that the number of cycle covers in $\vec{G}$ is the number of perfect matchings in H.

1.9 Solutions

Exercise 4. It is sufficient to show that for any regular d-simplex, the ratio of the radii of the inscribed and circumscribed hyperspheres is $\frac{1}{d}$. Let the coordinates of the regular d-simplex S be the unit vectors of the coordinate system of a $d+1$-dimensional space (that is, $(1,0,0,\ldots,0)$, $(0,1,0,0,\ldots,0)$, etc.). The inscribed hypersphere hits the surface of the simplex in the middle of the facets. The coordinates of these points are $(0,\frac{1}{d},\frac{1}{d},\ldots,\frac{1}{d})$, $(\frac{1}{d},0,\frac{1}{d},\ldots,\frac{1}{d})$, etc. Observe that these points are also vertices of a regular d-simplex S', whose circumscribed hypersphere is the inscribed hypersphere of S. Thus the ratio of the radii of the inscribed and circumscribed hyperspheres of S is the ratio of the edge lengths of S' and S. Since the edge length of S' is $\frac{\sqrt{2}}{d}$ and the edge length of S is $\sqrt{2}$, the ratio in question is indeed $\frac{1}{d}$.

Exercise 5. Let $D = \{d_1, d_2, \ldots, d_n\}$. The convex polytope is in $\mathbb{R}^{\binom{n}{2}}$. Let us denote the coordinates by the index pairs (i,j), where $i < j$. The linear inequalities defining the polytope are

$$0 \leq x_{(i,j)} \leq 1$$

and for each $i = 1, 2, \ldots, n$

$$\sum_{j=1}^{i-1} x_{(j,i)} + \sum_{j=i+1}^{n} x_{(i,j)} = d_i.$$

The bijection between the realizations $G = (V, E)$ and the integer points in the defined polytope is given by setting each $x_{(i,j)}$ to 1 if $(v_i, v_j) \in E$ and 0 otherwise.

Exercise 6. Repeat the given algorithm an appropriate number of times and take the majority answer. Use Chernoff's inequality to show that the appropriate number of repeats is indeed a polynomial function of n.

Exercise 7. Repeat the BPP algorithm an appropriate number of times, and take the majority answer. Use Chernoff's inequality to show that the appropriate number of repeats to achieve the prescribed properties is indeed a polynomial function of $-\log(\epsilon)$.

Exercise 8. This is a tricky question. One might think that the probability is $\frac{1}{2}$, since an RP algorithm says "yes" with $\frac{1}{2}$ probability if the correct answer is "yes". However, it answers "no" with probability 1 if the correct answer is "no", therefore a "yes" answer ensures that the correct answer is "yes". Therefore the probability that the RP algorithm gave the correct answer is actually 1.

Exercise 9. This is again a tricky question. We do not have information about the problem (whether or not the correct answer is "no"), so we cannot calculate exactly what the probability is that the answer was wrong. However, it is wrong with probability at most $\frac{1}{2}$ due to the definition of RP.

Exercise 10. Repeat the given algorithm an appropriate number of times, and answer "yes" if there is at least one "yes" answer, and "no" otherwise. Use the fact that for any $f(n)$ tending to infinity it holds that

$$\left(1 - \frac{1}{f(n)}\right)^{f(n)} \approx \frac{1}{e}.$$

Show that a polynomial number of repeats is sufficient to get an RP algorithm.

Exercise 11. Repeat the RP algorithm an appropriate number of times, and answer "yes" if there is at least one "yes" answer, and "no" otherwise. Basic algebraic considerations show that the appropriate number of repeats to achieve the prescribed properties is indeed a polynomial function of $-\log(\epsilon)$.

Exercise 13. Define set B as

$$B := \{x | p(x) \geq \pi(x)\}.$$

Observe that

$$\sum_{x \in B} |p(x) - \pi(x)| = \sum_{x \in \overline{B}} |p(x) - \pi(x)|.$$

Indeed,

$$\sum_{x \in B} (p(x) - \pi(x)) + \sum_{x \in \overline{B}} (p(x) - \pi(x)) = \sum_{x \in B \cup \overline{B}} p(x) - \sum_{x \in B \cup \overline{B}} \pi(x) = 0,$$

therefore

$$\sum_{x \in B} (p(x) - \pi(x)) = - \sum_{x \in \overline{B}} (p(x) - \pi(x)).$$

For any $x \in \overline{B}$, $p(x) - \pi(x)$ is negative, therefore,

$$-(p(x) - \pi(x)) = |p(x) - \pi(x)|.$$

What follows is that

$$d_{TV}(p, \pi) = \sum_{x \in B} (p(x) - \pi(x)).$$

We claim that B is a set on which the supremum

$$\sup_{A \subseteq X} (p(A) - \pi(A))$$

is realized. Indeed, for any $C \subseteq X$, the equation

$$p(C) - \pi(C) = p(B) - \pi(B) - (p(B \setminus C) - \pi(B \setminus C)) + (p(C \setminus B) - \pi(C \setminus B))$$

holds. However, $p(B \setminus C) - \pi(B \setminus C)$ cannot be negative and $p(C \setminus B) - \pi(C \setminus B)$ cannot be positive due to the definition of B. Therefore,

$$p(C) - \pi(C) \leq p(B) - \pi(B),$$

thus the supremum is indeed taken on B.

Exercise 15. Use the FPRAS algorithm for an estimation $\hat{f}$ and replace it with 1 or $c^{poly(|x|)}$ if the estimation is out of the boundaries. Set ε and δ in such a way that the prescribed inequalities for $\hat{f}$ and $\sigma_{\hat{f}}^2$ hold. Show that it is enough to set both $\frac{1}{\varepsilon}$ and $-\log(\delta)$ to some polynomial of $|x|$, thus, the overall running time of this procedure is only a polynomial function of $|x|$.

Exercise 16. Recall that the permanent of A is

$$per(A) := \sum_{\sigma \in S_n} \prod_{i=1}^{n} a_{i,\sigma(i)}.$$

A product is 1 if and only if for all i, $a_{i,\sigma(i)} = 1$, namely, $(i, \sigma(i))$ is a directed edge. We claim that there is a bijection between permutations for which the product is 1 and the cycle covers in $\vec{G}$. The bijection is given by the two representations of σ, the function representation, namely, for each i, $\sigma(i)$ is given, and the cycle representation of σ. Indeed, if a product is 1, then each $a_{i,\sigma(i)} = 1$, and these edges form a cycle cover in $\vec{G}$. On the other hand, each cycle cover indicates a permutation σ, such that each $a_{i,\sigma(i)} = 1$.

Part I

Computational Complexity
of Counting

Chapter 2

Algebraic dynamic programming and monotone computations

Dynamic programming is a powerful algorithmic technique to find the solution of a larger problem using solutions of smaller problems. It is mainly used in optimization problems, however, it is also applicable for counting problems. This chapter introduces a framework called *algebraic dynamic programming* which provides a unified framework of counting, sampling and optimizing.

    The name of algebraic dynamic programming was coined by Robert Giegerich and its theory was developed only for context-free grammars [76]. It was extended to further computational problems in bioinformatics [186].

In this book, we extend this theorem basically for any dynamic programming recursions. The main idea is to separate recursions building up combinatorial objects from using exact computations. To do this, we are going to introduce *yield algebras* that build the combinatorial objects (see Definition 19) and *evaluation algebras* that do the computations on these combinatorial objects (see Definition 21). Before giving the formal definitions, we introduce the concept via several well-known examples.

2.1   Introducing algebraic dynamic programming

2.1.1   Recursions, dynamic programming

Dynamic programming is one of the most fundamental methods in algorithmics. In a dynamic programming algorithm, larger problem instances are solved recursively using the solutions of smaller problem instances. In this way, any recursion can be considered as a dynamic programming. The relationship between algebraic dynamic programming and enumerative combinatorics is shown via a few examples.

Fibonacci numbers already appeared in Indian mathematics before Fibonacci as the number of k-unit-long sequences of patterns consisting of short syllables that are 1 unit of duration and long syllables that are 2 units of duration. Outside of India, the first appearance of the Fibonacci numbers is in the book *"Libre Abaci"* by Leonardo Bonacci, also known as Fibonacci. Fibonacci considered the growth of an idealized (biologically unrealistic) rabbit population. In the first month, a newborn couple of rabbits are put in a field. Newborn rabbits grow up in a month, never die, and after becoming adults, each couple produces a new couple of rabbits in each month. Therefore, if N_i denotes the number of newborn couples in the ith month and A_i denotes the number of adult couples in the ith month, then

$$A_{i+1} = A_i + N_i \tag{2.1}$$

and

$$N_{i+1} = A_i. \tag{2.2}$$

Using these recursions, it is easy to find the number of couples in month i. Indeed, let F_i denote the total number of couples in the ith month. Then

$$F_i = A_i + N_i = (A_{i-1} + N_{i-1}) + A_{i-1} =$$
$$(A_{i-1} + N_{i-1}) + (A_{i-2} + N_{i-2}) = F_{i-1} + F_{i-2} \tag{2.3}$$

which is the well-known recursion for the Fibonacci numbers. Assume also that we would like to count the children, grandchildren and grand-grandchildren of the founding couple, and in general, the rabbit couples of the kth generation in month i. Then we can assign the following characteristic polynomial to each subpopulation:

$$g_{\mathcal{A}_i} := \sum_{k=1}^{i} a_{i,k} x^k \tag{2.4}$$

$$g_{\mathcal{N}_i} := \sum_{k=1}^{i} n_{i,k} x^k \tag{2.5}$$

where $a_{i,k}$ is the number of adult couples of the kth generation in the ith month and $n_{i,k}$ is the number of newborn couples of the kth generation in the ith month. The first generation is the founding couple. The recursions for the characteristic polynomials are:

$$g_{\mathcal{A}_{i+1}} = g_{\mathcal{A}_i} + g_{\mathcal{N}_i} \tag{2.6}$$

$$g_{\mathcal{N}_{i+1}} = x g_{\mathcal{A}_i}. \tag{2.7}$$

The number of adult (newborn) rabbit couples of the kth generation in the ith month is the coefficient of x^k in $g_{\mathcal{A}_i}$ ($g_{\mathcal{N}_i}$). Let $r_{i,k}$ denote the total number of rabbit couples of the kth generation in the ith month. Then

$$r_{i,k} = a_{i,k} + n_{i,k} = (a_{i-1,k} + n_{i-1,k}) + a_{i-1,k-1}$$
$$(a_{i-1,k} + n_{i-1,k}) + (a_{i-2,k-1} + n_{i-2,k-1}) = r_{i-1,k} + r_{i-2,k-1}. \tag{2.8}$$

From this equation, we get that

$$r_{i,k} = \binom{i-k}{k-1}. \tag{2.9}$$

Indeed, $r_{i,k} = 0$ for any $i \leq 2k - 2$, and $r_{2k-1,k} = 1$, since each couple of rabbits has to grow up before giving birth, and thus, the kth generation appears first in the $2k - 1^{\text{st}}$ month. Therefore, $r_{2k-1,k}$ is indeed $\binom{k-1}{k-1} = 1$. For other coefficients, we get that

$$r_{i,k} = r_{i-1,k} + r_{i-2,k-1} = \binom{i-k-1}{k-1} + \binom{i-k-1}{k-2} = \binom{i-k}{k-1}. \tag{2.10}$$

From Equation (2.9) we get that

$$F_i = \sum_{k=1}^{\lfloor \frac{i+1}{2} \rfloor} r_{i,k} = \sum_{k=1}^{\lfloor \frac{i+1}{2} \rfloor} \binom{i-k}{k-1} = \sum_{k=0}^{\lfloor \frac{i-1}{2} \rfloor} \binom{i-k-1}{k}, \tag{2.11}$$

which is the well-known equality saying that the Fibonacci numbers are the sums of the "shallow" diagonals in Pascal's triangle.

There is an obvious similarity between Equations (2.1) and (2.2) and Equations (2.6) and (2.7). Both couples of recursions calculate *sums over sets*. Indeed, let $\mathcal{A}_i$ denote the set of adult rabbits in the ith month. Let f_1 assign the constant 1 to each rabbit couple, and let f_2 assign x^k to a rabbit couple in the kth generation. Then

$$A_i = \sum_{r \in \mathcal{A}_i} f_1(r) \tag{2.12}$$

and

$$g_{\mathcal{A}_i} = \sum_{r \in \mathcal{A}_i} f_2(r), \tag{2.13}$$

and similar equations hold for N_i and $g_{\mathcal{N}_i}$. We can also obtain recursions directly on the sets. Define two operations on rabbits, the o: "get one month older!" and the b: "give birth!" operators. Also define the action of these operators on sets of rabbit couples as

$$o * A := \{o * a | a \in A\}, \tag{2.14}$$

and similarly for b. Then the recursions for the subsets of populations are

$$\mathcal{A}_{i+1} = o * \mathcal{A}_i \cup o * \mathcal{N}_i \tag{2.15}$$
$$\mathcal{N}_{i+1} = b * \mathcal{A}_i. \tag{2.16}$$

In the algebraic dynamic programming approach, we will use such recursions to build up *yield algebras*, which describe how to build up sets subject to computations. The *evaluation algebras* describe what kind of computations have to be performed for each operator in the yield algebra. When we would like to count the rabbit couples, each operation has to be replaced by multiplication by 1, and the set union with the addition. When we would like to obtain the generating function to count the rabbit pairs of a given generation in a given month, then the o operation has to be replaced by multiplication by 1, and the b operation has to be replaced by multiplication by x. Such replacements indeed lead to the recursions in Equations (2.1)–(2.2) and (2.6)–(2.7).

Instead of applying operations on rabbits, we can assign a barcode to each rabbit couple in each month. The barcode in the ith month is an i-character-long sequence from the alphabet $\{0, 1\}$. It starts with a 1 and each o operation extends the sequence with a 0 and each b operation extends it with a 1 (so newborn rabbits get their parents' barcode extended with a 1). It is easy to see that each rabbit couple has a unique barcode, and the barcodes are exactly those sequences from the alphabet $\{0, 1\}$ which start with a 1 and there are no consecutive 1s. In such a sequence of length i, replace any 01 substring with a 2, then every 0 with 1, and then delete the starting 1. The resulting sequence contains 1s and 2s and the sum of the numbers is $i - 1$. Since the number of possible sequences consisting of k 2s and $i - 1 - 2k$ 1s is $\binom{i-1-k}{k}$, F_i is indeed the sum of the appropriate shallow diagonal in Pascal's triangle. The careful reader can also observe that the obtained sequences from the alphabet $\{1, 2\}$ decode the possible patterns of short and long syllables of a given total duration, thus the numbers of such patterns are also indeed the Fibonacci numbers.

As we can see, the Fibonacci numbers also appear as the number of certain legitimate code words or *regular expressions*. We can count any regular expressions using recursions. Consider the following example.

Example 2. *A code word from the alphabet $\{0, 1, 2, 3\}$ is said to be legitimate if it contains an odd number of 2s. How many code words of length n are legitimate? Give an efficient recursion to find the sum of the numbers in the legitimate code words of length n.*

Solution. The first question might be a standard exercise in any introductory combinatorics course, while the second one is somewhat unusual. Both questions can be answered if we first set up recursions concerning how to build up legitimate and illegitimate code words.

Let $\mathcal{L}_i$ denote the set of the legitimate code words of length i, and let $\mathcal{I}_i$ denote the set of illegitimate code words of length i. Let $\circ$ denote the concatenation operator and define this operator on a set as

$$\mathcal{A} \circ b := \{A \circ b \mid A \in \mathcal{A}\}. \tag{2.17}$$

Extending any legitimate code word with 0, 1 or 3 keeps it legitimate, while extending it with 2 makes it illegitimate. Similarly, extending an illegitimate code word with 2 makes it legitimate and extending it with 0, 1 or 3 keeps it illegitimate. Therefore, the recursions for the set of legitimate and illegitimate code words are

$$\mathcal{L}_{i+1} = (\mathcal{L}_i \circ 0) \cup (\mathcal{L}_i \circ 1) \cup (\mathcal{L}_i \circ 3) \cup (\mathcal{I}_i \circ 2) \tag{2.18}$$

and

$$\mathcal{I}_{i+1} = (\mathcal{I}_i \circ 0) \cup (\mathcal{I}_i \circ 1) \cup (\mathcal{I}_i \circ 3) \cup (\mathcal{L}_i \circ 2) \tag{2.19}$$

with the initial sets $\mathcal{L}_1 = \{2\}$ and $\mathcal{I}_1 = \{0, 1, 3\}$. To count the legitimate and illegitimate code words, we simply have to find the size of the sets appearing in Equations 2.18 and 2.19. Observe that each code word appears exactly once during the recursions, therefore

$$n_{\mathcal{L}_{i+1}} = 3n_{\mathcal{L}_i} + n_{\mathcal{I}_i} \tag{2.20}$$

and

$$n_{\mathcal{I}_{i+1}} = 3n_{\mathcal{I}_i} + n_{\mathcal{N}_i} \tag{2.21}$$

where $n_{\mathcal{L}_i}$ ($n_{\mathcal{I}_i}$) is the number of legitimate (illegitimate) code words of length i. The initial values are $n_{\mathcal{L}_1} = 1$ and $n_{\mathcal{I}_1} = 3$.

It is possible to find a closed form for such linear recurrences; such solutions can be found in any standard combinatorics textbook. From a computational

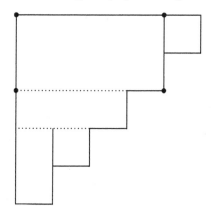

FIGURE 2.1: A stair-step shape of height 5 tiled with 5 rectangles. The horizontal lines inside the shape are highlighted as dotted. The four circles indicate the four corners of the rectangle at the top left corner that cuts the stair-step shape into two smaller ones. See text for details.

complexity point of view, the given recursions are sufficient to calculate $n_{\mathcal{L}_i}$ in a time proportional to some polynomial of i, therefore we are not going to provide closed forms here.

We also obtain recursions for the sum of the numbers in legitimate and illegitimate code words of a given length from the recursions in Equations (2.18) and (2.19). Extending with a number k each code word in a set of size m increases the sum of the numbers by mk. Therefore the recursions are

$$s_{\mathcal{L}_{i+1}} = s_{\mathcal{L}_i} + 4n_{\mathcal{L}_i} + s_{\mathcal{I}_i} + 2n_{\mathcal{I}_i} \tag{2.22}$$

and

$$s_{\mathcal{I}_{i+1}} = s_{\mathcal{I}_i} + 4n_{\mathcal{I}_i} + s_{\mathcal{L}_i} + 2n_{\mathcal{L}_i} \tag{2.23}$$

where $s_{\mathcal{L}_i}$ ($s_{\mathcal{I}_i}$) is the sum of the numbers in the legitimate (illegitimate) code words of length i. These recursions are sufficient to find the sum of the numbers in legitimate code words of length n using $O(poly(n))$ arithmetic operations. ∎

In both cases, the recursions were deducted from recursions in Equations (2.18) and (2.19). Namely, the recursions building the sets of objects of interest are fixed, and different calculations can be derived from these recursions. This is how to build different evaluation algebras from the same yield algebra in the algebraic dynamic approach.

Non-linear recursions also appear in enumerative combinatorics. The best known non-linear recursions are those that generate the Catalan structures, that is, those combinatorial objects that have C_n number of instances of size n, where C_n is the nth Catalan number. See the following example.

Example 3. *A stair-step shape of height n is the shape of the upper diagonal part of a square matrix of dimensions $(n + 1) \times (n + 1)$. Find the number of ways to tile a stair-step shape of height n with n rectangles. Rotate this stair-step shape in such a way that its corner is the top left corner (see Figure 2.1). Observe that different tilings might have a different number of horizontal lines inside the given shape. Compute the sum of the number of these horizontal lines in such tilings inside the stair-step shape of height n.*

Solution. Since the stair-step shape is tiled with n rectangles, each step must belong to a different rectangle. Therefore, one of the rectangles spans from a step to the top left corner, see Figure 2.1 for an example. Consider this rectangle in the top left corner in a tiling of a stair-step shape of height n. This rectangle has dimensions $i \times (n - i + 1)$ for some $i = 1, \ldots, n$. Removing this rectangle splits the tiling into tilings of stair-step shapes of height $n - i$ and $i - 1$. Here a stair-step shape of 0 height is an empty structure. The inverse of this operation is the merging of two stair-step shapes of height $n - i$ and $i - 1$ by inserting a rectangle with dimensions $i \times (n - i + 1)$ between the stair-step shapes. If $\circ$ denotes this inverse operation, $\mathcal{T}_i$ denotes the set of tilings of stair-step shapes of height i, and the $\circ$ operation is defined on sets as

$$\mathcal{A} \circ \mathcal{B} := \{a \circ b | a \in \mathcal{A} \text{ and } b \in \mathcal{B}\}, \tag{2.24}$$

then

$$\mathcal{T}_{n+1} = \cup_{i=0}^{n} \mathcal{T}_i \circ \mathcal{T}_{n-i} \tag{2.25}$$

where $\mathcal{T}_0$ is the set containing the empty structure of stair-step shape tiling of height 0. Furthermore, this is the initial case, therefore $|\mathcal{T}_0| = 1$. The careful reader might already have observed that $|\mathcal{T}_n| = C_n$, since

$$|\mathcal{T}_i \circ \mathcal{T}_j| = |\mathcal{T}_i||\mathcal{T}_j| \tag{2.26}$$

and the Catalan numbers also satisfy the recursion

$$C_{n+1} = \sum_{i=0}^{n} C_i C_{n-i} \tag{2.27}$$

with the initial condition $C_0 = 1$.

Any fixed tiling in $\mathcal{T}_{n-i}$ appears C_i times in the set $\mathcal{T}_i \circ \mathcal{T}_{n-i}$. Furthermore, any rectangle joining the tilings in $\mathcal{T}_i$ and in $\mathcal{T}_{n-1}$ has one horizontal line inside the shape except the rectangle with dimension $n \times 1$. Therefore the recursion for the number of horizontal lines, h_n, is

$$h_{n+1} = h_n + \sum_{i=1}^{n} (C_i h_{n-i} + h_i C_{n-i} + C_i C_{n-i}). \tag{2.28}$$

■

The careful reader might observe some similarity between Equation (2.28) and Equations (2.22) and (2.23). In both cases, the recursion finds the sum of an additive function over combinatorial objects, and the recursions use both these sums and the sizes of these smaller sets. We are going to introduce a commutative ring and will show that calculations involving these recursions can be naturally described in that ring.

In many textbooks, the introductory example for dynamic programming algorithms is the money change problem. We too introduce this problem together with its solution and show how it also fits into the algebraic dynamic programming approach.

Definition 18. *Let $C = \{c_1, c_2, \ldots, c_k\}$ be a finite set of positive integers called the* coin system, *and let x be a non-negative integer. The money change problem is to find the minimum number of coins necessary to change x. Each coin type has an infinite supply.*

To solve the money change problem, a function m is defined that maps the natural numbers $\mathbb{N}$ to $\mathbb{N} \cup \infty$; for each $x \in \mathbb{N}$, the value $m(x)$ is the minimum number of coins necessary to change x if changing of x is possible in this coin set C and ∞ otherwise. The following theorem is true for function m.

Theorem 18. *For $x = 0$, equation*

$$m(0) = 0 \tag{2.29}$$

holds, and for $x > 0$, equation

$$m(x) = \min_{\substack{i \in [1,k] \\ x - c_i \geq 0}} \{m(x - c_i) + 1\} \tag{2.30}$$

holds, where the minimum of the empty set is defined as ∞.

Proof. Equation (2.29) is obvious. To prove Equation (2.30), inequalities in both directions should be proved. If x cannot be changed, then $m(x) = \infty$ and naturally the inequality

$$m(x) \geq \min_{\substack{i \in [1,k] \\ x - c_i \geq 0}} \{m(x - c_i) + 1\} \tag{2.31}$$

holds. If x is changeable, then let $c_{i_1}, c_{i_2}, \ldots, c_{i_{m(x)-1}}, c'$ be a minimal change of x, that is, containing a minimum number of coins summing up to x. Then $c_{i_1}, c_{i_2}, \ldots c_{i_{m(x)-1}}$ is a change for $x - c'$ and thus cannot contain fewer coins than $m(x - c')$. Furthermore, a member of a set cannot be smaller than the minimal value of that set, thus

$$m(x) \geq m(x - c') + 1 \geq \min_{\substack{i \in [1,k] \\ x - c_i \geq 0}} \{m(x - c_i) + 1\}. \tag{2.32}$$

This means that for any x, the left-hand side of Equation (2.30) is greater than or equal to the right-hand side. If the set from which the minimum is taken on the right-hand side is empty for some x, then it is defined as ∞ and naturally inequality

$$m(x) \leq \min_{\substack{i \in [1,k] \\ x-c_i \geq 0}} \{m(x - c_i) + 1\} \tag{2.33}$$

holds. If the set is not empty, then let c' be the coin value for which the minimum is taken, and let $c_{i_1}, c_{i_2}, \ldots, c_{i_{m(x-c')}}$ be a minimal change for $x - c'$. Then the set of coins $c_{i_1}, c_{i_2}, \ldots, c_{i_{m(x-c')}}, c'$ is a change for x, it contains $m(x - c') + 1$ number of coins, and this number cannot be less than $m(x)$. Therefore

$$m(x) \leq m(x - c') + 1 = \min_{\substack{i \in [1,k] \\ x-c_i \geq 0}} \{m(x - c_i) + 1\}. \tag{2.34}$$

Since for all $x > 0$, inequalities in both directions hold, and therefore, Equation (2.30) also holds. $\square$

We can also build recursions on the possible coin sequences. Such recursions hold without stating what we want to calculate. After proving the correctness of the recursion on the possible coin sequences, we can solve different optimization and counting problems by replacing the operations in the given recursions. The construction is stated in the following theorem.

Theorem 19. *Let* $C = \{c_1, c_2, \ldots c_k\}$ *be a coin system,* $c_i \in \mathbb{Z}^+$, *and let* $S(x)$ *denote the set of coin sequences (repetitions are possible, order counts) summing up to* x. *Then*

$$S(0) = \{\epsilon\} \tag{2.35}$$

and for $x > 0$

$$S(x) = \sqcup_{\substack{i \in [1,k] \\ x-c_i \geq 0}} \{s \circ c_i \mid s \in S(x - c_i)\} \tag{2.36}$$

where ϵ *is the empty string,* $\circ$ *denotes string concatenation, and the use of disjoint union is to emphasize that each string in* $S(x)$ *is generated exactly once. Furthermore, the empty disjoint union (which is the case when for all* c_i, $x - c_i < 0$) *is defined as the empty set.*

Proof. Equation (2.35) is trivial: only the empty sum of the coin values has value 0. To prove Equation (2.36), it is sufficient to prove that

$$S(x) \subseteq \sqcup_{\substack{i \in [1,k] \\ x-c_i \geq 0}} \{s \circ c_i \mid s \in S(x - c_i)\} \quad \forall x > 0 \tag{2.37}$$

and

$$S(X) \supseteq \sqcup_{\substack{i \in [1,k] \\ x-c_i \geq 0}} \{s \circ c_i \mid s \in S(x - c_i)\} \quad \forall x > 0. \tag{2.38}$$

To prove Equation (2.37), consider the partitioning of the possible sequences of coin values summing up to x based on the last coin value. For each partition, and for each sequence in the partition, if the last coin c_i is removed, the remaining sequence will be in $S(x - c_i)$. Thus any sequence in $S(x)$ is in $\sqcup_{\substack{i \in [1,k] \\ x - c_i \geq 0}} \{s \circ c_i | s \in S(x - c_i)\}$.

To prove Equation (2.38), observe that each sequence in $\sqcup_{\substack{i \in [1,k] \\ x - c_i \geq 0}} \{s \circ c_i | s \in S(x - c_i)\}$ is in $S(x)$, since the sum of the coin values in each sequence is x. Furthermore, there is no multiplicity in $\sqcup_{\substack{i \in [1,k] \\ x - c_i \geq 0}} \{s \circ c_i | s \in S(x - c_i)\}$, since there is no multiplicity in any of the $S(x - c_i)$ sets and two sequences cannot be the same if their last coin values are different.

$\square$

Theorem 19 immediately provides a recursion to count the number of coin sequences summing up to a given amount: simply the size of $S(x)$ is to be calculated. Due to Equation (2.36), the following equality is true:

$$|S(x)| = \sum_{\substack{i \in [1,k] \\ x - c_i \geq 0}} |S(x - c_i)|. \tag{2.39}$$

Furthermore, the initial value

$$S(0) = 1 \tag{2.40}$$

can be read out from Equation (2.35). This calculation can be formalized in the following way. Let $f(s) := 1$ for any coin sequence s, and let

$$F(S(x)) := \sum_{s \in S(x)} f(s). \tag{2.41}$$

Then $F(S(x))$ will be the size of the set $S(x)$ and it holds that

$$F(S(x)) = \sum_{\substack{i \in [1,k] \\ x - c_i \geq 0}} F(S(x - c_i)). \tag{2.42}$$

This is the evaluation algebra, which can be changed without changing the underlying yield algebra. For example, let $g(s) := z^{|s|}$ (here z is an indeterminate) and let

$$G(S(x)) := \sum_{s \in S(x)} g(s). \tag{2.43}$$

Then $G(S(x))$ is a polynomial in which the coefficient of z^k is the number of those coin sequences of length k in which the coin values sum up to x. This polynomial is called the *partition polynomial*. It is easy to see that

$$g(s \circ c_i) = g(s)z \quad \forall s, c_i \tag{2.44}$$

thus

$$G(S(x)) = \sum_{\substack{i \in [1,k] \\ x - c_i \geq 0}} G(S(x - c_i))z. \tag{2.45}$$

In this way, it is possible to count the number of coin sequences of length k summing up to x for arbitrary k and x. Observe that once $G(S(X))$ is calculated, the optimization problem can also be solved. Indeed, the minimum number of coins necessary to change x is the degree of the minimum-degree monomial in $G(S(x))$.

2.1.2 Formal definition of algebraic dynamic programming

Below we give the formal description of algebraic dynamic programming. As mentioned, two algebras, the yield algebra and the evaluation algebra, have to be constructed. A yield algebra describes how to build up the space subject to calculations. First, we give its definition.

Definition 19. *A yield algebra is a tuple $\{A, (\Theta, \leq), p, \mathcal{O}, \mathcal{R}\}$. A is a set of objects, $(\Theta, \leq)$ is a partially ordered set called parameters. Let $B \subseteq \Theta$ be the subset of minimal elements. B is required to be non-empty and each $\theta \in \Theta$ must be accessible from it in a finite number of steps. Both A and Θ might be countably infinite. The function p maps A to Θ. $\mathcal{O}$ is a finite set of partial operators possibly with different arities. Each operator is defined on A. If $\circ$ is an m-ary operator, the expression*

$$a_1 \circ a_2 \circ \ldots \circ a_m$$

is also written as

$$\circ \left((a_j)_{j=1}^m \right).$$

The set of recursions, $\mathcal{R}$, contains a recursion for each $\theta \in \Theta \setminus B$ in the form

$$S(\theta) = \sqcup_{i=1}^n \left(\circ_i \left((S(\theta_{i,j}))_{j=1}^{m_i} \right) \right) \tag{2.46}$$

where

$$S(\theta) := \{a \in A \,|\, p(a) = \theta\} \tag{2.47}$$

and

$$\circ_i \left((S(\theta_j))_{j=1}^{m_i} \right) := \left\{ \circ_i \left((a_j)_{j=1}^{m_i} \right) \,|\, a_j \in S(\theta_{i,j}) \right\} \tag{2.48}$$

where $\circ_i \in \mathcal{O}$ is an m_i-ary operator. These operators might not be defined on some arguments, however, they must be defined on all possible arguments appearing in the recursions in $\mathcal{R}$. Furthermore, for all i and j, $\theta_{i,j} < \theta$.

For example, in the case of the population dynamics of rabbits following the Fibonacci numbers, A contains the set of adult and newborn rabbits in given years. Θ contains the (i, l) pairs, where i indicates the year and l is a label, which is either "adult" or "newborn". Parameter $(i_1, l_1) \leq (i_2, l_2)$ if $i_1 < i_2$ or $i_1 = i_2$ and $l_1 = l_2$. The operators are the already introduced "give birth!" and "get older!" operators, and the recursions have been given in Equations (2.15) and (2.16).

In the stairstep-shape tiling problem, A is the set of possible tilings. The parameter n tells the height of the tiled shape. The ordering of these parameters is the natural ordering of the natural numbers. The binary operator $\circ$ merges two tilings of shapes of heights k and l via a $(k+1) \times (l+1)$ rectangle, thus forming a tiling of a shape of height $k + l + 1$. The recursion is given in Equation (2.25).

In the money change problem, A is the set of coin sequences of finite length, $(\Theta, \leq)$ is the natural numbers with the usual ordering, and p assigns the sum of the coin values to each coin sequence. $\mathcal{O}$ contains a unary operation for each coin value such that $\circ_{c_i}(s) = s \circ c_i$. Then $\mathcal{R}$ indeed contains the recursions in Equation (2.36).

We are going to define the evaluation algebra, which describes what to calculate in the dynamic programming algorithms. Each evaluation algebra contains an algebraic structure called semiring. For readers not familiar with abstract algebra, the definition of semiring is given.

Definition 20. *A semiring* $(R, \oplus, \odot)$ *is an algebraic structure with two operations. The operation* $\oplus$ *is called addition, and* $(R, \oplus)$ *is a commutative monoid, namely, a commutative semigroup with a unit, that is,* $\oplus$ *is an associative and commutative operator on* R. *The operation* $\odot$ *is called multiplication, and* $(R, \odot)$ *is a semigroup, that is,* $\odot$ *is an associative operator on* R. *The distributive rule connects the two operations: for any* $a, b, c \in R$, *equations*

$$a \odot (b \oplus c) = (a \odot b) \oplus (a \odot c) \tag{2.49}$$

and

$$(b \oplus c) \odot a = (b \odot a) \oplus (c \odot a) \tag{2.50}$$

hold. The additive unit is usually denoted by 0.

We would like to emphasize that any ring is also a semiring in which the addition can be inverted (that is, there is also a subtraction). Readers not familiar with abstract algebra might consider the integer ring as an example of semirings with the additional rule that subtraction is forbidden. Later on, we will see that subtraction has very high computational power. Some computational problems can be solved in polynomial time when subtraction is allowed and otherwise those computational problems have exponential lower bound for their running time when only addition and multiplication are allowed. In this chapter, we would like to study the computational problems efficiently solvable using only additions and multiplications. This is why we restrict evaluation algebras to semiring computations.

Definition 21. *An* evaluation algebra *is a tuple* $\{Y, R, f, \mathcal{T}\}$, *where* Y *is a* *yield algebra, and f is a function mapping A (the set of objects in the yield algebra) to some semiring R. $\mathcal{T}$ is a set of functions T_i mapping $R^{m_i} \times \Theta^{m_i}$ to R where $\circ_i$ is an m_i-ary operator , with the property that for any operands of $\circ_i$*

$$f\left(\circ_i\left((a_j)_{j=1}^{m_i}\right)\right) = T_i\left(f(a_1),\ldots,f(a_{m_i}); p(a_1),\ldots,p(a_{m_i})\right). \qquad (2.51)$$

In many cases, the operator T_i does not depend on the parameters (i.e., $p(a_1),\ldots,p(a_{m_i})$). In those cases, the parameters will be omitted. We also require that each T_i should be expressed with operations in the algebraic structure R. When T_i depends on the parameters, the expression is given via a hidden function h mapping Θ^{m_i} to some R^{n_i} and then T_i is rendered as an algebraic expression of $m_i + n_i$ indeterminates (the m_i values of $f(a_j)$s and the n_i values coming from the image of h). Each operation $T_i \in \mathcal{T}$ must satisfy the distributive rule with respect to the addition in the semiring, that is, for any $\theta_{i,1},\ldots,\theta_{i,m_i}$

$$T_i\left(\sum_{a_{i,1} \in S(\theta_{i,1})} f(a_{i,1}),\ldots,\sum_{a_{i,m_i} \in S(\theta_{i,m_i})} f(a_{i,m_i}); \theta_{i,1},\ldots,\theta_{i,m_i}\right) =$$
$$\sum_{a_{i,1} \in S(\theta_{i,1})} \cdots \sum_{a_{i,m_i} \in S(\theta_{i,m_i})} f\left(\circ_i\left((a_{i,j})_{j=1}^{m_i}\right)\right). \qquad (2.52)$$

Our aim in the evaluation algebra is to calculate

$$F(S(\theta)) := \sum_{a \in S(\theta)} f(a). \qquad (2.53)$$

Thus, the evaluation algebra tells us what to calculate in the recursions in the yield algebra given in Equation (2.46). Due to the properties of the evaluation algebra, the following theorem holds.

Theorem 20. *Let $Y = \{A, (\Theta, \leq), p, \mathcal{O}, \mathcal{R}\}$ be a yield algebra, and let $E = \{Y, R, f, \mathcal{T}\}$ be an evaluation algebra. If for some parameter θ, the recursion*

$$S(\theta) = \sqcup_{i=1}^{n} \circ_i\left((S(\theta_{i,j}))_{j=1}^{m_i}\right) \qquad (2.54)$$

holds in the yield algebra, then

$$F(S(\theta)) = \sum_{i=1}^{n} T_i(F(S(\theta_{i,1})),\ldots,F(S(\theta_{i,m_i})), \theta_{i,1},\ldots,\theta_{i,m_i}) \qquad (2.55)$$

also holds. Namely, it is sufficient to know the values $F(S(\theta_{i,j}))$ for each $S(\theta_{i,j})$ to be able to calculate $F(S(\theta))$.

Proof. Equation (2.55) is the direct consequence of the distributive property of the T_i functions. Indeed,

$$F(S(\theta)) = \sum_{i=1}^{n} \left(\sum_{a_{i,1} \in S(\theta_{i,1})} \cdots \sum_{a_{i,m_i} \in S(\theta_{i,m_i})} f\left(\circ_i \left((a_{i,j})_{j=1}^{m_i} \right) \right) \right). \qquad (2.56)$$

From Equation (2.52), we get that

$$F(S(\theta)) = \sum_{i=1}^{n} \left(T_i \left(\sum_{a_{i,1} \in S(\theta_{i,1})} f(a_{i,1}), \dots, \right. \right.$$

$$\left. \left. \sum_{a_{i,m_i} \in S(\theta_{i,m_i})} f(a_{i,m_i}); \theta_{i,1}, \dots, \theta_{i,m_i} \right) \right). \qquad (2.57)$$

We can write back the definition of the F function (that is, Equation (2.53)) into the arguments of the T_i function, so we get Equation (2.55). $\qquad \square$

The amount of computation necessary to calculate $F(S(\theta))$ depends on how many $\theta' < \theta$ exist in Θ, how big n is in Equation (2.55) and how much time it takes to calculate each T_i. This is stated in the theorem below.

Theorem 21. *A computational problem X (X might be a decision problem, optimization problem or a counting problem) can be solved with algebraic dynamic programming in polynomial running time, if a pair of yield algebra, $Y = \{A, (\Theta, \leq), p, \mathcal{O}, \mathcal{R}\}$ and evaluation algebra $E = \{Y, R, f, \mathcal{T}\}$ exists such that for any problem instance $x \in X$, the following holds:*

(a) *a polynomial time computable θ exists such that the solution of x is $F(S(\theta))$,*

(b) *the size of $\theta_{\downarrow} := |\{\theta'|\theta' \leq \theta\}|$ is $O(poly(|x|))$, furthermore, for any $\theta' \in \theta_{\downarrow}$, the set of parameters covered by θ' can be calculated in polynomial time,*

(c) *for any $\theta' \leq \theta$, each T_i in Equation (2.55) can be calculated in $O(poly(|x|))$ time, and*

(d) *for any $\theta' \in B \cap \theta_{\downarrow}$, $F(S(\theta'))$ can be calculated in $O(poly(|x|))$ time.*

2.1.3 The power of algebraic dynamic programming: Variants of the money change problem

To show the power of algebraic dynamic programming, we consider several variants of the money change problem. We fix the following yield algebra. A is the set of possible coin sequences, Θ is the natural numbers with the usual ordering, and p maps the sequences to their summed coin values. The unary operator $\circ_i$ extends a coin sequence with the coin c_i. The recursions are given in Equations (2.35) and (2.36). For this yield algebra, we give several evaluation algebras that solve the following computational problems.

2.1.3.1 Counting the coin sequences summing up to a given amount

When the number of coin sequences summed up to a given value is calculated, R is the integer number ring and function f is the constant 1 function. Each T_i is the identity function (recall that each $\circ_{c_i}$ operation and thus, each T_i operation is unary). Then indeed Equation (2.55) turns into Equation (2.42).

2.1.3.2 Calculating the partition polynomial

How do we change the evaluation algebra when $f(s) = z^{|s|}$? Then $R = \mathbb{Z}[z]$. Each T_i is the multiplication with z. It is easy to see that Equation (2.55) turns into Equation (2.45).

2.1.3.3 Finding the recursion for optimization with algebraic dynamic programming

To show the power of algebraic dynamic programming, an example is given for solving optimization problems using algebraic dynamic programming. For this, first the *tropical semiring* is defined.

Definition 22. *The tropical semiring* $(R, \oplus, \odot)$ *is a semiring where* $R = \mathbb{R} \cup \infty$, $\odot$ *is the usual addition of the reals, extended by the symbol of infinity with the identity relations*

$$a \odot \infty = \infty \odot a = \infty \quad \forall a \in R \tag{2.58}$$

and $a \oplus b$ *is the minimum of* a *and* b, *extended by the symbol of infinity with the identity relations*

$$a \oplus \infty = \infty \oplus a = a \quad \forall a \in R. \tag{2.59}$$

It is easy to see that the tropical semiring is indeed a commutative semiring, that is, both $(R, \odot)$ and $(R, \oplus)$ are commutative semigroups and the distributive rule

$$a \odot (b \oplus c) = (a \odot b) \oplus (a \odot c) \tag{2.60}$$

holds for all $a, b, c \in R$. Note also that ∞ is the additive unit, therefore the empty tropical summation is ∞.

The tropical semiring can be utilized to solve the money change problem using algebraic dynamic programming. In the evaluation algebra, the algebraic structure R is the tropical semiring. For a coin sequence s, we set $f(s) = |s|$.

Each T_i operator increases the current value by 1, so it is the tropical multiplication by 1 in the tropical semiring. The F function takes the minimum of the given values, namely, it is the tropical addition. Then

$$F(S(x)) = \oplus_{\substack{i \in [1,k] \\ x - c_i \geq 0}} F(S(x - c_i)) \odot 1 \tag{2.61}$$

which is exactly Equation (2.30) with the notation of the operators in the tropical semiring and $m(x)$ is denoted by $F(S(x))$.

We are going to introduce three variants of the tropical semiring which are also used in optimization problems.

Definition 23. *The* dual tropical semiring $(R, \oplus, \odot)$ *is a semiring, where* $R = \mathbb{R} \cup \{-\infty\}$, $a \oplus b$ *is the maximum of a and b, and $\odot$ is the usual addition. The dual tropical semiring can be obtained from the tropical semiring by multiplying each element by -1.*

The exponentiated tropical semiring $(R, \oplus, \odot)$ *is a semiring, where* $R = \mathbb{R}^+ \cup \{\infty\}$, $a \oplus b$ *is the minimum of a and b, and $\odot$ is the usual multiplication. The exponentiated tropical semiring can be obtained from the tropical semiring by taking the exponent of each element.*

The dual exponentiated tropical semiring $(R, \oplus, \odot)$ *is a semiring, where* $R = \mathbb{R}^+ \cup \{0\}$, $a \oplus b$ *is the maximum of a and b, and $\odot$ is the usual multiplication. We will denote $\mathbb{R}^+ \cup \{0\}$ by $\mathbb{R}^{\geq 0}$. The dual exponentiated tropical semiring can be obtained from the tropical semiring by first multiplying each element by -1 and then taking the exponent of it.*

2.1.3.4 Counting the total sum of weights

It is also possible to count the sum of the coin weights in all possible coin sequences that sum up to a certain value. Assume that there is a weight function $w : C \to \mathbb{R}^+$. The weights might be arbitrary, even irrational numbers, thus it would be computationally intractable to count the sequences with each possible total sum. Instead, we can define the commutative ring $((\mathbb{N}, \mathbb{R}), \oplus, \odot)$ where

$$(n_1, w_1) \odot (n_2, w_2) := (n_1 n_2, w_1 n_2 + w_2 n_1) \tag{2.62}$$

and

$$(n_1, w_1) \oplus (n_2, w_2) := (n_1 + n_2, w_1 + w_2). \tag{2.63}$$

It is easy to see that both operations are commutative and associative and the distributive rule

$$r_1 \odot (r_2 \oplus r_3) = (r_1 \odot r_2) \oplus (r_1 \odot r_3) \tag{2.64}$$

holds for any $r_1, r_2, r_3 \in (\mathbb{N}, \mathbb{R})$. Here the first coordinate is the number of coin sequences and the second coordinate is the total sum of weights. For any two sets of coin sequences, both the number of sequences and the sum of the weights are added in the union of the two sets. For the operation $\circ_{c_i}$, the associated function T_i is the multiplication with $(1, w_i)$, where w_i is the weight of c_i. Indeed, if A is a set of coin sequences then the total sum of weights in $\circ_{c_i}(A)$ is $w + nw_i$, where w is the total sum of weights in A and n is the number of sequences in A. Furthermore, the number of sequences in $\circ_{c_i}(A')$ is still n'. Therefore the appropriate recursion is

$$F(S(x)) = \oplus_{\substack{i \in [1,k] \\ x - c_i \geq 0}} F(S(x - c_i)) \odot (1, w_i). \tag{2.65}$$

The empty sum in this ring is $(0, 0)$, since $(0, 0)$ is the additive unit.

2.1.3.5 Counting the coin sequences when the order does not count

When the order of the coins does not count, the base set A must be changed (and so the yield algebra), since several coin sequences contain the same coins, just in different order. In this case, the base set A must contain only the non-decreasing coin sequences. $(\Theta, \leq)$ contains the pairs $\mathbb{N} \times C$, where C is the set of coin values, and the partial ordering is coordinatewise, that is, for any two members of Θ, $(n_1, c_1) \leq (n_2, c_2)$ if and only if $n_1 \leq n_2$ and $c_1 \leq c_2$. The p function is

$$p(c_1 c_2 \ldots c_n) := \left(\sum_{i=1}^{n} c_i, c_n \right). \tag{2.66}$$

The operator set $\mathcal{O}$ contains the unary operators $\circ_{c_i}$ which still concatenates c_i to the end of a coin sequence. However, this operator can be applied only on the sequences that end with a coin whose value is at most c_i. It is guaranteed in the recursion of the yield algebra, since the recursion is

$$S(x, c_i) = \sqcup_{\substack{c_j \geq c_i \\ x - c_j \geq 0}} \circ_{c_i} (S(x - c_i, c_j)). \tag{2.67}$$

Once the yield algebra is obtained, several evaluation algebras can be associated to it. For example, to count the number of set of coins that sum up to a given value, the f function is the constant 1 in the evaluation algebra and each T_i operator is the identity function.

2.2 Counting, optimizing, deciding

In this section, an algebraic description is given explaining the relationship amongst counting, optimization and decision problems. We saw in previous examples that it is useful to introduce generating functions in algebraic dynamic programming. A generating function is a polynomial coming from a polynomial ring. We first define an algebraic structure that is a natural generalization of polynomial rings.

Definition 24. *Let G be a monoid, with its operation written as multiplication and let R be a semiring. The monoid semiring of G over R denoted by $R[G]$, is the set of mappings $h : G \to R$ with finite support. The addition of h_1 and h_2 is defined as $(h_1 + h_2)(x) \to h_1(x) + h_2(x)$ while the multiplication is defined as $(h_1 h_2)(x) = \sum_{uv=x} h_1(u)h_2(v)$.*

It is easy to see that $R[G]$ is indeed a semiring. If both R and G are commutative, then $R[G]$ is a commutative one. The additive unit is the constant 0 function, that is, the function that maps each element in the monoid G to the additive unit of R. The multiplicative unit is the function that maps 1_G, the unit of the monoid to 1_R, the multiplicative unit of the semiring, and any other member of G is mapped to 0.

An equivalent notation of the mappings is the formal summation

$$\sum_{g \in G} h(g)g \tag{2.68}$$

where only those members of the summation are indicated for which $h(g) \neq 0$. Both notations (mappings and formal summations) are used below.

An example for a monoid semiring is the integer polynomial ring $\mathbb{Z}[x]$. Here the monoid is the one variable free monoid generated by x. The semiring is the integer numbers. Although the integer numbers with the usual addition and multiplication form a ring, we do not use the subtractions of this ring in algebraic dynamic programming algorithms. Another example for a monoid semiring is the natural number polynomial semiring, $\mathbb{N}[x]$, that is, the integer polynomials with non-negative coefficients. 0 is considered to be a natural number to get a semiring for the usual addition and multiplication. In fact, $\mathbb{N}[x]$ is a sub-semiring of the $\mathbb{Z}[x]$ semiring, and that is the semiring what we use in algebraic dynamic programming algorithms, even if $\mathbb{Z}[x]$ is given as the algebraic structure in the evaluation algebra.

Monoid semirings over the natural number semiring can be used to build evaluation algebras if the combinatorial objects are scored by a multiplicative function based on some monoid. This is precisely stated in the following theorem.

Theorem 22. *Let $Y = (A, \Theta, \mathcal{O}, \mathcal{R})$ be a yield algebra, G a (commutative) monoid, and let $f : A \to G$ be a function, such that for any m-ary operator $\circ \in \mathcal{O}$, a function $h_\circ : \Theta^m \to G$ exists, for which the equality*

$$f\left(\circ\left((a_i)_{i=1}^m\right)\right) = h_\circ(\theta_1, \theta_2, \ldots, \theta_m) \prod_{i=1}^m f(a_i) \tag{2.69}$$

holds, where θ_i is the parameter of a_i. Then $(Y, \mathbb{N}[G], f', \mathcal{T})$ is an evaluation algebra, where

$$f'(a) := 1f(a) \tag{2.70}$$

and for each operator $\circ_i$, the corresponding function T_i is

$$T_i(f'(a_1), \ldots, f'(a_{i_m}), \theta_1, \ldots, \theta_{i_m}) = h_{\circ_i}(\theta_1, \ldots, \theta_{i_m}) \prod_{j=1}^{i_m} f'(a). \qquad (2.71)$$

Proof. It is the direct consequence of the fact that $\mathbb{N}[G]$ is a semiring, and the distributive rule holds. Indeed,

$$T_i \left(\sum_{a_{i_1} \in S(\theta_1)} f'(a_{i_1}), \ldots, \sum_{a_{i_{m_i}} \in S(\theta_{m_i})} f'(a_{i_{m_i}}); \theta_1, \ldots, \theta_{m_i} \right) =$$

$$h_{\circ_i}(\theta_1, \ldots, \theta_{m_i}) \left(\sum_{a_{i_1} \in S(\theta_1)} f'(a_{i_1}) \right) \cdots \left(\sum_{a_{i_{m_i}} \in S(\theta_{m_i})} f'(a_{i_{m_i}}) \right) =$$

$$\sum_{a_{i_1} \in S(\theta_1)} \cdots \sum_{a_{i_{m_i}} \in S(\theta_{m_i})} h_{\circ_i}(\theta_1, \ldots, \theta_{m_i}) f'(a_{i_1}) \ldots f'(a_{i_{m_i}}) =$$

$$\sum_{a_{i_1} \in S(\theta_1)} \cdots \sum_{a_{i_{m_i}} \in S(\theta_{m_i}) \in S(\theta_{m_i})} f' \left(\circ_i \left((a_{i_j})_{j=1}^{m_i} \right) \right). \qquad (2.72)$$

$\square$

Definition 25. *The evaluation algebra given in Theorem 22 is called the statistics evaluation algebra.*

In algebraic dynamic programming, two monoid semirings are of central interest, $\mathbb{N}[\mathbb{R}^+]$ and $\mathbb{N}[\mathbb{R}^\times]$, where $\mathbb{R}^+$ is the additive group of real numbers and $\mathbb{R}^\times$ is the multiplicative group of the positive numbers. Computation in these semirings might be intractable, since the number of terms in the formal summation in Equation (2.68) representing the semiring elements might grow exponentially during the recursion. However, some homomorph images might be easy to calculate. Indeed, any homomorph image indicates an evaluation algebra, stated in the following theorem.

Theorem 23. *Let $E = \{Y, \mathbb{N}[G], f, \mathcal{T}\}$ be a statistics evaluation algebra, with functions*

$$T_i(f(a_1), \ldots, f(a_{i_m}), \theta_1, \ldots, \theta_{i_m}) = h_{\circ_i}(\theta_1, \ldots, \theta_{i_m}) \prod_{j=1}^{i_m} f(a)$$

and let $\varphi : \mathbb{N}[G] \to R'$ be a semiring homomorphism. Then $E' = \{Y, R', f', \mathcal{T}'\}$ is an evaluation algebra, where

$$f'(a) := \varphi(f(a))$$

and for each operator $\circ_i$,

$$T_i'(f'(a_1),\ldots,f'(a_{i_m}),\theta_1,\ldots,\theta_{i_m}) := \varphi(h_{\circ_i}(\theta_1,\ldots,\theta_{i_m}))\prod_{j=1}^{i_m} f'(a_j).$$

Furthermore, E' calculates $F'(S(\theta)) = \varphi(F(S(\theta)))$.

Proof. It is the direct consequence of the definition of homomorphism. Indeed, if

$$a = \circ_i\left((a_j)_{j=1}^{m_i}\right)$$

then

$$f'(a) = \varphi(f(a)) = \varphi\left(h_{\circ_i}(\theta_1,\ldots,\theta_{i_m})\prod_{j=1}^{i_m} f(a_j)\right) =$$

$$\varphi(h_{\circ_i}(\theta_1,\ldots,\theta_{i_m}))\prod_{j=1}^{i_m}\varphi(f(a_j)) = \varphi(h_{\circ_i}(\theta_1,\ldots,\theta_{i_m}))\prod_{j=1}^{i_m} f'(a) \quad (2.73)$$

and the distributive rule holds, since

$$T_i\left(\sum_{a_{i_1}\in S(\theta_1)} f'(a_{i_1}),\ldots,\sum_{a_{i_{m_i}}\in S(\theta_{m_i})} f'(a_{i_{m_i}});\theta_1,\ldots,\theta_{m_i}\right) =$$

$$\varphi(h_{\circ_i}(\theta_1,\ldots,\theta_{m_i}))\left(\sum_{a_{i_1}\in S(\theta_1)} f'(a_{i_1})\right)\cdots\left(\sum_{a_{i_{m_i}}\in S(\theta_{m_i})} f'(a_{i_{m_i}})\right) =$$

$$\sum_{a_{i_1}\in S(\theta_1)}\cdots\sum_{a_{i_{m_i}}\in S(\theta_{m_i})}\varphi(h_{\circ_i}(\theta_1,\ldots,\theta_{m_i}))f'(a_{i_1})\ldots f'(a_{i_{m_i}}) =$$

$$\sum_{a_{i_1}\in S(\theta_1)}\cdots\sum_{a_{i_{m_i}}\in S(\theta_{m_i})} f'\left(\circ_i\left((a_{i_j})_{j=1}^{m_i}\right)\right). \quad (2.74)$$

It is also clear that

$$F'(S(\theta)) = \sum_{a\in S(\theta)} f'(a) = \sum_{a\in S(\theta)} \varphi(f(a)) =$$

$$\varphi\left(\sum_{a\in S(\theta)} f(a)\right) = \varphi(F(S(\theta))). \quad (2.75)$$

$\square$

Below we define three homomorphisms. These homomorphisms construct evaluation algebras, and the constructed evaluation algebras solve the counting, minimizing and decision problems for the same yield algebra.

(a) $(Y, \mathbb{N}[G], f, \mathcal{T})$ is a statistics evaluation algebra, $\varphi : \mathbb{N}[G] \to \mathbb{N}$, and

$$\varphi(h) = \sum_{g \in G} h(g).$$

(Recall that $h \in \mathbb{N}[G]$ is a mapping from G to $\mathbb{N}$.) It is indeed a semiring homomorphism, since

$$\varphi(h_1 + h_2) = \sum_{g \in G} (h_1 + h_2)(g) =$$

$$\sum_{g \in G} h_1(g) + \sum_{g \in G} h_2(g) = \varphi(h_1) + \varphi(h_2) \qquad (2.76)$$

and

$$\varphi(h_1 h_2) = \sum_{g \in G} (h_1 h_2)(g) = \sum_{g \in G} \sum_{g_1 g_2 = g} h_1(g_1) h_2(g_2) =$$

$$\left(\sum_{g_1 \in G} h_1(g_1) \right) \left(\sum_{g_2 \in G} h_2(g_2) \right) = \varphi(h_1) \varphi(h_2). \qquad (2.77)$$

The homomorph image calculates $\varphi(F(S(\theta)))$, that is, the size of $S(\theta)$.

(b) $(Y, \mathbb{N}[\mathbb{R}^+], f, \mathcal{T})$ is a statistics evaluation algebra, $\varphi : \mathbb{N}[\mathbb{R}^+] \to \mathcal{R}$, where $\mathcal{R}$ is the tropical semiring, and

$$\varphi(h) = \min\{supp(h)\}$$

where
$$supp(h) := \{g \in \mathbb{R} | h(g) \neq 0\}.$$

(Recall that $h \in \mathbb{N}[\mathbb{R}^+]$ is a mapping from $\mathbb{R}^+$ to $\mathbb{N}$.) When the support of h is the empty set, $\varphi(h) = +\infty$, the additive unit of $\mathcal{R}$. It is indeed a semiring-homomorphism, since

$$\varphi(h_1 + h_2) = \min\{supp(h_1 + h_2)\} =$$
$$\min\{supp(h_1)\} \oplus \min\{supp(h_2)\} = \varphi(h_1) \oplus \varphi(h_2) \qquad (2.78)$$

and

$$\varphi(h_1 h_2) = \min\{supp(h_1 h_2)\} =$$
$$\min\{supp(h_1)\} \odot \min\{supp(h_2)\} = \varphi(h_1) \odot \varphi(h_2). \qquad (2.79)$$

This homomorph image calculates $\varphi(F(S(\theta)))$, that is, the minimal value in $S(\theta)$.

Similar construction exists with evaluation algebra $\{Y, \mathbb{N}[\mathbb{R}^\times] f, \mathcal{T}\}$.

(c) $(Y, \mathrm{N}[G], f, \mathcal{T})$ is a statistics evaluation algebra, $\varphi : \mathrm{N}[G] \rightarrow (\{0, 1\}, \vee, \wedge)$, where the constant 0 function (the additive unit of $\mathrm{N}[G]$) is mapped to 0 and all other members of $\mathrm{N}[G]$ are mapped to 1. Here $(\{0, 1\}, \vee, \wedge)$ is the Boolean algebra with two elements. φ is a homomorphism, in fact, any mapping from any semiring to the two element Boolean algebra is a homomorphism if the additive unit is mapped to 0 and all other elements are mapped to 1.

The homomorph image calculates $\varphi(F(S(\theta)))$, which is 0 if $S(\theta) = \emptyset$ and 1 otherwise. Therefore, with this evaluation algebra, we can decide if an object with parameter θ exists.

As can be seen from these examples, if a yield algebra $\{A, \Theta, p, \mathcal{O}, \mathcal{R}\}$ exists on some objects A, and f is an additive function on reals or a multiplicative function on the positive real numbers in the sense of Equation (2.69), then counting, minimizing and deciding if an object exists with parameter θ are similar computational problems, in the sense that they can be calculated with evaluation algebras that are homomorph images of the same statistics evaluation algebra.

If the function of the objects is a multiplicative function on real numbers, then calculating the weighted sum (instead of just counting the objects in some set $S(\theta)$) can be obtained with a homomorphism easily. The homomorphism $\varphi : \mathrm{N}[\mathbb{R}^\times] \rightarrow \mathbb{R}$ is simply

$$\varphi(h) = \sum_{g \in \mathbb{R}} gh(g).$$

It is indeed a homomorphism, since

$$\varphi(h_1 + h_2) = \sum_{g \in \mathbb{R}} g(h_1 + h_2)(g) =$$

$$\sum_{g \in \mathbb{R}} gh_1(g) + \sum_{g \in \mathbb{R}} gh_2(g) = \varphi(h_1) + \varphi(h_2) \tag{2.80}$$

and

$$\varphi(h_1 h_2) = \sum_{g \in \mathbb{R}} g(h_1 h_2)(g) = \sum_{g \in \mathbb{R}} g \sum_{g_1 g_2 = g} h_1(g_1) h_2(g_2) =$$

$$\left(\sum_{g_1 \in \mathbb{R}} g_1 h_1(g_1) \right) \left(\sum_{g_2 \in \mathbb{R}} g_2 h_2(g_2) \right) = \varphi(h_1) \varphi(h_2) \tag{2.81}$$

The homomorph image calculates $\varphi(F(S(\theta)))$, which is indeed the weighted sum of the objects.

However, if the monoid semigroup is $\mathrm{N}[\mathbb{R}^+]$, then

$$\varphi(h) = \sum_{g \in \mathbb{R}} gh(g)$$

is not a homomorphism. Instead, the ring introduced in Subsection 2.1.3.4 can be used. Let R denote the ring defined in Subsection 2.1.3.4, and define $\varphi : \mathbb{N}[\mathbb{R}^+] \to R$ as

$$\varphi(h) = \left(\sum_{g \in \mathbb{R}} h(g), \sum_{g \in \mathbb{R}} gh(g) \right).$$

It is indeed a homomorphism, since

$$\varphi(h_1 + h_2) = \left(\sum_{g \in \mathbb{R}} (h_1 + h_2)(g), \sum_{g \in \mathbb{R}} g(h_1 + h_2)(g) \right) =$$

$$\left(\sum_{g \in \mathbb{R}} h_1(g) + \sum_{g \in \mathbb{R}} h_2(g), \sum_{g \in \mathbb{R}} gh_1(g) + \sum_{g \in \mathbb{R}} gh_2(g) \right) =$$

$$\varphi(h_1) + \varphi(h_2) \tag{2.82}$$

and

$$\varphi(h_1 h_2) = \left(\sum_{g \in \mathbb{R}} (h_1 h_2)(g), \sum_{g \in \mathbb{R}} g(h_1 h_2)(g) \right) =$$

$$\left(\sum_{g \in \mathbb{R}} \sum_{g_1 + g_2 = g} h_1(g_1) h_2(g_2), \sum_{g \in \mathbb{R}} g \sum_{g_1 + g_2 = g} h_1(g_1) h_2(g_2) \right) =$$

$$\left(\left(\sum_{g_1 \in \mathbb{R}} h_1(g_1) \right) \left(\sum_{g_2 \in \mathbb{R}} h_2(g_2) \right), \sum_{g \in \mathbb{R}} \sum_{g_1 + g_2 = g} (g_1 + g_2) h_1(g_1) h_2(g_2) \right) =$$

$$\left(\left(\sum_{g_1 \in \mathbb{R}} h_1(g_1) \right) \left(\sum_{g_2 \in \mathbb{R}} h_2(g_2) \right), \sum_{g_1 \in \mathbb{R}} \sum_{g_2 \in \mathbb{R}} (g_1 + g_2) h_1(g_1) h_2(g_2) \right) =$$

$$\left(\sum_{g_1 \in \mathbb{R}} h_1(g_1), \sum_{g_1 \in \mathbb{R}} g_1 h_1(g_1) \right) \left(\sum_{g_2 \in \mathbb{R}} h_2(g_2), \sum_{g_2 \in \mathbb{R}} g_2 h_2(g_2) \right) =$$

$$\varphi(h_1)\varphi(h_2). \tag{2.83}$$

The homomorph image calculates

$$\varphi(F(S(\theta)) = \left(|S(\theta)|, \sum_{a \in S(\theta)} f(a) \right)$$

thus, weighted sums can also be calculated when the function f is additive on the objects.

This idea can be extended to an arbitrary commutative ring, and even higher-order moments can be calculated. A commutative ring $(R^m, \oplus, \odot)$ is introduced. R is an arbitrary commutative ring with a multiplicative unit, $\oplus$ is the coordinate wise addition and $\odot$ is the multiplication defined in the following way. When two elements $(a_0, a_1, \ldots, a_{m-1})$ and $(b_0, b_1, \ldots, b_{m-1})$ are multiplied, the product at the kth coordinate contains

$$\sum_{i=0}^{k} \binom{k}{i} a_{k-i} b_i. \tag{2.84}$$

This multiplication is associative since

$$\sum_{i=0}^{k} \binom{k}{i} a_{k-i} \sum_{j=0}^{i} \binom{i}{j} b_{i-j} c_j =$$

$$= \sum_{0 \le j \le i \le k} \binom{k}{j, i-j, k-i} a_{k-i} b_{i-j} c_j =$$

$$= \sum_{i=0}^{k} \binom{k}{i} \left(\sum_{j=0}^{i} \binom{i}{j} a_{i-j} b_j \right) c_{k-i}. \tag{2.85}$$

It is easy to see that the distributive rule also holds. This commutative ring can be used to calculate the moments of an additive function over an ensemble of combinatorial objects, A, on which a yield algebra exists where all operations are binary. An additive function is a function g such that for all binary operations $\circ_i$,

$$g(a \circ_i b) = g(a) + g(b) \tag{2.86}$$

where the addition is in the ring R. If $x = (p^0, p^1, \ldots, p^{m-1})$ and $y = (q^0, q^1, \ldots, q^{m-1})$, then indeed

$$x \odot y = \left((p+q)^0, (p+q)^1, \ldots, (p+q)^{m-1} \right). \tag{2.87}$$

Hence, if $f(a) = \left(g(a)^0, g(a)^1, \ldots, g(a)^{m-1} \right)$ then

$$f(a \circ_i b) = f(a) \odot f(b) \tag{2.88}$$

namely, each T_i function is the $\odot$ multiplication, which is naturally distributive with respect to the $\oplus$ addition. The evaluation algebra calculates

$$F(S(\theta)) = \left(\sum_{a \in S(\theta)} g(a)^0, \sum_{a \in S(\theta)} g(a)^1, \ldots, \sum_{a \in S(\theta)} g(a)^{m-1} \right). \tag{2.89}$$

An example for computing higher-order moments is given at the end of Subsection 2.3.3.

Optimization problems can also be extended to further cases. The usual multiplication with non-negative real numbers can be considered as tropical powering, and tropical powering also satisfies the distributive rule. Indeed, if for an operator $\circ_i$, the f function satisfies the equation

$$f\left(\circ_i\left((a_j)_{j=1}^{m_i}\right)\right) = c_{i,0} + \sum_{j=1}^{m_i} c_{i,j} f(a_j)$$

$$[= T_i(f(a_1), \ldots, f(a_{m_i}), p(a_1), \ldots, p(a_{m_i}))] \tag{2.90}$$

where each $c_{i,j}$ is non-negative (and might depend on parameters $p(a_j)$), then it also holds that

$$T_i(\min_{a_1 \in S(\theta_1)} f(a_1), \ldots, \min_{a_{m_i} \in S(\theta_{m_i})} f(a_{m_i}), \theta_1, \ldots, \theta_{m_i}) =$$

$$\min_{a_1 \in S(\theta_1)} \cdots \min_{a_{m_i} \in S(\theta_{m_i})} f\left(\circ_i\left((a_j)_{j=1}^{m_i}\right)\right). \tag{2.91}$$

Another interesting case is when R is a distributive lattice, and for each operator $\circ_i$, $f\left(\circ_i\left((a_j)_{j=1}^{m_i}\right)\right)$ can be described with operations in R. An evaluation algebra can be built in this case, since the distributive rules

$$\bigvee_{k,l}(f(a_k) \vee f(a_l)) = \bigvee_k f(a_k) \vee \bigvee_l f(a_l) \tag{2.92}$$

and

$$\bigvee_{k,l}(f(a_k) \wedge f(a_l)) = \bigvee_k f(a_k) \wedge \bigvee_l f(a_l) \tag{2.93}$$

as well as the dual equalities hold. A special case is when R is the set of real numbers, $\vee$ is the maximum and $\wedge$ is the minimum. The so-obtained (min,max)-semiring can also be used in optimization problems, for example, finding the highest vehicle that can travel from some point A to point B on a network of roads containing several bridges with different height (see Example 6).

2.3   The zoo of counting and optimization problems solvable with algebraic dynamic programming

In this section, we give a large ensemble of combinatorial objects on which evaluation algebras can be constructed. We also define several computational problems and provide evaluation algebras to solve them.

2.3.1   Regular grammars, Hidden Markov Models

Transformational grammars have been invented by Noam Chomsky [41]. Regular grammars are one of the simplest transformational grammars as they are in the lowest level of the Chomsky hierarchy [42]. Stochastic versions of regular grammars are related to Hidden Markov Models [163, 15].

Definition 26. *A regular grammar is a tuple* $(T, N, \mathbb{S}, R)$, *where* T *is a finite set called terminal characters,* N *is a finite set called non-terminal characters. The sets* T *and* N *are disjoint.* $\mathbb{S} \in N$ *is a special non-terminal character, called starting non-terminal.* R *is a finite set of rules, each in one of the following forms*

$$W \quad \rightarrow \quad xW' \tag{2.94}$$
$$W \quad \rightarrow \quad x \tag{2.95}$$
$$W \quad \rightarrow \quad \epsilon \tag{2.96}$$

where $W, W' \in N$, $x \in T$, ϵ *denotes the empty sequence. The shorthand for the rules (2.94)-(2.96) is*

$$W \rightarrow xW' \mid x \mid \epsilon.$$

A generation is a finite series of transformations

$$\mathbb{S} = X_0 \rightarrow X_1 \rightarrow \ldots \rightarrow X_k \in T^* \tag{2.97}$$

where for each $i = 1, \ldots, k - 1$, *there exists a rule* $W \rightarrow \beta$ *and a word* $X_p \in T^*$ *such that* $X_i = X_p W$ *and* $X_{i+1} = X_p \beta$. *(Here* β *might be any sequence appearing at the right-hand side of a rewriting rule.)* T^* *denotes the finite sequences from* T. *The language* $L_G \subseteq T^*$ *contains those sequences that can be generated by the grammar. A grammar is said to be* unambiguous *if any sequence* $X \in L_G$ *can be generated in exactly one way. An* ambiguous *grammar contains at least one sequence in its language that can be generated in at least two different ways.*

A generation is possible if there is a non-terminal in the intermediate sequence. Once the sequence contains only terminal characters, the generation is terminated. This is the rationale behind the naming of terminals and non-terminals.

Given a sequence $X \in T^*$, the following questions can be asked:

(a) Is $X \in L_G$?

(b) How many generations are there which produce X?

(c) Given a function $w : R \rightarrow \mathbb{R}^{\geq 0}$, which generation $\mathcal{G}$ maximizes

$$\prod_{i=0}^{k-1} w(W_i \rightarrow \beta_i)$$

where the rewriting rule $W_i \rightarrow \beta_i$ is applied in the ith step of the generation $\mathcal{G}$ generating X in k steps?

(d) Given a function $w : R \to \mathbb{R}^{\geq 0}$, compute

$$\sum_{\mathcal{G}_i} \prod_{j=0}^{k_i-1} w(W_{i,j} \to \beta_{i,j})$$

where the rewriting rule $W_{i,j} \to \beta_{i,j}$ is applied in the jth step of the generation $\mathcal{G}_i$ generating X in k_i steps.

These questions can be answered using the same yield algebra and different evaluation algebras. The yield algebra builds the possible generations of intermediate sequences. Note that in any generation of any regular grammar, each intermediate sequence appearing in the sequence of generations is in form YW, where $Y \in T^*$ and $W \in N$. The yield algebra is the following. The set A contains the possible generations of intermediate sequences $X_i W$, where X_i denotes the prefix of X of length i. The parameters are pairs (i, W) denoting the length of the prefix and the current non-terminal character. For $i = |X|$, a parameter (i, ϵ) is also considered. This parameter describes the set of possible generations of X. In the partial ordering of the parameters, $(i_1, W_1) \leq (i_2, W_2)$ if $i_1 \leq i_2$. For each rewriting rule $W \to \beta$, there is a unary operation $\circ_{W \to \beta}$ extending the rewriting with a new rewriting $W \to \beta$. The recursions are

$$S((i, W)) \quad = \quad \bigsqcup_{W' | (W' \to x_i W) \in R} \circ_{W' \to x_i W} \left(S((i - 1, W')) \right) \qquad (2.98)$$

$$S((i, \epsilon)) \quad = \quad \left(\bigsqcup_{W | (W \to x_i) \in R} \circ_{W \to x_i} \left(S((i - 1, W)) \right) \right) \sqcup$$
$$\bigsqcup_{W | (W \to \epsilon) \in R} \circ_{W \to \epsilon} \left(S((i, W)) \right) \qquad \text{for } i = |X| \quad (2.99)$$

with the initial condition $S((0, \mathbb{S})) = \{\mathbb{S}\}$, namely, the set contains $\mathbb{S}$ as the rewriting sequence containing no rewriting steps, and $S((0, W)) = \emptyset$ for all $W \neq \mathbb{S}$.

For the given problems, the following evaluation algebras can be constructed.

(a) The semi-ring is the Boolean semiring $(\{0, 1\}, \vee, \wedge)$. The function f is the constant 1. Each function $T_{\alpha \to \beta}$ is the identity. The answer for the decision question is "yes" if $F(S((|X|, \epsilon))) = 1$, and "no" if $F(S((|X|, \epsilon))) = 0$. This latter can happen if $S((|X|, \epsilon)) = \emptyset$, since the empty sum in the Boolean semiring is 0, being the additive unit.

(b) The semi-ring is $\mathbb{Z}$. The f function is the constant 1. Each function $T_{\alpha \to \beta}$ is the identity function. The number of possible generations is $F(S((|X|, \epsilon)))$.

(c) The semiring is the dual exponentiated tropical semiring, $(\mathbb{R}^{\geq 0}, \max, \cdot)$.

The f function for a generation $\mathcal{G}$ is

$$\prod_{i=1}^{k-1} w(W_i \to \beta_i).$$

Each $T_{W \to \beta}$ is the multiplication with $w(W \to \beta))$. The maximum score is $F(S((|X|, \epsilon)))$.

(d) The semiring is $(\mathbb{R}^{\geq 0}, +, \cdot)$. The f function for a generation $\mathcal{G}$ is

$$\prod_{i=1}^{k-1} w(W_i \to \beta_i).$$

Each $T_{W \to \beta}$ is the multiplication with $w(W \to \beta))$. The sum of the scores over all possible generations is $F(S((|X|, \epsilon)))$.

If the grammar is unambiguous, then the following counting problem can be also solved: Given a regular grammar $G = \{T, N, \mathbb{S}, R\}$ and a series $\tau_1, \tau_2, \ldots, \tau_n$, where $\forall \tau_i \subseteq T$, how many sequences $X = x_1 x_2 \ldots x_n$ exist in the language of G, such that $\forall x_i \in \tau_i$? The yield algebra builds the possible generations of intermediate sequences YW such that $\forall y_i \in \tau_i$ and $W \in N$. The same parameters and operations can be used, and the recursions are

$$S((i, W)) = \bigsqcup_{x \in \tau_i} \bigsqcup_{W' | (W' \to xW) \in R} {}^{\circ} W' \to xW \left(S((i-1, W')) \right) \quad (2.100)$$

$$S((n, \epsilon)) = \left(\bigsqcup_{x \in \tau_n} \bigsqcup_{W | (W \to x) \in R} {}^{\circ} W \to x \left(S((n-1, W)) \right) \right) \sqcup$$
$$\bigsqcup_{W | (W \to \epsilon) \in R} {}^{\circ} W \to \epsilon \left(S((n, W)) \right) \quad (2.101)$$

with the same initial conditions as above. The evaluation algebra is the standard one for counting the size of the sets (case (b) above). $F(S((n, \epsilon)))$ counts the number of possible *generations* that produce a sequence X, such that $\forall x_i \in \tau_i$. However, since the grammar is unambiguous, this number is also the number of sequences satisfying the prescribed conditions.

On the other hand, the same counting problem is #P-complete for ambiguous grammars. This will be proven in Chapter 4.

When each $\tau_i = T$, the given algebraic dynamic programming algorithm counts the number of sequences of length n in the language that the grammar generates. This is demonstrated with the following example.

Example 4. *Compute the number of sequences of length n from the alphabet $\{a, b\}$ that contains an even number of a's and an odd number of b's.*

Solution. The following unambiguous regular grammar generates those sequences. $T = \{a, b\}$, $N = \{W_{ee}, W_{eo}, W_{oe}, W_{oo}\}$, $\mathbb{S} = W_{ee}$, and the rewriting rules are

$$W_{ee} \to aW_{oe} \mid bW_{eo} \quad (2.102)$$
$$W_{eo} \to aW_{oo} \mid bW_{ee} \mid \epsilon \quad (2.103)$$
$$W_{oe} \to aW_{ee} \mid bW_{oo} \quad (2.104)$$
$$W_{oo} \to aW_{eo} \mid bW_{oe}. \quad (2.105)$$

Indeed, here e stands for even, and o stands for odd, and the two characters in the index of the nonterminals tell the parity of the number of a's and b's generated so far. For example, W_{oe} denotes that so far an odd number of a's and an even number of b's have been generated, etc. W_{ee} is indeed the starting non-terminal, since at the beginning, 0 number of characters has been generated and 0 is an even number. The generation can be stopped when an even number of a's and an odd number of b's have been generated, as indicated by the $W_{eo} \to \epsilon$ rule. ∎

As can be seen, non-terminals play the role of "memory" in generating sequences. Above the non-terminals at the end of the intermediate sequences, the generation is memoryless. We can consider the stochastic version of regular grammars, and the memoryless property provides that the stochastic versions of regular grammars are Markov processes. The stochastic regular grammars can be defined in the following way.

Definition 27. *A stochastic regular grammar is a tuple* $(T, N, \mathbb{S}, R, \pi)$, *where* $(T, N, \mathbb{S}, R)$ *is a regular grammar and* $\pi : R \to \mathbb{R}^+$ *is a probability distribution for each non-terminal, that is, for each* $W \in N$, *the equality*

$$\sum_{\beta \mid (W \to \beta) \in R} \pi(W \to \beta) = 1 \qquad (2.106)$$

holds. (Recall that β *is any of the sequences that might appear in the right-hand side of a rewriting rule of a regular grammar, including the empty sequence.) A stochastic regular grammar makes random generations*

$$S = X_1 \to X_2 \to \dots$$

where in each rewriting step, the rewriting rule $W \to \beta$ *is chosen randomly following the distribution* π.

The random generation in a stochastic regular grammar can be viewed as a random process, in which the states are the intermediate sequences. This process indeed has the Markovian property. That is, what the intermediate sequence X_{i+1} is depends only on X_i and does not depend on any $X_j, j < i$. One can ask for a given sequence $X \in T^*$ what the most likely generation and the total probability of the generation are. This latter is the sum of the probabilities of the possible generations. Both questions can be answered by the algebraic dynamic programming algorithms using the appropriate evaluation grammars described above for a general w function.

Stochastic regular grammars are closely related to Hidden Markov Models.

Definition 28. *A Hidden Markov Model is a tuple* $(\vec{G}, START, END, \Gamma, T, e)$, *where* $\vec{G} = (V, E)$ *is a directed graph in which loops are allowed but parallel edges are not, START and END are distinguished vertices in* $\vec{G}$, *START has 0 in-degree, END has 0 out-degree,* Γ *is a finite set of symbols, called an* alphabet, $T : E \to \mathbb{R}^+$ is the transition probability function satisfying for all $u \neq END$

$$\sum_{v|(u,v)=e\in E} T(e) = 1.$$

and $e : \Gamma \times (V \setminus \{START, END\}) \to \mathbb{R}^{\geq 0}$ is the emission probability function satisfying for all $v \in V \setminus \{START, END\}$

$$\sum_{x\in\Gamma} e(x, v) = 1$$

The vertices of $\vec{G}$ are called states. A random walk on the states is defined by $\vec{G}$ and the transition probabilities. The random walk starts in the state $START$ and ends in the state END. During such a walk, states emit characters according to the emission distribution e. In case of loops, the random walk might stay in one state for several steps. A random character is emitted in each step. The process is hidden in the sense that an observer can see only the emitted characters and cannot observe the random walk itself. An emission path is a random walk together with the emitted characters. The probability of an emission path is the product of its transition and emission probabilities.

If (u, v) is an edge, then the notation $T(v|u)$ is also used, emphasizing that T is a conditional distribution. Indeed $T(v|u)$ is the probability that the Markov process will be in state v in the next step given that it is in the state u in the current step.

One can ask, for an $X \in \Gamma^*$, what the most likely emission path and the total emission probability are. This latter is the sum of the emission path probabilities that emit X. These questions are equivalent with those that can be asked for the most likely generation and the total generation probability of a sequence in a stochastic regular grammar, stated by the following theorem.

Theorem 24. *For any Hidden Markov Model $H = (\vec{G}, START, END, \Gamma, T, e)$, there exists a stochastic regular grammar $G = (T, N, \mathbb{S}, R, \pi)$ such that $\Gamma = T$, L_G is exactly the set of sequences that H can emit, and for any $X \in L_G$, the probability of the most likely generation in G is the probability of the most likely emission path in H, and the total generation probability of X in G is the total emission probability of X in H. Furthermore, the running time needed to construct G is a polynomial function of the size of $\vec{G}$.*

Proof. For $\vec{G} = (V, E)$, let the non-terminals of the regular grammar correspond to V. That is, for any $v \in V$, there is a non-terminal W_v. The starting non-terminal $\mathbb{S} = W_{START}$. For each $(u, v) \in E$, $v \neq END$ and $x \in \Gamma$ such that $e(v, x) \neq 0$, construct a rewriting rule

$$W_u \to xW_v$$

with probability $T(v|u)e(v, x)$ and for each $(u, END) \in E$ construct a rewriting rule

$$W_u \to \epsilon$$

with probability $T(END|u)$. This is indeed a probability distribution, since for any u

$$\sum_{\beta | W_u \to \beta} \pi(W_u \to \beta) = \sum_{\substack{v \neq END, \\ (u,v) \in E}} \sum_{x | e(x,v) \neq 0} \pi(W_u \to xW_v) + \pi(W_u \to \epsilon)$$

$$= \sum_{\substack{v \neq END, \\ (u,v) \in E}} \sum_{x | e(x,v) \neq 0} T(v|u)e(v,x) + T(END|u)$$

$$= \sum_{\substack{v \neq END, \\ (u,v) \in E}} T(v|u) \sum_{x | e(x,v) \neq 0} e(v,x) + T(END|u)$$

$$= \sum_{\substack{v \neq END, \\ (u,v) \in E}} T(v|u) + T(END|u) = 1 \qquad (2.107)$$

where $\pi(W_u \to \epsilon) = T(END|u) = 0$ if there is no edge going from u to END. There is a bijection mapping an emission path

$$START \to u_1 \to u_2 \to \ldots \to u_n \to END$$

emitting sequence $X = x_1 x_2 \ldots x_n$ to the generation

$$W_{START} \to x_1 W_{u_1} \to x_1 x_2 W_{u_2} \to \ldots \to x_1 x_2 \ldots x_n W_{u_n} \to x_1 x_2 \ldots x_n.$$

This bijection proves that the set of emittable sequences is indeed L_G. Furthermore, it is easy to check that the bijection keeps the probabilities, therefore, the most likely generation of a sequence X is the most likely emission of X in H and the total probability of generating X is the total emission probability of X in H. □

From this proof, it is also clear that there is a yield algebra building the possible emission paths of a sequence X, furthermore, for any series of $\tau_1, \tau_2, \ldots, \tau_n$, ($\tau_i \subseteq \Gamma$), there is a yield algebra building the emission paths generating such $Y = y_1 y_2 \ldots y_n$ sequences that for each i, $y_i \in \tau_i$. Therefore, similar problems will be tractable for HMMs than for stochastic regular grammars. The algorithms solving these problems are well known in the scientific literature. The algorithm finding the most likely emission path is known as the Viterbi algorithm [180, 69] and the algorithm summing the probabilities of possible emission paths is called the Forward algorithm [14, 16]. It is also well-known in the Hidden Markov Model literature that "the Viterbi algorithm is similar [...] in implementation to the forward calculation" [145].

2.3.2   Sequence alignment problems, pair Hidden Markov Models

The sequence alignment problem was first considered by two biologists, Saul B. Needleman and Christian D. Wunch [137]. They developed a dynamic programming algorithm to find an optimal alignment of two sequences and used this algorithm to infer the relationship between two protein sequences. The mathematically rigorous description of sequence alignment was published by Peter H. Shellers [153]. Sequence alignment methods have been the central procedures of bioinformatics.

Definition 29. *A sequence alignment of sequences* $X, Y \in \Gamma^*$ *is a* $2 \times L$ *table filled in with characters from* $\Gamma \cup \{-\}$ *where* $- \notin \Gamma$, *called the gap character, satisfies the following rules:*

(a) there is no column containing two gap characters and

(b) the non-gap characters in the first row form sequence X, *and the non-gap characters in the second row form sequence* Y.

The $\genfrac{}{}{0pt}{}{-}{y}$ *columns are called* insertions, *the* $\genfrac{}{}{0pt}{}{x}{-}$ *columns are called* deletions, *and the* $\genfrac{}{}{0pt}{}{x}{y}$ *columns are called* matches *if* $x = y$ *and* mismatches *or substitutions if* $x \neq y$. *The joint name of insertions and deletions is* gap.

The minimum length of a sequence alignment of sequences X and Y is $\max\{|X|, |Y|\}$ and the maximum length is $|X| + |Y|$. It is easy to show that the number of alignments of two sequences with length n and m is

$$\sum_{i=0}^{\min\{n,m\}} \frac{(n + m - i)!}{i!(m - i)!(n - i)!}, \tag{2.108}$$

see also Exercise 13. A subset of alignments called substitution-free alignments consists of those in which for any alignment column containing characters x and y, if x is not a gap symbol and y is not a gap symbol then $x = y$. It is easy to build a yield algebra of these alignments. The base set A contains such alignments for all possible prefixes of the sequences X and Y. The parameters (i, j) indicate the length of the prefixes. In the partial ordering of the parameters, $(i_1, j_1) \leq (i_2, j_2)$ if $i_1 \leq i_2$ and $j_1 \leq j_2$. The unary operators $\circ_{\substack{x\\x}}, \circ_{\substack{x\\-}}$ and $\circ_{\substack{-\\y}}$ extend the alignment with the alignment column $\genfrac{}{}{0pt}{}{x}{x}$, $\genfrac{}{}{0pt}{}{x}{-}$ and $\genfrac{}{}{0pt}{}{-}{y}$, respectively. The recursions are

$$S((i, j)) = \circ_{\substack{x_i\\-}} (S((i - 1, j))) \sqcup \circ_{\substack{-\\y_j}} (S((i, j - 1))) \quad \text{if } x_i \neq y_j \tag{2.109}$$

$$S((i, j)) = \circ_{\substack{x_i\\y_j}} (S((i - 1, j - 1))) \sqcup$$

$$\circ_{\substack{x_i\\-}} S((i - 1, j)) \sqcup \circ_{\substack{-\\y_j}} S((i, j - 1)) \quad \text{if } x_i = y_j \tag{2.110}$$

with the initial condition $S((0, 0)) = \left\{ \genfrac{}{}{0pt}{}{\epsilon}{\epsilon} \right\}$, where $\genfrac{}{}{0pt}{}{\epsilon}{\epsilon}$ is the empty alignment.

This yield algebra can be used to solve the following optimization problems with appropriate evaluation algebras:

(a) **Longest common subsequence** The subsequences of a sequence $X = x_1 x_2 \ldots x_n$ are the sequences $x_{i_1} x_{i_2} \ldots x_{i_m}$ where for all $k = 1, \ldots, m-1$, $i_k < i_{k+1}$. Sequence Z is a common subsequence of X and Y if Z is a subsequence of X and a subsequence of Y. Any substitution-free alignment corresponds to a common subsequence: the subsequence containing the characters of the $\frac{x}{x}$ columns (one character from each column, obviously), and *vice versa*, any common subsequence can be represented with a substitution-free alignment. There is no bijection between the common subsequences and the substitution-free alignments, since there might be more than one substitution-free alignment indicating the same common subsequence. On the other hand, the mapping of substitution-free alignments onto the common subsequences is a surjection. The length of the longest common subsequence can be found by finding the substitution-free alignment with the maximum number of $\frac{x}{x}$ columns. The semiring in the evaluation algebra must be the dual tropical semiring (with maximum instead of minimum), and the function f assigns the number of $\frac{x}{x}$ columns to each substitution-free alignment. The $T_{\frac{x}{x}}$ function for operator $o_{\frac{x}{x}}$ is the tropical multiplication by 1 (that is, the usual adding of 1), while the functions $T_{\frac{x}{-}}$ and $T_{\frac{-}{y}}$ are the identity functions. The length of the longest common subsequence is $F(S((n, m)))$.

(b) **Edit distance** The edit distance of sequences X and Y is the minimum number of insertion and deletion operations necessary to transform X to Y. Any substitution-free alignment corresponds to a series of insertion and deletion operations (although the order of these operations is not specified by the alignment). The semiring in the evaluation algebra is the tropical one, and the f function maps the number of insertions and deletions to a substitution-free alignment. The function $T_{\frac{x}{x}}$ is the identity function, while both the $T_{\frac{x}{-}}$ and $T_{\frac{-}{y}}$ functions are both the tropical multiplication by 1. The edit distance is $F(S((n, m)))$.

(c) **Shortest common supersequence**. Sequence Z is a supersequence of X if X is a subsequence of Z. A shortest common supersequence of X and Y is a common supersequence with minimal length. Any substitution-free alignment corresponds to a common supersequence. Indeed, just read the non-gap characters in each insertion and deletion column and one copy of the common characters in each match column. *Vice versa*, any common supersequence can be represented with a substitution-free alignment. Similar to the longest common subsequence problem, there is no bijection between substitution-free alignments and common supersequences, however, the mapping of substitution-free alignments to common supersequences is a surjection. The semiring in the evaluation algebra is the tropical one, and the f function assigns its length to a substitution-free alignment. Each T function is the tropical multiplication by 1. The length of the shortest common supersequence is $F(S((n, m)))$.

There are natural corresponding counting problems to these optimization problems. However, we can count the *ways* that common subsequences or common supersequences can be represented with substitution-free alignments, and therefore the number of these ways and not the number of *different* common subsequences and supersequences can be found with algebraic dynamic programming. On the other hand, if the sequences are permutations, then any common subsequence can be represented in exactly one way. Thus, the longest common subsequences of two permutations can be counted using algebraic dynamic programming. The algebraic structure in the evaluation algebra should be $\mathbb{Z}[z]$, f assigns z^k to a substitution-free alignment with k match columns. The function T_x is the multiplication with z, and both T_x and T_- are the identity functions. The number of longest common subsequences of two permutations of length n is the coefficient of the largest monomial of $F(S((n, n)))$.

For permutations, yet another question arises, based on the following definition.

Definition 30. *A permutation $\sigma = \sigma_1 \sigma_2 \ldots \sigma_k$ is a subpermutation of $\pi = \pi_1 \pi_2 \ldots \pi_n$ if π contains a subsequence $\pi_{i_1} \pi_{i_2} \ldots \pi_{i_k}$ such that $\forall l, m$, $\pi_{i_l} < \pi_{i_m} \iff \sigma_l < \sigma_m$*

It is clear that the longest common subsequences of two permutations correspond to the longest common subpermutations. However, two longest common subsequences might be the same subpermutations. For example, there are 4 longest common subsequences of the permutations 2143 and 1234: 24, 23, 14, 13. However, they are the same subpermutations, 12. Counting the longest common subpermutations is known to be #P-complete. Actually, deciding if a permutation contains a given subpermutation is already NP-complete, and the corresponding counting problem is #P-complete [21].

Regarding the edit distance, it is natural to count the ways that a sequence X can be transformed into sequence Y using insertion and deletion operations. When X is transformed into Y with the total number of p insertion and deletion operations, there are $p!$ ways to perform these operations. A natural attempt is to count the substitution-free alignments of two sequences containing p insertions and deletions, then multiply this number by $p!$. However, different alignments might contain the same insertions and deletions, just in different order. For example

```
AB-C
A-DC
```

and

$$\text{A-BC}$$
$$\text{AD-C}$$

both contain the deletion of B and insertion of D. The problem can be eliminated if insertions are not allowed after deletions, however, deletions are allowed after insertions. This needs a modification of the yield algebra, since it must build only these substitution-free alignments. The parameter set also should be modified. The triplet (i, j, t), with $t \in \{I, D, M\}$ indicates the length of the prefixes and whether the last alignment column is an insertion, deletion or a match. The operators are the same, and the recursions are

$$S((i, j, M)) \quad = \quad \sqcup_{t \in \{I, D, M\}} \circ_{y_j}^{x_i} (S((i-1, j-1, t))) \quad \text{if } x_i = y_j \tag{2.111}$$

$$S((i, j, M)) \quad = \quad \emptyset \text{ if } x_i \neq y_j \tag{2.112}$$

$$S((i, j, I)) \quad = \quad \sqcup_{t \in \{I, M\}} \circ_{y_j}^{-} (S((i, j-1, t))) \tag{2.113}$$

$$S((i, j, D)) \quad = \quad \sqcup_{t \in \{I, D, M\}} \circ_{-}^{x_i} (S((i-1, j, t))) \tag{2.114}$$

with initial conditions $S((0, 0, M)) \left\{ \begin{array}{c} \epsilon \\ \epsilon \end{array} \right\}$ and $S((0, 0, I)) = S((0, 0, D)) = \emptyset$. Technically, the match here means neither insertion nor deletion.

In the evaluation algebra, the algebraic structure is $\mathbb{Z}[z]$, and f assigns z^k to a substitution-free alignment with k insertions and deletions. The function $T_{\frac{x}{-}}$ is the identity function, while both the $T_{\frac{x}{-}}$ and $T_{\frac{-}{y}}$ functions are multiplication by z. If

$$F(S((n, m))) = \sum_{i=\min\{n,m\}}^{n+m} c_i z^i \tag{2.115}$$

then the number of ways that X can be transformed into Y is

$$\sum_{i=\min\{n,m\}}^{n+m} c_i i! \tag{2.116}$$

and the number of ways that X can be transformed into Y with a minimum number of insertion and deletion operations is $c_k k!$ where k is the smallest index such that $c_k \neq 0$.

When substitutions are allowed, the yield algebra building all possible alignments might be given. The parameters (i, j) and operators $\circ_{y_j}^{x_i}$, $\circ_{-}^{x_i}$ and $\circ_{y_j}^{-}$ are as above. The recursions are simply

$$S((i, j)) \quad = \quad \circ_{y_j}^{x_i} (S((i-1, j-1))) \sqcup$$
$$\circ_{-}^{x_i} (S((i-1, j))) \sqcup \circ_{y_j}^{-} (S((i, j-1))) \tag{2.117}$$

with the initial condition $S((0,0)) = \left\{ \begin{array}{c} \epsilon \\ \epsilon \end{array} \right\}$.

A weight function w mapping from the operators to reals or integers can be introduced. The following questions can be asked, which are solvable with the appropriate evaluation algebras:

(a) **The alignment minimizing/maximizing the sum/product of the weights.** The semiring in the evaluation algebra is the ordinary/dual ordinary/exponentiated tropical semiring. In the exponentiated tropical semiring, the weights might be only non-negative. The function f assigns the appropriate score to each alignment, the operators are the tropical multiplications with the given weight. The score of the optimal alignment is $F(S((n, m)))$.

(b) **The number of alignments with a given score.** This can be calculated easily if the weights are integers, small in absolute value, and the score of the alignments is additive. Then the algebraic structure in the evaluation algebra is $\mathbb{Z}[z]$, and f assigns z^s to an alignment with score s. Each T function is the multiplication with z^w, where w is the weight of the operator. The number of alignments with a prescribed score s is the coefficient of z^s in $F(S((n, m)))$.

Biologists would like to see sequence alignments in which insertions and deletions are aggregated; therefore they define different gap penalty functions. Recall that the joint name of insertions and deletions is "gap" and a gap penalty function is a scoring of gaps such that it depends on only the length of the gaps, that is, how many insertions and deletions are aggregated. The most commonly used gap penalty function is the *affine gap penalty* [81], in which a k-long run of insertions or deletions gets a $g_o + (k - 1)g_e$ score, where g_o is the gap opening penalty and g_e is the gap extension penalty. Gaps must be the same type in a run of gaps, that is, the score of a run of insertions of length k followed by the run of deletions of length l gets a score $g_o + (k - 1)g_e + g_o + (l - 1)g_e$ and not $g_o + (k + l - 1)g_e$. When the score is to be minimized, g_o is set larger than g_e. This causes alignments in which gaps are aggregated to get a smaller score than the alignments having the same number of gaps and the same number of matches and mismatches of the same type, just the gaps are scattered. For example, the alignment

AACTAT
ACC--T

has a smaller score than the alignment

AACTAT
A-C-CT

under the affine gap penalty scoring scheme, although the two alignments are alignments of the same sequences containing the same type of alignment columns, just in different order.

Under this scoring scheme, the scoring of an alignment is no longer additive in the strict sense that the score depends on only the individual alignment columns. However, the score of an insertion or deletion only depends on whether or not the previous alignment column is of the same type. If the yield algebra is built up separating the different types of alignments by extending the parameters with the different indicator variables, then the scoring can be done appropriately in the evaluation algebra. That is, the yield algebra is exactly the same as introduced above with recursions in Equations (2.111)–(2.114). Recall that in this yield algebra, only those alignments are built that do not contain insertions after deletions, however, deletions might occur after insertions.

If the aim is to find the smallest possible score, then the semiring in the evaluation algebra is the tropical one. The f function assigns the score to each alignment. The $T_{\substack{x \\ y}}$ function for operator $\circ_{\substack{x \\ y}}$ is the tropical multiplication with the weight of the alignment column $\substack{x \\ y}$. The functions $T_{\substack{x \\ -}}$ and $T_{\substack{- \\ y}}$ depend on the parameters. $T_{\substack{x \\ -}}$ is the tropical multiplication with g_o if the parameter is (i, j, M) or (i, j, I) and the tropical multiplication with g_e if the parameter is (i, j, D). Similarly, $T_{\substack{- \\ y}}$ is the tropical multiplication with g_o if the parameter is (i, j, M) and the tropical multiplication with g_e if the parameter is (i, j, I) (recall that insertions cannot follow deletions).

Counting problems can be solved with the appropriate modification of the evaluation algebra, see for example, Exercise 20.

The parameters (i, j, x), where i and j denote the length of the prefixes and x is a member of a finite set are very similar to the parameters (i, W) used in regular grammars and Hidden Markov Models. It is natural to define pair Hidden Markov Models and to observe that the introduced yield algebras for sequence alignments (for example, the ones defined by recursions in Equations (2.111)–(2.114)) are special cases of the class of yield algebras that can be constructed based on pair Hidden Markov Models.

Definition 31. *A pair Hidden Markov Model is a tuple $(\vec{G}, START, END, \Gamma, T, e)$, where $\vec{G} = (V, E)$ is a directed graph with two distinguished vertices, $START$ and END. Vertices are called states. Loops are allowed, however, parallel edges are not. The in-degree of the $START$ and the out-degree of the END states is 0. Γ is a finite set of symbols, called an alphabet, and $T : E \to \mathbb{R}^+$ are the transition probability function satisfying for all $v \neq END$*

$$\sum_{u \mid (v,u)=e \in E} T(e) = 1.$$

The function $e : (\Gamma \cup \{-\}) \times (\Gamma \cup \{-\}) \times (V \setminus \{START, END\}) \to \mathbb{R}^{\geq 0}$ is the *emission probability function satisfying for all $v \in V \setminus \{START, END\}$*

$$\sum_{(x,y)\in(\Gamma\cup\{-\})\times(\Gamma\cup\{-\})} e((x,y),v) = 1$$

where $- \notin \Gamma$. Furthermore, for all $v \in V$ and $x, y \in \Gamma$ the following implications also hold

$$e((x,-),v) = 0 \implies \forall x' \in \Gamma \ e((x',-),v) = 0$$
$$e((-,y),v) = 0 \implies \forall y' \in \Gamma \ e((-,x'),v) = 0$$
$$e((x,y),v) = 0 \implies \forall x',y' \in \Gamma \ e((x',y'),v) = 0.$$

Depending on which emission probabilities are not 0, the states are called insertion, deletion and match states. A random walk on the states is defined by $\vec{G}$ and the transition probabilities. The random walk starts in the state $START$ and ends in the state END. During such a walk, states emit characters or pair of characters according to the emission distribution e. In case of loops, the random walk might stay in one state for several consecutive steps. In each step, a random character or a pair of characters is emitted. The emitted characters generate two strings, X and Y. The process is hidden in the sense that an observer can see only the emitted sequences and cannot observe the random walk itself. The observer cannot even observe which characters are emitted together and which ones individually, that is, the observer cannot see the so-called co-emission pattern. An emission path is a random walk together with the two emitted sequences. The probability of an emission path is the product of its transition and emission probabilities.

Given two sequences, X and Y, and a pair Hidden Markov Model H, a yield algebra can be constructed whose base set, A, is the partial emission paths that emit prefixes of X and Y. The parameter set is (i, j, W), where i and j are the length of the prefixes and W is the current state of the emission path, assuming that W has already emitted a character or pair of characters. In the partial ordering of the parameters, $(i, j, W) \leq (i', j', W')$ if $i \leq i'$ and $j \leq j'$. The operators $\circ_{W_M}$, $\circ_{W_I}$ and $\circ_{W_D}$ extend the emission path with one step and the emission of the new state. Here, the indices M, I and D denote the type of the states. The recursions are

$$S((i,j,W_M)) =$$
$$\begin{cases} \sqcup_{W|(W,W_M)\in E} \circ_{W_M} (S((i-1,j-1,W))) & \text{if } e((x_i,y_j),W_M) \neq 0 \\ \emptyset & \text{if } e((x_i,y_j),W_M) = 0 \end{cases}$$
$$S((i,j,W_I)) =$$
$$\begin{cases} \sqcup_{W|(W,W_I)\in E} \circ_{W_I} (S((i,j-1,W))) & \text{if } e((-,y_j),W_I) \neq 0 \\ \emptyset & \text{if } e((-,y_j),W_I) = 0 \end{cases}$$
$$S((i,j,W_D)) =$$
$$\begin{cases} \sqcup_{W|(W,W_D)\in E} \circ_{W_D} (S((i-1,j,W))) & \text{if } e((x_i,-),W_D) \neq 0 \\ \emptyset & \text{if } e((x_i,-),W_D) = 0 \end{cases}.$$

Similar evaluation algebras can be combined with this yield algebra like the ones prescribed earlier in this section. The method can be extended to triple-wise and multiple alignments as well as triple and multiple Hidden Markov Models. However, the size of the parameter set grows exponentially with the number of sequences. Indeed, the number of parameters below a parameter $(i_1, i_2, \ldots, i_k, W)$ in the partial ordering of the set of the parameters is $\Omega(\prod_{j=1}^{k} i_j)$. Finding the minimum scored multiple alignment is proven to be NP-hard [104, 183].

2.3.3 Context-free grammars

Context-free grammars are on the second level of the Chomsky hierarchy [42]. They are widely used in computational linguistics, computer program compilers, and also in bioinformatics.

Definition 32. *A* context-free grammar *is a tuple $(T, N, \mathbb{S}, R)$, where T is a finite set of symbols called* terminals, *N is a finite set of symbols called* non-terminals, *$T \cap N = \emptyset$, $\mathbb{S} \in N$ is a distinguished non-terminal called the* starting non-terminal, *and R is a finite set of rewriting rules in the form*

$$W \to \beta$$

where W is a non-terminal and $\beta \in (T \cup N)^*$.
   A generation *is a finite series*

$$\mathbb{S} = X_1 \to X_2 \to \ldots \to X_k \in T^* \tag{2.118}$$

where for each $i = 1, \ldots, k - 1$, there exists a rule $W \to \beta$ and words $X_p, X_s \in T^*$ such that $X_i = X_p W X_s$ and $X_{i+1} = X_p \beta X_s$. T^* denotes the finite sequences from T.
   The language $L_G \subseteq T^*$ contains those sequences that can be generated by the grammar.
   Each generation can be described with a parse tree. *A parse tree is a vertex-labeled, rooted tree in which the children of internal nodes are naturally ordered. The root of the tree is labeled with $\mathbb{S}$, and for any internal node v labeled with W, the number of descendants of v is $|\beta|$, where the rewriting rule $W \to \beta$ is applied on W during the generation. If the kth character of β is a terminal character a, the kth child is a leaf that is labeled with a; if the kth character is a non-terminal W', then the kth child is an internal node W' and its descendant will be determined by the rewriting rule $W' \to \beta'$ applied on W' later in the generation. A grammar is said to be* unambiguous *if any sequence $X \in L_G$ has exactly one parse tree. An* ambiguous *grammar contains at least one sequence in its language that has at least two different parse trees.*

Note that several generations might have the same parse tree; however, these generations apply the same rewriting rules on the same non-terminals, just in different order.

What happens with a non-terminal W depends on only W and not how W has been generated. This is again a Markov property, and indeed, random generations in a context-free grammar constitute a branching process, which is indeed a Markov process. Therefore, similar to the stochastic regular grammars, stochastic context-free grammars can be defined.

Definition 33. *A stochastic context-free grammar is a tuple* $(T, N, \mathbb{S}, R, \pi)$, *where* $(T, N, \mathbb{S}, R)$ *is a context-free grammar and* $\pi : R \to \mathbb{R}^+$ *is a probability distribution for each non-terminal, that is, for each* $W \in N$, *the equality*

$$\sum_{\beta | W \to \beta \in R} \pi(W \to \beta) = 1 \tag{2.119}$$

holds. A stochastic context-free grammar makes random generations. A random generation can be described with its parse tree. The probability of a parse tree is the product of the probabilities of rewriting rules indicated by its internal nodes. Each rewriting rule has the appropriate multiplicity in the product. The probability of a sequence in the grammar is the sum of the probabilities of the parse trees that can generate it.

Remark 1. *Unlike stochastic regular grammars, stochastic context-free grammars might not define a distribution over the language they generate. This is because the branching process described by a context-free grammar might not end with probability 1. It can happen that with some probability separated from 0, a stochastic context-free grammar generates more and more non-terminals in the intermediate sequences and the generation never stops. However, it is a probability theory question in the theory of branching processes whether or not a branching process halts with probability 1, and not discussed in this book.*

An important class of context-free grammars which are in the so-called Chomsky normal form [42] is defined below.

Definition 34. *A context-free grammar is in Chomsky normal form if each rewriting rule is one of the following*

$$W \to W_1 W_2 \mid a$$

where $W, W_1, W_2 \in N$ *and* $a \in T$.

It is a well-known theorem that any context-free grammar G can be rewritten into a context-free grammar G' such that G' is in Chomsky normal form and $L_G = L_{G'}$ [159]. Here an equivalent theorem is proved for stochastic context-free grammars.

Theorem 25. *For any context-free grammar* $G = \{T, N, \mathbb{S}, R, \pi\}$ *there exists a context-free grammar* $G' = \{T, N', \mathbb{S}', R', \pi'\}$ *such that* G' *is in Chomsky normal form, and the following holds:*

(a) *There is a surjective function* $g : \mathcal{T}_G \to \mathcal{T}_{G'}$, *where* $\mathcal{T}_G$ *is the set of possible parse trees of* G, *such that* g *keeps the probabilities, that is, for any* $\tau' \in \mathcal{T}_{G'}$

$$\pi'(\tau') = \pi(g^{-1}(\tau'))$$

where

$$\pi(g^{-1}(\tau')) = \sum_{\tau \mid g(\tau) = \tau'} \pi(\tau).$$

Specifically, $L_G = L_{G'}$, *and the total generation probability of a sequence* X *in* G *is the total generation probability of* X *in* G'.

(b) *The set of rewriting rules* R' *satisfy the inequality*

$$|R'| \leq 128|R|^3 + 4|R|^2$$

where $|R|$ *is defined as the total sum of the lengths of the* β *sequences in the rewriting rules in* R.

(c) *There is a polynomial running time algorithm that constructs* G' *from* G.

Proof. The proof is constructive, and G' is constructed in the following three phases. Each phase consists of several steps, so G is transformed into G' via a series of steps, $G = G_1, G_2, \ldots G_k = G'$ such that for all $i = 1, \ldots, k - 1$, it is proved that there is a surjective function from $\mathcal{T}_{G_i}$ to $\mathcal{T}_{G_{i+1}}$ that keeps the probability. Finally, it will be shown that the entire construction can be done in polynomial time and indeed $|R'| \leq 128|R|^3 + 4|R|^2$.

In the first phase, those rewriting rules $W \to \beta$ are considered for which $|\beta| > 2$. If the current grammar is G_i in which there is a rewriting rule $W \to \beta$ such that $\beta = b_1, b_2 \ldots b_k$, then the new non-terminals W_1 and W_2 are added to the nonterminals, and thus we get $N_{i+1} = N_i \cup \{W_1, W_2\}$ together with the following rewriting rules and probabilities

$$\begin{aligned}
\pi_{i+1}(W \to W_1 W_2) &= \pi_i(W \to \beta) \\
\pi_{i+1}(W_1 \to b_1 b_2) &= 1 \\
\pi_{i+1}(W_2 \to b_3 \ldots b_k) &= 1.
\end{aligned}$$

The rule $W \to \beta$ is removed from R_i and the above rules with the given probabilities are added to get R_{i+1} and π_{i+1}. The g_i function replaces each rule $W \to \beta$ with the above three rules. It is easy to see that g_i is a bijection between $\mathcal{T}_{G_i}$ and $\mathcal{T}_{G_{i+1}}$ that keeps the probabilities.

The first phase finishes in $O(|R|)$ time, since the integer value

$$\sum_{\beta \mid \exists W, W \to \beta \in R_i \,\wedge\, |\beta| > 2} |\beta|$$

is strictly monotonously decreasing. At the end of the first phase, a grammar

G_I is constructed in which for each rewriting rule $W \to \beta$ in R_I, $|\beta| \leq 2$. Furthermore it is clear that $|R_I| \leq 2|R|$, and the construction runs in polynomial time with $|R|$.

In the second phase, those rewriting rules $W \to \beta$ are considered in which $|\beta| = 2$, but $\beta \neq W_1 W_2$ for some non-terminals W_1 and W_2. If the current grammar is G_i containing a rewriting rule $W \to a_1 a_2$, where $a_1, a_2 \in T$, then the new non-terminals W_1, W_2 are added to the non-terminals N_i to get N_{i+1} together with the following rewriting rules and probabilities

$$
\begin{aligned}
\pi_{i+1}(W \to W_1 W_2) &= \pi_i(W \to a_1 a_2) \\
\pi_{i+1}(W_1 \to a_1) &= 1 \\
\pi_{i+1}(W_2 \to a_2) &= 1.
\end{aligned}
$$

The rule $W \to a_1 a_2$ is removed from R_i and the above rules are added with the prescribed probabilities to get R_{i+1} and π_{i+1}. The g_i function replaces each rule $W \to a_1 a_2$ with the above three rules. It is easy to see that g_i is a bijection between $\mathcal{T}_{G_i}$ and $\mathcal{T}_{G_{i+1}}$ that keeps the probabilities. Rewriting rules $W \to a_1 W_2$ and $W \to W_1 a_2$ are handled in a similar way. However, observe that it is sufficient to introduce only one new non-terminal.

The second phase finishes in $O(|R_I)|$ time since each rewriting rule is modified at most once, and the new rules are not modified further. At the end of the second phase, a grammar G_{II} is constructed in which each rewriting rule is either in Chomsky normal form or it is $W \to W'$ for some $W, W' \in N_{II}$. Furthermore it is clear that $|R_{II}| \leq 2|R_I|$, thus $|R_{II}| \leq 4|R|$, and the construction runs in polynomial time with $|R|$.

In the third phase, those rules $W \to W'$ are considered which are the only rules that are not in Chomsky normal form in G_{II}. If G_i is a grammar in which a W' exists so that for some W, $W \to W' \in R$, then we do the following:

(a) For all W, such that $W \to W' \in R$, for each $W' \to \beta$ the following rewriting rules with the given probabilities are added

$$
\pi_{i+1}(W \to \beta) = \pi_i(W \to \beta) + \pi_i(W \to W')\pi_i(W \to \beta). \tag{2.120}
$$

Simultaneously, the rewriting rule $W \to W'$ is removed.

(b) If a $W \to W$ rule appeared, remove it from the rules and adjust the probabilities of all other rewriting rules $W \to \beta$

$$
\pi_{i+1}(W \to \beta) := \frac{\pi_{i+1}(W \to \beta)}{1 - \pi_{i+1}(W \to W)}. \tag{2.121}
$$

(c) If there is no $W \neq W'$ such that W' appears in the sequence β in a rewriting rule $W \to \beta \in R_{i+1}$, then remove W' from N_{i+1} and all of its rewriting rules.

The g_i function is constructed such that any pair of rewriting rules $W \to W'$ and $W' \to \beta$ is replaced by $W \to \beta$. In this way, several parse trees might be mapped onto the same parse tree. However, the definition of the new probabilities in Equation (2.120) together with the distributivity of multiplication over addition provides that the mapping keeps the probability.

If the $W \to W$ rules appear, then there might be $0, 1, \ldots W \to W$ rules before any $W \to \beta$ rule. The infinite sum of the geometric series, that is,

$$1 + \pi_{i+1}(W \to W) + \pi_{i+1}(W \to W)^2 + \ldots = \frac{1}{1 - \pi_{i+1}(W \to W)}$$

and the definition of the new probabilities in Equation (2.121) together provide that g_i keeps the probabilities.

The third phase finishes in $O(|N_{II}|)$ number of steps, since in each step one non-terminal is eliminated on the right-hand side, and no rule added in which a non-terminal would appear on the right-hand side that was eliminated earlier. At the end of the third phase, the grammar G' is in Chomsky normal form.

In any context-free grammar with non-terminals N and terminals T in Chomsky normal form, the number of rewriting rules cannot exceed $2|N|^3 + |N||T|$. Notice that $|N'| \leq |N_{II}|$ and $T' = T$. Furthermore, a very rough upper bound is $|N_{II}| \leq |R_{II}|$. Since the number of terminal characters cannot be more than the sum of the length of the rewriting rules, it follows that

$$|R'| \leq 2|N|^3 + |N||T| \leq 2|R_{II}|^3 + |R_{II}||R| \leq 128|R|^3 + 4|R|^2. \qquad (2.122)$$

$\square$

The importance of rewriting the grammar in Chomsky normal form is that there is a polynomially sized yield algebra for those grammars which are in Chomsky normal form. Given a context-free grammar G and a sequence X, the yield algebra builds those parse trees that generate a substring of X starting with an arbitrary non-terminal W. The parameters are (i, j, W), indicating the first and the last index of a substring and the non-terminal labeling of the root of the parse tree. In the partial ordering of the parameters, $(i_1, j_1, W_1) \leq (i_2, j_2, W_2)$ if $i_1 \geq i_2$ and $j_1 \leq j_2$. The binary operator $\circ_{W \to W_1 W_2}$ takes two parse trees whose roots are labeled with W_1 and W_2 and connects them to a larger parse tree with a root labeled with W. The recursions are

$$S((i, j, W)) = \sqcup_{W_1} \sqcup_{W_2} \sqcup_{k=i}^{j-1} S((i, k, W_1)) \circ_{W \to W_1 W_2} S((k+1, j, W_2)) \quad (2.123)$$

with the following initial conditions. If $W \to x_i \notin R$, then $S((i, i, W)) = \emptyset$. If $W \to x_i \in R$, then $S((i, i, W))$ is the set containing the parse tree generating x_i from W in one step. Similar yield algebras can be constructed that for a given series of sets $\tau_1, \tau_2, \ldots \tau_n, \forall \tau_i \subseteq T$ build all parse trees generating all possible sequences X in which all $x_i \in \tau_i$. Note that the corresponding evaluation algebra counting the size of the sets with a given parameter counts all possible parse trees and not just the possible sequences that can be generated with the given condition. These two numbers are equal only if the grammar is unambiguous. For ambiguous grammars it is also #P-complete to count the number of sequences that can be generated with a given constraint since any regular grammar is also context-free, and the counting problem in question is already #P-complete for regular grammars.

The reason why a context-free grammar must be rewritten into Chomsky normal form is that the binary operation $\circ_{W \to W_1 W_2}$ indicates only $O(n)$ disjoint union operations in Equation (2.123), where n is the length of the generated sequence. If there were a rewriting rule $W \to W_1 W_2 W_3$, it would require a ternary operator $\circ_{W \to W_1 W_2 W_3}$, and the recursion in the yield algebra would require $\Omega(n^2)$ disjoint union in form

$$
\begin{aligned}
S((i,j,W)) &= \sqcup_{k_1=i}^{j-2} \sqcup_{k_2=k_1+1}^{j-1} \circ_{W \to W_1 W_2 W_3} (S((i,k_1,W_1)), \\
&\quad S((k_1+1,k_2,W_2)), S((k_2+1,j,W_3)) \sqcup \\
&\quad \sqcup [\text{further disjoint unions}]
\end{aligned}
\tag{2.124}
$$

and generally, a rewriting rule with k non-terminals on the right-hand side would require $\Omega(n^{k-1})$ disjoint unions in the yield algebra (and thus, that many operations in the corresponding evaluation algebra). The fact that the recursion in Equation (2.124) can be split into recursions

$$
\begin{aligned}
S((i,j,W)) &= \sqcup_{k_2=i+1}^{j-1} S((i,k_2,W')) \circ_{W \to W' W_3} S((k_2+1,j,W_3)) \sqcup \\
&\quad \sqcup [\text{further disjoint unions}] \\
S((i,k_2,W')) &= \sqcup_{k_1=i}^{k_2-1} S((i,k_1,W_1)) \circ_{W' \to W_1 W_2} S((k_1+1,k_2,W_2))
\end{aligned}
$$

clearly highlights that the new non-terminals introduced in the first phase of rewriting a grammar into Chomsky normal form provide additional "memory" with which the calculations might be speeded up. It is also clear that efficient algorithms are available for all context-free grammars in which the rewriting rules are not in Chomsky normal form. However, there are at most 2 non-terminals on the right-hand side in each rewriting rule, furthermore, there are no rewriting rules in the form $W \to W'$.

One family of the most known combinatorial structures that might be described with context-free grammars are the Catalan structures. Catalan structures with a parameter k are combinatorial structures whose number is the Catalan number C_k. An example for them is the set of Dyck words.

Definition 35. *A Dyck word is a finite sequence from the alphabet $\{x, y\}$, such that in any prefix, the number of x characters is greater than or equal to the number of y characters, and the number of x and y characters is the same in the whole sequence.*

Dyck words can be generated by context-free languages, as stated in the following theorem.

Theorem 26. *The following unambiguous context-free grammar* $G =$ $(T, N, \mathbb{S}, R)$ *generates the Dyck words.* $T = \{x, y\}$, $N = \{\mathbb{S}\}$, *and the recursions are*

$$\mathbb{S} \rightarrow xy \mid x\mathbb{S}y \mid x\mathbb{S}y\mathbb{S} \mid xy\mathbb{S}. \qquad (2.125)$$

Proof. It is clear that G generates Dyck words, since in each rewriting rule, there is one x and one y, and each x precedes its corresponding y. Therefore it is sufficient to prove that each Dyck word is generated by this grammar in exactly one way. The proof is inductive. The only Dyck word with one x and one y is xy, which can be generated, and there is only one parse tree generating it.

Assume that $D = d_1 d_2 \ldots d_n$ is a Dyck word, and $n > 2$. Let i be the smallest index such that in the prefix D_i, the number of x characters equals the number of y characters.

If $i = n$, then $d_2 d_3 \ldots d_{n-1}$ is also a Dyck word. The first rewriting rule is $\mathbb{S} \rightarrow x\mathbb{S}y$, and $\mathbb{S}$ can generate $d_2 d_3 \ldots d_{n-1}$ by induction.

If $i < n$, then $d_{i+1} d_{i+2} \ldots d_n$ is also a Dyck word. If $i = 2$, then the first rewriting rule is $\mathbb{S} \rightarrow xy\mathbb{S}$, and $\mathbb{S}$ can generate $d_{i+1} d_{i+2} \ldots d_n$. If $2 < i < n$, then both $d_2 d_3 \ldots d_{i-1}$ and $d_{i+1} d_{i+2} \ldots d_n$ are Dyck words, The first rewriting rule is $\mathbb{S} \rightarrow x\mathbb{S}y\mathbb{S}$, and the two $\mathbb{S}$ can generate $d_2 d_3 \ldots d_{i-1}$ and $d_{i+1} d_{i+2} \ldots d_n$.

To prove that the grammar is unambiguous, first observe that the first character x of the Dyck word is generated in the first rewriting. Namely, the first x character is a child of the root in the parse tree. Assume that its corresponding y is at a position j which is not the first index such that in D_j, the number of x characters equals the number of y characters. Then $j \neq 2$, and the substring $d_2 d_3 \ldots d_{j-1}$ is not a Dyck word. However, it should be generated by $\mathbb{S}$, but any sequence generated by $\mathbb{S}$ is a Dyck word, a contradiction. □

Although the recursions in Equation (2.125) are not in Chomsky normal form, there are at most two non-terminals on the left-hand side, therefore, a polynomial yield algebra can be constructed based on the grammar building the Dyck words. The parameters p in the yield algebra are the positive integers, indicating the number of x characters in the word. The ordering is the natural ordering of the integers. Each rewriting rule has an appropriate operator. Due to brevity, it is indexed with the right-hand side of the rewriting rule. The recursions are

$$S(i) = \circ_{x\mathbb{S}y} \left(S(i-1) \right) \sqcup \circ_{xy\mathbb{S}} \left(S(i-1) \right) \sqcup$$
$$\left(\sqcup_{j=1}^{i-2} S(j) \circ_{x\mathbb{S}y\mathbb{S}} S(i-j-1) \right) \qquad (2.126)$$

with the initial condition $S(1) = \{xy\}$.

This yield algebra can be combined with several evaluation algebras. One example is given below.

Example 5. *Dyck words can be represented as monotonic lattice paths along the edges of a grid with* $n \times n$ *square cells, going from the left bottom corner*

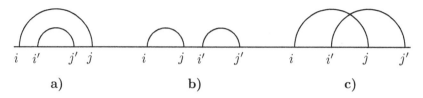

FIGURE 2.2: a) Nested, b) separated, c) crossing base pairs. Each base pair is represented by an arc connecting the index positions of the base pair.

to the top right corner not stepping above the diagonal. Each x is represented by a horizontal step, and each y is represented by a vertical step. The area of a Dyck word is the area below the lattice path representing it. Give a dynamic programming algorithm that calculates the average area of a Dyck word of length n.

Solution. The yield algebra is the same as defined above. Since area is an additive function, the algebraic structure in the evaluation algebra is the ring $R = (\mathbb{R}^2, \oplus, \odot)$, where the addition is coordinatewise and multiplication is defined by

$$(x_1, y_1) \odot (x_2, y_2) = (x_1 x_2, x_1 y_2 + y_1 x_2).$$

The f function assigns $(1, a)$ to each Dyck word D where a is the area of D. The T functions for the operators depend on the parameters. The $T_{xy\mathbb{S}}$ function is the multiplication with $(1, k)$ when the unary operator $c_{xy\mathbb{S}}$ is applied on a Dyck word of length k. The function $T_{x\mathbb{S}y}$ is the identity. Finally, the $T_{x\mathbb{S}y\mathbb{S}}(a, b, k, l)$ function is $a \odot b \odot (1, (k+1)l)$, where $a, b \in R$ and k and l are parameters. Indeed, adding xy at the beginning of a Dyck word of length $2k$ (thus, with parameter k) increases its area by k. Adding an x to the beginning and a y to the end of a Dyck word do not change its area. Finally, the area of a Dyck word $xDyD'$ is the area of D plus the area of D' plus $(k+1)l$ if the parameters of the Dyck words are $p(D) = k$ and $p(D') = l$.

If $F(S(n)) = (X, Y)$, then the average area is $\frac{Y}{X}$. ∎

Context-free grammars are also used in bioinformatics, since the pseudo-knot-free RNA structures can be described with these grammars.

Definition 36. *An RNA sequence is a finite long string from the alphabet $\{a, u, c, g\}$. A secondary structure is a set of pair of indexes (i, j), $i + 2 < j$ such that each index is in at most one pair. For each pair of indexes, the pair of characters might only be (a, u), (c, g), (g, c), (u, a), (g, u) or (u, g). These pairs are called base pairs. A secondary structure is pseudo-knot free if for all pair of indexes, (i, j) and (i', j'), $i < i'$, it holds that either $j' < j$ or $j < i'$. Namely, any pair of indexes are either nested or separated, and there are no crossing base pairs, see also Figure 2.2.*

The four characters represent the possible nucleic acids building the RNA molecules. An RNA molecule is a single stranded polymer, the string can be folded, and the nucleotides can form hydrogen bonds stabilizing the structure of the RNA. The pair of indexes represent the nucleic acids making hydrogen bonds. Due to spherical constraints, it is required that $i + 2 < j$. From now on, any RNA secondary structure is considered to be pseudo-knot-free, and the adjective "pseudo-knot-free" will be omitted.

Theorem 27. *The following grammar (also known as the Knudsen-Hein grammar, [115]) can generate all possible RNA secondary structures.* $T = \{a, c, g, u\}$, $N = \{\mathbb{S}, L, F\}$, *and the rewriting rules are*

$$\mathbb{S} \rightarrow L\mathbb{S} \mid L \tag{2.127}$$

$$L \rightarrow a \mid c \mid g \mid u \mid aFu \mid cFg \mid gFc \mid uFa \mid gFu \mid uFg \tag{2.128}$$

$$F \rightarrow aFu \mid cFg \mid gFc \mid uFa \mid gFu \mid uFg \mid L\mathbb{S} \tag{2.129}$$

The base pairs are the indexes of those pair of characters that are generated in one rewriting step. Furthermore, each pseuknot-free RNA structure that a given RNA sequence might have has exactly one parse tree in the grammar.

Proof. It is clear that the grammar generates RNA structures, so it is sufficient to show that each possible RNA structure that a given RNA sequence might have can be generated with exactly one parse tree.

In a given RNA secondary structure of a sequence X, let those base pairs (i, j) be called *outermost*, for which no base pair (i', j') exists such that $i' < i$ and $j < j'$. If (i, j) is an outermost base pair, then the base pairs inside the substring X' from index $i + 1$ till index $j - 1$ form also an RNA structure on the substring. Furthermore, either there is an outermost base pair $(i+1, j-1)$ on the substring X' or at least one of the following is true:

(a) there are at least two outermost base pairs or

(b) there is an outermost base pair (i', j') and a character which is not base-paired and outside of the base pair (i', j'), that is, if it has index k, then $k < i'$ or $k > j'$ or

(c) there are at least two characters in X' which are not base-paired and are outside of any outermost base pair.

This comes from the fact that the length of X' is at least 2, according to the definition of RNA secondary structure, i.e., $i + 2 < j$. Having said this, a given RNA secondary structure can be generated in the following way. If the outermost base pairs have indexes (i_1, j_1), (i_2, j_2), $\ldots (i_k, j_k)$, then first an intermediate sequence containing

$$i_1 + \sum_{l=2}^{k}(i_l - j_{l-1}) + n - j_k$$

number of Ls is generated, where n is the length of the sequence, applying an appropriate number of times the $\mathbb{S} \to L\mathbb{S}$ rule, then the $\mathbb{S} \to L$ rule. If there is no outermost base pair (and thus, there are no base pairs at all), n number of Ls are generated. Then each character which is not base-paired is generated using the appropriate rewriting rule from the possibilities

$$L \to a \mid c \mid g \mid u$$

and the outermost base pairs are generated using the appropriate rewriting rule from the possibilities

$$L \to aFu \mid cFg \mid gFc \mid uFa \mid gFu \mid uFg.$$

Then each intermediate string from the index $i_l + 1$ till the index $j_l - 1$ must be generated. If $(i_l + 1, j_l - 1)$ is an outermost base pair, then it is generated by the appropriate rule from the possibilities

$$F \to aFu \mid cFg \mid gFc \mid uFa \mid gFu \mid uFg,$$

otherwise the rule $F \to L\mathbb{S}$ is applied, and the appropriate number of Ls are generated if $x_{i_l+1} x_{i_l+2} \ldots x_{j_l-1}$ was the entire sequence.

It is easy to prove that each secondary structure can be generated by only one parse tree. If there are no base pairs, then no non-terminals F can be generated, and then the only possibility is to generate n number of Ls and rewrite them to the appropriate characters. If there are base pairs, and thus some of them are outermost ones, they can be generated only from a non-terminal L, which was generated by a starting nonterminal $\mathbb{S}$. Then the substring from index $i_l + 1$ till the index $j_l - 1$ should be generated from a non-terminal F. This is the same as generating a secondary structure from $\mathbb{S}$, just at least two characters must be generated. However, this is required by the definition of RNA secondary structures. $\qquad \square$

Although the grammar is not in Chomsky normal form, there are at most 2 non-terminals on the right-hand side of each rewriting rule. Furthermore, although there is a rewriting rule $\mathbb{S} \to L$, there is no possibility of a circular generation

$$W \to W_1, \quad W_1 \to W_2, \quad \ldots \quad W_k \to W$$

in this grammar. Therefore, the yield algebra building possible RNA structures based on this grammar is computationally efficient. A stochastic version of this grammar might be used to predict RNA secondary structures: the one which is generated by a most likely parse tree is a natural prediction for the secondary structure.

Better predictions might be available by extending the grammar with further non-terminals. Recall that non-terminals play the role of constant size memory, thus the generation might "remember" what happened in previous rewriting steps. For example, physical chemists measured that consecutive *pair of base pairs* rather than just base pairs stabilizes an RNA secondary structure [168, 169]. Two base pairs (i, j) and (i', j') are consecutive if $i' = i + 1$ and $j' = j - 1$. This can be emphasized via introducing a new non-terminal F'. The non-terminal F represents the first base pair, and F' represents the fact that there was a base pair in the previous rewriting step in the parse tree. That is, the rewriting rules are modified as

$$L \ \rightarrow \ a \mid c \mid g \mid u \mid aFu \mid cFg \mid gFc \mid uFa \mid gFu \mid uFg \quad (2.130)$$
$$F \ \rightarrow \ aF'u \mid cF'g \mid gF'c \mid uF'a \mid gF'u \mid uF'g \quad\quad\quad (2.131)$$
$$F' \ \rightarrow \ aF'u \mid cF'g \mid gF'c \mid uF'a \mid gF'u \mid uF'g \mid L\mathbb{S}. \quad (2.132)$$

With this modification, any base pair must be in a consecutive pair of base pairs. After further thermodynamic considerations, a rather complicated context-free grammar is constructed with many non-terminals and rewriting rules. A thermodynamic free energy is assigned to each rewriting rule. The free energy of the secondary structure is additive, that is, the free energy of a secondary structure is the sum of free energies assigned to each rewriting rule. In this model, the following questions might be asked.

(a) **The secondary structure with minimum free energy.** This can be obtained by choosing the tropical semiring in the evaluation algebra together with the appropriate functions not detailed here.

(b) **The probability of the minimum free energy secondary structure.** In thermodynamics, the probability of a secondary structure is given by the Boltzmann distribution

$$P_T(S) = \frac{1}{Z} e^{-\frac{\Delta G(S)}{RT}}.$$

Here $P_T(S)$ denotes the probability of the structure S in the Boltzmann distribution at temperature T, $\Delta G(S)$ denotes the free energy of S, and R is the Regnault or universal gas constant making the exchange between temperature measured in Kelvin and free energy measured in J/mol. Z is the so-called partition function defined as

$$Z := \sum_{S \in \mathcal{S}} e^{-\frac{\Delta G(S)}{RT}}$$

where $\mathcal{S}$ is the set of all possible RNA secondary structures that an RNA sequence might have. To be able to calculate the probability of the minimum free energy structure, the partition function must be calculated. Since the free energies are additive in this model, the values

$$e^{-\frac{\Delta G(S)}{RT}}$$

are multiplicative. Therefore the real numbers can be chosen as the algebraic structure in the evaluation algebra, and the T functions for binary operators are multiplications (possibly with multiplications with constants, not detailed here), and the T functions for unary operators are multiplications with constants.

(c) **Moments of the Boltzmann distribution.** It is also important to know what the average value of the free energy in the Boltzmann distribution is and what its variance is, since they indicate how extreme the minimum free energy is in the distribution. The average free energy is defined as

$$\frac{1}{Z} \sum_{S \in \mathcal{S}} \Delta G(S) e^{-\frac{\Delta G(S)}{RT}}$$

and the variance as

$$\frac{1}{Z} \sum_{S \in \mathcal{S}} \Delta G^2(S) e^{-\frac{\Delta G(S)}{RT}} - \left(\frac{1}{Z} \sum_{S \in \mathcal{S}} \Delta G(S) e^{-\frac{\Delta G(S)}{RT}} \right)^2.$$

To calculate the moments of the Boltzmann distribution, it is necessary to calculate

$$Z = \sum_{S \in \mathcal{S}} e^{-\frac{\Delta G(S)}{RT}},$$

$$\sum_{S \in \mathcal{S}} \Delta G^{(}S) e^{-\frac{\Delta G(S)}{RT}} \quad \text{and}$$

$$\sum_{S \in \mathcal{S}} \Delta G^2(S) e^{-\frac{\Delta G(S)}{RT}}.$$

If $\circ$ is a binary operator and $\mathcal{S}_1$ and $\mathcal{S}_2$ are secondary structures such that for all $S_1 \in \mathcal{S}_1$ and $S_2 \in \mathcal{S}_2$, $\Delta G(S_1 \circ S_2) = \Delta G(S_1) + \Delta G(S_2)$, then the following equalities hold:

$$\sum_{S_1 \in \mathcal{S}_1} \sum_{S_2 \in \mathcal{S}_2} \Delta G(S_1 \circ S_2) e^{-\frac{\Delta G(S_1 \circ S_2)}{RT}} =$$

$$\sum_{S_1 \in \mathcal{S}_1} \sum_{S_2 \in \mathcal{S}_2} (\Delta G(S_1) + \Delta G(S_2)) e^{-\frac{\Delta G(S_1)}{RT}} e^{-\frac{\Delta G(S_2)}{RT}} =$$

$$\sum_{S_1 \in \mathcal{S}_1} \Delta G(S_1) e^{-\frac{\Delta G(S_1)}{RT}} \sum_{S_2 \in \mathcal{S}_2} e^{-\frac{\Delta G(S_2)}{RT}} +$$

$$\sum_{S_1 \in \mathcal{S}_1} e^{-\frac{\Delta G(S_1)}{RT}} \sum_{S_2 \in \mathcal{S}_2} \Delta G(S_2) e^{-\frac{\Delta G(S_2)}{RT}} \qquad (2.133)$$

$$\sum_{S_1 \in \mathcal{S}_1} \sum_{S_2 \in \mathcal{S}_2} \Delta G(S_1 \circ S_2)^2 e^{-\frac{\Delta G(S_1 \circ S_2)}{RT}} =$$

$$\sum_{S_1 \in \mathcal{S}_1} \sum_{S_2 \in \mathcal{S}_2} (\Delta G(S_1) + \Delta G(S_2))^2 e^{-\frac{\Delta G(S_1)}{RT}} e^{-\frac{\Delta G(S_2)}{RT}} =$$

$$\sum_{S_1 \in \mathcal{S}_1} \Delta G(S_1)^2 e^{-\frac{\Delta G(S_1)}{RT}} \sum_{S_2 \in \mathcal{S}_2} e^{-\frac{\Delta G(S_2)}{RT}} +$$

$$2 \sum_{S_1 \in \mathcal{S}_1} \Delta G(S_1) e^{-\frac{\Delta G(S_1)}{RT}} \sum_{S_2 \in \mathcal{S}_2} \Delta G(S_2) e^{-\frac{\Delta G(S_2)}{RT}} +$$

$$\sum_{S_1 \in \mathcal{S}_1} e^{-\frac{\Delta G(S_1)}{RT}} \sum_{S_2 \in \mathcal{S}_2} \Delta G(S_2)^2 e^{-\frac{\Delta G(S_2)}{RT}}. \tag{2.134}$$

Therefore, knowing the values

$$\sum_{S_i \in \mathcal{S}_i} e^{-\frac{\Delta G(S_i)}{RT}},$$

$$\sum_{S_i \in \mathcal{S}_i} \Delta G^{(}S_i) e^{-\frac{\Delta G(S_i)}{RT}} \quad \text{and}$$

$$\sum_{S_i \in \mathcal{S}_i} \Delta G^2(S_i) e^{-\frac{\Delta G(S_i)}{RT}}$$

for each $\mathcal{S}_i$ is sufficient to calculate the moments of the Boltzmann distribution.

These algorithms are also well known in the scientific literature; here we just gave a unified description of them. Michail Zucker and David Sankoff gave the first dynamic programming algorithm for computing the minimum free-energy RNA structure [187]. John S. McCaskill described the dynamic programming algorithm computing the partition function of RNA structures [126]. István Miklós, Irmtraud Meyer and Borbála Nagy gave an algebraic dynamic programming algorithm to compute the moments of the Boltzmann distribution [132].

2.3.4   Walks on directed graphs

Shortest path problems are typical introductory problems in lectures on dynamic programming. From the algebraic dynamic programming point of view, it should be clear that they are shortest walk problems. The difference between path and walk is given in the definition below.

Definition 37. *Let $\vec{G} = (V, E)$ be a directed graph. A* walk *is a series $v_0, e_1, v_1, e_2, \ldots, e_k, v_k$ of vertices $v_i \in V$ and edges $e_i \in E$, such that for all $1 \leq i \leq k$, e_i is incident to v_{i-1} and v_i. A* path *is a walk with no repeated vertices (and thus, no repeated edges). A* closed walk *is a walk in which the first and the last vertex are the same.*

Yield algebra can be built for walks in the following way. Given a directed graph $\vec{G} = (V, E)$, let A be the set of walks on $\vec{G}$. The parameters are triplets (i, j, k), and the walk from vertex v_i to vertex v_j containing k edges is assigned to the parameter (i, j, k). The partial ordering of the parameters is the natural ordering of their last index. The unary operator $\circ_{(v_l, v_j)}$ concatenates the edge $e = (v_l, v_j)$ and vertex v_j to the end of a walk. The recursions are

$$S((i, j, k)) = \bigsqcup_{(v_l, v_j) \in E} \circ_{(v_l, v_j)} (S((i, l, k-1))). \qquad (2.135)$$

The base cases are $S((i, i, 0)) = \{v_i\}$ and $S((i, j, 0)) = \emptyset$ for all $i \neq j$.

Given a weight function $w : E \to \mathbb{R}$, define

$$f(v_0, e_1, \ldots, e_k, v_k) := \sum_{i=1}^{k} w(e_i).$$

An evaluation algebra can be built in which R is the tropical semiring, and each $T_{(v_l, v_j)}$ is the tropical multiplication (the usual addition) with $w((v_l, v_j))$. $F(S((i, j, k)))$ calculates the weight of the smallest weight (= shortest) walk from v_i to v_j in k steps. If $\vec{G}$ does not contain a negative cycle (that is, a closed walk, $v_0, e_1, \ldots e_k, v_0$ such that $\sum_{i=1}^{k} w(e_i) < 0$), then

$$\min_{k=1,\ldots,|V|-1} \{F(S((i, j, k)))\}$$

is the weight of the shortest *path* from v_i to v_j. Indeed, any path cannot contain more edges than $|V| - 1$, and the shortest walk must be a shortest path in case of no negative cycles.

On the other hand, finding the longest path is an NP-hard optimization problem. If each weight is the constant 1, then the longest path from v_i to v_j is exactly $|V| - 1$ if and only if there is a Hamiltonian path from v_i to v_j. Equivalently, it is NP-hard to find the shortest path in a given number of steps in case of negative cycles. To see this, set all weights to -1, and ask for the shortest path from v_i to v_j in $|V| - 1$ steps.

Since the shortest walks are all shortest paths when there are no negative cycles, it is also possible to count the number of shortest paths in polynomial time on graphs without negative cycles. To do this, first build up the statistics evaluation algebra using the monoid semiring $\mathbb{N}[\mathbb{R}^+]$, then take the homomorph image

$$\varphi(h) := g_{minsupp} h(g_{minsupp})$$

where $g_{minsupp}$ is the smallest element in the support of h. It is easy to see that this is indeed a semiring homomorphism. Then $\varphi(F(S((i, j, k))))$ calculates the number of the shortest walks from v_i to v_j in k steps. Summing this for all $k = 1, \ldots, |V| - 1$ in the homomorph image gives the number of shortest walks from v_i to v_j which coincides with the number of shortest paths.

The (max,min)-semiring can be utilized in the following example.

Example 6. *The* $\vec{G} = (V, E)$ *be a directed graph, and let* $w : E \rightarrow \mathbb{R}$ *be a weight function. Define*

$$f(v_0, e_1, \ldots, e_k, v_k) := \min_i \{w(e_i)\}$$

Find the path π *from* v_i *to* v_j *that maximizes* $f(\pi)$.

Solution. Build the yield algebra for the possible walks on $\vec{G}$ in the way described above. Build the following evaluation algebra. Let $R = (\mathbb{R} \cup \{-\infty\}, \max, \min)$ be the semiring in the yield algebra, in which the addition is the maximum and the multiplication is the minimum of the two operands. The f function is the one given in the exercise, and each $T_{(v_l,v_j)}$ is the multiplication with $w((v_l, v_j))$ in this semiring, that is, taking the minimum of the argument and $w((v_l, v_j))$. Then $F(S((i, j, k)))$ calculates

$$\max_{\pi|p(\pi)=(i,j,k)} f(\pi)$$

namely, the maximum value of the walks from v_i to v_j in k steps. The value of the maximum value path from v_i to v_j is

$$\max_{k=1,\ldots,|V|-1} F(S((i, j, k))).$$

Indeed, any path from v_i to v_j contains at most $|V| - 1$ edges. Furthermore, for any walk with a cycle $v_l, e_{l+1}, \ldots, v_l$:

$$\pi = v_i, \ldots, e_l, v_l, e_{l+1}, \ldots, v_l, e_{l'}, \ldots, v_j$$

it holds that $f(\pi) \leq f(\pi')$, where

$$\pi' = v_i, \ldots, e_l, v_l, e_{l'}, \ldots, v_j.$$

∎

It is clear that counting the number of walks that maximizes the prescribed f function in Example 6 can be calculated with algebraic dynamic programming in polynomial time. Some of them might not be paths, and counting the number of such *paths* is a hard computational problem. Indeed, deciding that there is a Hamiltonian path in a graph is NP-complete. Observe that the same blowing-up technique used in the proof of Theorem 16 can be applied to decide whether or not there is a path with $n - 1$ edges between two vertices in a graph, where n is the number of vertices.

However, the number of shortest (= having minimum number of edges) paths that maximizes the f function in Example 6 can be calculated in polynomial time. Indeed, the shortest walks that maximize f are shortest paths.

The set of walks that minimizes the function f in Example 6 might not contain any path. Interestingly, this optimization problem is NP-hard, since the edge-disjoint pair of paths problem can be easily reduced to it [70].

To conclude, algebraic dynamic programming algorithms might solve several counting and optimization problems on walks of directed graphs. When the solutions coincide with the solutions of the corresponding problems on paths of directed graphs, then naturally, these problems can be solved for paths. However, whenever the optimal walk might not be a path, the path versions of these problems always seem to be NP-hard.

2.3.5 Trees

Removing an internal node from a tree splits the tree into several subtrees. This gives the possibility to build a yield algebra on trees and to solve vertex coloring and related problems on trees. To do this, the trees should be rooted as defined below.

Definition 38. *Let $G = (V, E)$ be a tree, and $v \in V$ and arbitrary vertex. Rooting the tree at vertex v means an orientation of the edges such that the orientation of an edge (v_i, v_j) is from v_i to v_j if v_j is closer to v than v_i. In such a case, vertex v_i is a* child *of v_j and v_j is the* parent *of v_i. Vertex v is called the* root *of the tree. The* subtree rooted in vertex v_j *is the tree that contains v_j and all of its descendants (its children, the children of its children, etc.).*

Every node in a rooted tree except its root has exactly one parent. Any internal node has at least one child. The central task here is to solve coloring problems on trees. First, the definition of r-proper coloring is given.

Definition 39. *Let $G = (V, E)$ be a rooted tree, C is a finite set of colors, and $r \subseteq C \times C$ is an arbitrary relation. A coloring $c : V \to C$ is an r-proper coloring if for all oriented edges $(v_i, v_j) \in E$, $(c(v_i), c(v_j)) \in r$.*

The r-proper coloring is an extension of the usual proper coloring definition. Indeed, if

$$r = (C \times C) \setminus (\cup_{c \in C} \{(c, c)\})$$

then the r-proper coloring coincides with the usual definition of proper coloring.

The yield algebra of the r-proper coloring of a tree rooted in v can be constructed based on the subtrees of the tree and its proper colorings. Let $G = (V, E)$ be a tree rooted in vertex v, let C be a finite set of colors, and let $r \subseteq C \times C$ be a relation. Let A be the set of r-proper colorings of all the subtrees of G. The parameters (u, c) describe the root of the subtree, u, and its color, c. The partial ordering of the parameters is based on the root of the subtrees, $(u_1, c_1) \leq (u_2, c_2)$ if the path from u_1 to v contains u_2. The arity of the operator $\circ_{u,c}$ is the number of children of vertex u. It takes a subtree for each child, and connects them together with their common parent node u colored by color c. The recursions are

$$S((u, c)) = \circ_{u,c} \left(\left(\sqcup_{c_i | (c_i, c) \in r} S((w_i, c_i)) \right)_{i=1}^{k} \right) \tag{2.136}$$

where the w_i nodes are the children of u. The initial condition for a leaf u is that $S((u, c))$ is the set containing the trivial tree consisting only of the vertex u, colored by color c. There are problems where the colors are given for the leaves of the tree and the task is to give r-proper coloring of the entire tree. For those problems, the initial conditions prescribe that $S((u, c))$ be the empty set if leaf u is not colored by c.

Combining this yield algebra with different evaluation algebras provides algebraic dynamic programming algorithms to find, for example,

(a) the number of r-proper colorings of a tree with a given number of colors,

(b) the number of independent vertex sets of a tree,

(c) the size of the largest independent set of a tree,

(d) the coloring that minimizes

$$\sum_{(u_1, u_2) \in E} w(c(u_1), c(u_2))$$

for some weight function $w : C \times C \to \mathbb{R}$ (known as the Sankoff-Rousseau algorithm [150]),

(e) the sum

$$\sum_{c \in C} \pi(c(v)) \left(\prod_{(u_1, u_2) \in E} P_{w((u_1, u_2))}(c(u_1), c(u_2)) \right)$$

where π is a function mapping C to the (non-negative) real numbers, w is an edge weight function assigning a non-negative weight for each edge, and P_x is a parameterized family of functions, where the parameter x is a non-negative real number, and the function itself maps from $C \times C$ to the (non-negative) real numbers (known as Felsenstein's algorithm [68]).

2.4 Limitations of the algebraic dynamic programming approach

There is an algebraic operation and an algorithmic step that the algebraic dynamic programming approach does not use. We already emphasized that subtraction is forbidden in algebraic dynamic programming. There is also no branching in algebraic dynamic programming: the operations to be performed later in a recursion do not depend on the result in previous computations.

Subtractions might improve the running time. Consider the problem of counting the different subsequences of a sequence. Let X be a finite string from a finite alphabet, and let A be the set of subsequences of X. If Y is a subsequence of X, then let the parameter of Y be the minimal i such that Y is a subsequence of X_i, the i-character-long prefix of X.

Let Y be a subsequence of X with parameter i. Let k denote the length of Y. Let $g(i)$ be the index of the previous occurrence of x_i in X, if x_i also appears in X_{i-1}, otherwise let $g(i)$ be 0. It is easy to see that the parameter of Y_{k-1} is equal to or larger than $g(i)$. Therefore we can set up a yield algebra on the subsequences. The parameter is as described above, and its ordering is the natural one of the integers. The unary operator $\circ_a$ concatenates character a to the end of a sequence. The recursion is

$$S(i) \quad = \quad \sqcup_{j=g(i)}^{i-1} \circ_{x_i} (S(j)) \tag{2.137}$$

with the initial condition $S(0) = \{\varepsilon\}$, where ε is the empty string. To count the number of subsequences, we simply have to replace each disjoint union with the addition and each $\circ_a$ operation with multiplication by 1. Since the number of indices between $g(i)$ and $i-1$ is comparable with the length of X, this algorithm runs in $O(n^2)$ time.

However, if subtraction is allowed, the number of different subsequences can be counted in linear time. Let $m(i)$ denote the number of different subsequences of the prefix X_i, and let $g(i)$ be defined as above. Then

$$m(i) = 2m(i-1) - m(g(i)). \tag{2.138}$$

Indeed, we can extend any subsequence of X_{i-1} with the character x_i. This yields some overcounting as some of the subsequences could appear twice in this way. However, only those subsequences can appear that are also subsequences of $X_{g(i)}$. Clearly, the recursion in Equation (2.138) takes constant time for each index, and therefore, the number of different subsequences can be computed in linear time if subtractions are allowed.

Schnorr showed that matrix multiplication of two $n \times n$ matrices requires $\Omega(n^3)$ arithmetic operations if only multiplications and additions are allowed [151]. However, matrix multiplications can be done in $O(n^{\log_2(7)})$ time if subtractions are allowed [162], or even faster [47, 51]. Subtractions might have even more computational power. We are going to discuss it further at the end of Chapter 3.

Branching (test and branch instructions in an algorithm) might also speed up calculations. It is well known that a greedy algorithm can find a minimum spanning tree in an edge-weighted graph in polynomial time [116]. On the other hand, Mark Jerrum and Marc Snir proved that computing the spanning tree polynomial needs an exponential number of additions and multiplications [101]. Let $G = (V, E)$ be a graph, and assign a formal variable x_i to each edge e_i of G. The spanning tree polynomial is defined as

$$\sum_{T \in \mathcal{T}(G)} \prod_{x_i \in g(T)} x_i \tag{2.139}$$

where $\mathcal{T}(G)$ denotes the set of spanning trees of G, and for each spanning tree T, $g(T)$ denotes the set of formal variables assigned to the edges of T. It is easy to see that the evaluation of the spanning tree polynomial in the tropical semiring is the score (that is, the sum of the weights of the edges) of the minimum spanning tree. This score can be calculated with a simple greedy algorithm but not with algebraic dynamic programming, at least not in polynomial time. Also, the number of spanning trees, as well as the number of minimum spanning trees can be computed in polynomial time, however, such a computation needs subtraction. See also Chapter 3 for more details. Thus, spanning trees are combinatorial objects for which optimization and counting problems both can be computed in polynomial time, however, these algorithms cannot be described in the algebraic dynamic programming framework, and they are quite different.

2.5 Exercises

1. Show that the variants of the tropical semiring given in Definition 23 are all isomorphic to the tropical semiring.

2. Let $a_{i,k}$ and $n_{i,k}$ be the coefficients of the polynomials defined in Equations (2.4) and (2.5). Show that

$$a_{i,k} = a_{i-1,k} + n_{i-1,k}$$

and

$$n_{i,k} = a_{i-1,k-1}.$$

3. * Let $S = a_1, a_2, \ldots, a_n$ be a series of real numbers. Define the score of any subsequence $a_{i_1}, a_{i_2}, \ldots a_{i_k}$ to be $\prod_{j=1}^{k}(-1)^j a_{i_j}$. Give a dynamic programming algorithm that

   (a) calculates the maximum score of the subsequences,

   (b) calculates the sum of the scores over all possible subsequences.

4. Let $A = \{a_1, a_2, \ldots, a_n\}$ be a set of real numbers. Give a dynamic programming algorithm which

   (a) calculates

   $$\max_{S \subseteq A, |S|=k} \prod_{a_i \in S} a_i \quad \text{and}$$

   (b) calculates

   $$\sum_{S \subseteq A, |S|=k} \prod_{a_i \in S} a_i.$$

5. A subsequence of a sequence is *isolated* if does not contain consecutive characters. Let $a_1, a_2, \ldots a_n$ be a series of real numbers, and treat each number as a character. Give a dynamic programming algorithm that calculates for this series

   (a) the largest sum of isolated subsequences and

   (b) the sum of the products of isolated subsequences of length k.

6. ○ Give a dynamic programming algorithm which computes the number of sequences of length n from the alphabet $\{a, b\}$ that do not contain the substring aba.

7. ○ Give a dynamic programming algorithm that computes the number of sequences of length n from the alphabet $\{a, b, c, d, e\}$ that contain an odd number of a's and the sum of the numbers of c's and d's is even.

8. Give a dynamic programming algorithm that computes the number of sequences of length n from the alphabet $\{a, b\}$ in which the number of a's can be divided by 3 and does not contain 3 consecutive b's.

9. Give a dynamic programming algorithm that for a given Hidden Markov Model, $(\vec{G}, START, END, \Gamma, T, e)$ and a series of sets, $\tau_1, \tau_2, \ldots, \tau_n$, $\forall \tau_i \subseteq \Gamma$,

   (a) calculates the probability that the HMM emits a sequence $Y y_1 y_2 \ldots y_n$ such that for all i, $y_i \in \tau_i$ and

   (b) counts the number of possible emission paths that emit a sequence Y such that for all i, $y_i \in \tau_i$.

10. * A fast food chain is considering opening some restaurants along a highway. There are n possible locations at distances $d_1, d_2, \ldots d_n$ from a given starting point. At most one restaurant can be opened at a location, and the expected profit from opening a restaurant at location i is p_i. Due to some regulations, any two restaurants should be at least d miles apart from each other. Give a dynamic programming algorithm that

    (a) calculates the maximum expected total profit,

    (b) calculates the number of legal plans of opening restaurants, and

    (c) calculates the variance of the expected total profit of the uniform distribution of legal opening plans.

    cf. Dasgupta-Papadimitriou-Vazirani: Algorithms, Exercise 6.3.

11. ○ A checkerboard of 4 rows and n columns is given. An integer is written on each square. Also $2n$ pebbles are given and some or all of them can be put on squares such that no horizontal or vertical neighbors have both pebbles. Furthermore, we require that there be at least one pebble

in each column. The score of a placement is the sum of the integers on the occupied squares. Give a dynamic programming algorithm that

(a) calculates the maximum score in $O(n)$ time,

(b) calculates the number of possible placements in O(n) time,

(c) calculates the number of possible placements of k pebbles for each k in $O(n^2)$ time, and

(d) calculates the maximum score of placements with k pebbles for each k in $O(n^2)$ time.

cf. Dasgupta-Papadimitriou-Vazirani: Algorithms, Exercise 6.5.

12. Give a dynamic programming algorithm that computes the number of ways to tile a $3 \times n$ table with dominoes.

13. * Prove that the number of alignments of two sequences of length n and m is

$$\sum_{i=0}^{\min\{n,m\}} \frac{(n+m-i)!}{i!(m-i)!(n-i)!}.$$

14. Prove that two parse trees, both in Chomsky normal form, generating the same sequence contain the same number of internal nodes, thus the same number of rewriting rules.

15. Give a dynamic programming algorithm that counts all possible generations of a sequence by a context-free grammar in Chomsky normal form. Here two generations

$$S = A_1 \to A_2 \to \ldots \to A_n = A$$

and

$$S = B_1 \to B_2 \to \ldots \to B_n = A$$

are considered to be different even if they have the same parse tree.

Note that although all parse trees generating the same sequence contain the same number of rewriting rules, different parse trees might represent different numbers of generations.

16. Suppose that in a Dyck word, each x is replaced by (123) and each y is replaced with (12), and the Dyck word is evaluated as the product of the defined cycles in the permutation group S_3. For example, $xxyy$ becomes

$$(123)(123)(12)(12) = (123)(123) = (132)$$

on the other hand, $xyxy$ becomes

$$(123)(12)(123)(12) = (23)(23) = id.$$

Give a dynamic programming algorithm that computes how many Dyck word of length n exist that is evaluated as a given member of the permutation group S_3.

17. * The following operation on symbols a, b, c is defined according to the following table:

	a	b	c
a	b	b	a
b	c	b	a
c	a	c	c

Notice that the operation defined by the table is neither associative nor commutative. Give a dynamic programming algorithm that counts how many parenthesizations of a given sequence of symbols $\{a, b, c\}$ there are that yield a.

cf. Dasgupta-Papadimitriou-Vazirani: Algorithms, Exercise 6.6.

18. Let B be a Boolean expression containing words from $V = \{\text{"TRUE"}, \text{"FALSE"}\}$ and from $O = \{\text{"and"}, \text{"or"}, \text{"xor"}\}$. The expression is legitimate, that is, it starts and ends with a word from V and contains words from V and O alternatively. Give a dynamic programming algorithm that counts how many ways there are to parenthesize the expression such that it will evaluate to "$TRUE$".

19. An algebraic expression contains positive integer numbers, and addition and multiplication operations. Give a dynamic programming algorithm that

    (a) calculates the maximum value that a parenthesization might take and

    (b) calculates the average value of a random parenthesization.

20. * Give a dynamic programming algorithm that counts how many sequence alignments of two sequences, X and Y, there are which contain s substitutions, i insertions and d deletions.

21. * Assume that breaking a string of length l costs l units of running time in some string-processing programming language. Notice that different breaking scenarios using the same breaking positions might have different running time. For example, if a 30-character string is broken at positions 5 and 20, then making the first cut at position 5 has a total running time of $30 + 25 = 55$, on the other hand, making the first break at position 20 has a total running time of $30 + 20 = 50$.

Give a dynamic programming algorithm that counts

(a) how many ways there are to break a string into $m + 1$ pieces at m prescribed points such that the total running time spent on breaking the substrings into smaller pieces is exactly w units and

(b) how many ways there are to break a string of length n into $m + 1$ pieces such that the total running time spent on breaking the substrings into smaller pieces is exactly w units.

Two breaking scenarios are not distinguished if they use the same breaks of the same substrings just in different order. For example, breaking a string at position 20 and then at position 10 and then at position 30 is not distinguished from the scenario which breaks a string at position 20 then at position 30 and then at position 10.

cf. Dasgupta-Papadimitriou-Vazirani: Algorithms, Exercise 6.6.

22. Give a dynamic programming algorithm that takes a sequence of numbers and calculates the total sum of sums that can be obtained by inserting + symbols into the text. For example, if the input is 123, then the number to be calculated is

$$168 = 123 + (1 + 23) + (12 + 3) + (1 + 2 + 3).$$

23. There are n biased coins, and the coin with index i has probability p_i for a head. Give a dynamic programming algorithm that computes the probability that there will be exactly k heads when all the coins are tossed once.

cf. Dasgupta-Papadimitriou-Vazirani: Algorithms, Exercise 6.10.

24. * Given a convex polygon P of n vertices in the Euclidian plane (by the coordinates of the vertices). A triangulation of P is a collection of $n - 3$ diagonals such that no two diagonals cross each other. Notice that a triangulation breaks a polygon into $n - 2$ triangles. Give a dynamic programming algorithm that calculates the maximum and average score of a triangulation if it is defined by

(a) the sum of the edge lengths, and

(b) the product of the edge lengths.

cf. Dasgupta-Papadimitriou-Vazirani: Algorithms, Exercise 6.12.

25. * There is an $n \times m$ chocolate bar that should be broken into 1×1 pieces. During a breaking scenario, any rectangular part in itself can be broken along any horizontal or vertical lines. Each 1×1 piece has a unique label, thus breaking two rectangles can be distinguished even if they have the same dimensions. Give a dynamic programming recursion that computes the number of possible breaking scenarios if

(a) two breaking scenarios are equivalent if they contain the same breakings in different order, and

(b) the order of the breakings count.

Question for fun: What is the minimum number of breakings necessary to get nm number of 1×1 pieces?

26. There is a rectangular piece of cloth with dimensions $n \times m$, where n and m are positive integers. Also given a list of k products, for each product a triplet (a_i, b_i, c_i) is given, such that the product needs a rectangle of cloth of dimensions $a_i \times b_i$ and can be sold at a price c_i. Assume that a_i, b_i and c_i are all positive integers. Any rectangular piece of cloth can be cut either horizontally or vertically into two pieces with integer dimensions. Give a dynamic programming algorithm that counts how many ways there are to cut the clothes into smaller pieces maximizing the total sum of prices. The order of the cuts does not count, but on the other hand, it does count where the smaller pieces are located on the $n \times m$ rectangle.

   cf. Dasgupta-Papadimitriou-Vazirani: Algorithms, Exercise 6.14.

27. * A *binary search tree* is a vertex-labeled, rooted uni-binary tree with the following properties. The labels are objects with a total ordering, and for each internal node v, the labels on the left subtree of v are all smaller than the label of v and the labels on the right subtree of v are all bigger than the label of v. It is allowed that an internal node does not have a right or a left child; however, any internal node must have at least one child.

   Let $T = (V, E)$ be a rooted uni-binary tree, let the bijection $g : V \to L$ label the vertices, and let $\pi : L \to \mathbb{R}^+$ be a probability distribution of the labels. Let $d : V \to \mathbb{Z}^+$ be the distance function of vertices measuring the number of vertices of the path to the root. That is, $d(root) = 1$, the distance of the children of the root is 2, etc. This distance measures the number of comparisons necessary to find a given object (label) in T. The score of a binary search tree is the expected number of comparisons necessary to find a random object following the distribution π, namely,

$$\sum_{v \in V} d(v) \pi(g(v))$$

   Give a dynamic programming algorithm that for a given distribution of objects, $\pi : L \to \mathbb{R}^+$,

   (a) finds a binary search tree with the minimum cost, and

   (b) calculates the expected score of a uniformly distributed binary search tree.

28. * Give a dynamic programming algorithm that in a binary tree T

    (a) computes the number of matchings,
    (b) calculates the size of a maximal matching,
    (c) calculates the number of maximal matchings, and
    (d) calculates the average size of a matching.

29. A vertex cover of a graph $G = (V, E)$ is a subset of vertices that in-
    cludes at least one endpoint of each edge. Give a dynamic programming
    algorithm that for a binary tree

    (a) computes the number of vertex covers,
    (b) computes the average number of vertices in a vertex cover, and
    (c) computes the number of minimal vertex covers.

30. Give a dynamic programming algorithm that calculates for each k how
    many increasing subsequences of length k are in a given permutation.

31. A sequence of numbers is called *zig-zag* if the differences of the consec-
    utive numbers are positive and negative alternatively. Give a dynamic
    programming algorithm that computes for each k how many zig-zag
    subsequences of length k there are in a given permutation.

32. * Give a dynamic programming algorithm that for an input text, an
    upper bound t, and for each k and x, calculates how many ways there
    are to break the text into k lines such that each line contains at most t
    characters, and the squared number of spaces at the end of the lines for
    each line has a total sum x.

33. A sequence is palindromic if it is equal to its reverse. Give a dynamic
    programming algorithm that in a given sequence

    (a) finds the longest palindromic subsequence,
    (b) calculates the number of longest palindromic subsequences, and
    (c) calculates the average length of palindromic subsequences.

34. * Give a dynamic programming algorithm that computes the number
    of ways to insert a minimum number of characters into a sequence to
    make it a palindrome.

35. ○ There are n types of boxes with dimensions (a_i, b_i, c_i), and the dimen-
    sions might be arbitrary positive real numbers. A stable stack of boxes is
    such that for each consecutive pair of boxes, the horizontal dimensions
    of the bottom box are strictly larger than those of the top box. The
    boxes might be rotated in any direction, and thus, the same type of box
    might be used more than once. Give a dynamic programming algorithm
    that

(a) finds the height of the largest possible stack,

(b) computes the number of possible stacks, and

(c) calculates the average height of the stacks.

36. * Give a dynamic programming algorithm that for a given rooted binary tree

   (a) finds the number of subsets of vertices not containing any neighboring vertices,

   (b) computes the number of proper colorings of vertices with $k \geq 3$ colors (recall that no neighboring vertices should have the same color in a proper coloring of the vertices of a graph), and

   (c) computes the number of ways that the vertices can be partially colored with $k \geq 2$ colors such that no neighboring vertices have the same colors, however, one or both might be uncolored.

37. Generalize Exercise 36 for arbitrary trees.

38. * There is an $n \times m$ checkerboard, and each square contains some coins. A tour of the checkerboard consists of two paths, one from the top left corner to the bottom right one, containing only right and downward steps; the other path is from the bottom right corner to the top left one, containing only left and upward steps. The score of the tour is the sum of coins on the visited squares, and coins on squares visited twice count only once. Give a dynamic programming algorithm that

   (a) computes the number of tours,

   (b) calculates the maximum possible scores, and

   (c) calculates the average score of a uniformly distributed tour.

39. ○ Generalize Exercise 38 such that diagonal steps are possible on both paths.

40. * Let $A = \{a_1, a_2, \ldots a_n\}$ be a set of positive integers. Give a dynamic programming algorithm that calculates

   (a) the number of ways A can be split into 3 disjoint subsets, X, Y and Z such that

   $$\sum_{a_i \in X} a_i = \sum_{a_j \in Y} a_j = \sum_{a_k \in Z} a_k$$

   (b) the average value

   $$\left(\sum_{a_i \in X} a_i - \sum_{a_j \in Y} a_j \right)^2 + \left(\sum_{a_i \in X} a_i - \sum_{a_j \in Z} a_j \right)^2 + \left(\sum_{a_i \in Y} a_i - \sum_{a_j \in Z} a_j \right)^2$$

   of a random tripartition $X \sqcup Y \sqcup Z = A$.

41. Given a directed graph $\vec{G} = (V, E)$ and a weight function $w : E \rightarrow \mathbb{R}$, for any path π, define

$$f(\pi) := \min_{e \in \pi} w(e).$$

Give a dynamic programming that computes the number of shortest paths maximizing $f(\pi)$.

42. * A decreasing pair of non-negative integer sequences (bidegree sequences) $D = \{(b_1, b_2, \ldots, b_n), (c_1, c_2, \ldots, c_m)\}$ is called *graphical* if there exists a bipartite graph whose degrees are exactly D. The Gale-Ryser theorem states that a bidegree sequence is graphical iff

$$\sum_{i=1}^{n} b_i = \sum_{j=1}^{m} c_j$$

and for all $k = 1, \ldots, n$,

$$\sum_{i=1}^{k} b_i \leq \sum_{j=1}^{k} c_j^*$$

where c^* is defined as

$$c_j^* := |\{c_i | f_i \geq j\}|.$$

Give a dynamic programming algorithm that computes the number of graphical bidegree sequences.

2.6   Solutions

Exercise 3. Construct the following yield algebra. Let A be the subsequences of prefixes of the given string. Technically, these subsequences can be represented as sequences from $\{0, 1\}$, where 0 means that the corresponding character is not part of the subsequence and 1 means that the corresponding character is part of the subsequence. For example, if $A = 1, -5, 2, 4, -3, 2, 1, 1, 3$, then 010110 denotes $-5, 4, -3$. Any subsequence is parameterized with the length of the prefix and the parity indicator of the length of the subsequence. That is, Θ contains the (i, p) pairs, $i \in [0, \ldots n]$, $p \in \{0, 1\}$. The parameters are partially ordered based on their first coordinate. The operation $\circ_1$ concatenates 1 to the end of the sequence representing the subsequence, and the operation $\circ_0$ concatenates 0 to the end of the sequence representing the subsequence. The recursion of the yield algebra is

$$S((i, p)) = \circ_0 (S((i - 1, p))) \sqcup \circ_1 (S((i - 1, p + 1 (\mathrm{mod}\ 2))))$$

with initial values $S((0,0)) = \{\epsilon\}$ and $S((0,1)) = \emptyset$, where ϵ denotes the empty sequence.

When the largest product is to be calculated, the evaluation algebra is the following. $R = ((\mathbb{R}^{\geq 0} + \cup\{-\infty\}) \times (\mathbb{R}^{\geq 0} \cup \{-\infty\}), \oplus, \odot)$ semiring, $\oplus$ is the coordinatewise maximum with the rule

$$\max\{-\infty, a\} = a$$

and

$$(x_1, y_1) \odot (x_2, y_2) = (\max\{x_1 x_2, y_1 y_2\}, \max\{x_1 y_2, y_1 x_2\})$$

where the multiplication of $-\infty$ with itself is defined as $-\infty$. The rationale is that the first coordinate stores the largest available maximum and the second coordinate stores the absolute value of the largest possible negative value, if it exists. Here $-\infty$ stands for "non defined". If X is a subsequence $\{x_1, x_2, \ldots x_k\}$, then

$$f(X) = \begin{cases} \left(\prod_{i|x_i=1}(-1)^i a_i, -\infty\right) & \text{if } \prod_{i|x_i=1}(-1)^i a_i > 0 \\ \left(-\infty, \left|\prod_{i|x_i=1}(-1)^i a_i\right|\right) & \text{if } \prod_{i|x_i=1}(-1)^i a_i < 0 \\ (0,0) & \text{if } \prod_{i|x_i=1}(-1)^i a_i = 0 \end{cases}.$$

The T_1 function for the operator $\circ_1$ depends on the parameter (i,p). If $(-1)^i a_i > 0$, then T_1 is the multiplication with $((-1)^i a_i, -\infty)$, and if $(-1)^i a_i < 0$, then T_1 is the multiplication with $(-\infty, ((-1)^i a_i))$. If $a_i = 0$, then T_1 is the multiplication with $(0,0)$. Finally, the function T_0 for the operator $\circ_0$ is the identity function.

When the sum of the score of all possible subsequences is to be calculated, the evaluation algebra is the following. R is the real field; if X is a subset, then

$$f(X) = \prod_{a_i \in X} (-1)^i a_i.$$

The function T_1 for operation $\circ_1$ depends on the parameter (i,p); it is the multiplication with $(-1)^i a_i$. The function T_0 for the operation $\circ_0$ is the identity function.

Exercise 6. The following unambiguous regular grammar generates the possible sequences. $T = \{a, b\}$, $N = \{S, A, B, X\}$ and the rewriting rules are

$$
\begin{aligned}
S &\rightarrow aA \mid bB \mid a \mid b \\
A &\rightarrow aA \mid bX \mid a \mid b \\
B &\rightarrow aA \mid bB \mid a \mid b \\
X &\rightarrow bB \mid b.
\end{aligned}
$$

The nonterminal A means that the last generated character was a, B means that the last generated character was b and the next to last character was not a, and X means that the last generated character was b and the next to last character was a. The usual yield algebra and the corresponding evaluation algebra as described in Subsection 2.3.1 gives the recursion on the sets of possible generations and counts them. Since the grammar is unambiguous, the number of possible generations is the number of possible sequences.

Exercise 7. The following unambiguous grammar generates the possible sequences. $T = \{a, b, c, d, e\}$, $N = \{S, W_1, W_2, W_3\}$, and the rewriting rules are

$$
\begin{aligned}
S &\rightarrow aW_2 \mid a \mid bS \mid cW_1 \mid dW_1 \mid eS \\
W_1 &\rightarrow aW_3 \mid bW_1 \mid cS \mid dS \mid eW_1 \\
W_2 &\rightarrow aS \mid bW_2 \mid b \mid cW_3 \mid dW_3 \mid eW_2 \mid e \\
W_3 &\rightarrow aW_1 \mid bW_3 \mid cW_2 \mid c \mid dW_2 \mid d \mid eW_3.
\end{aligned}
$$

The non-terminal S means that in the so-far generated sequence, the number of generated a's is even, and the sum of the number of c's and d's is even. Similarly, W_1 stands for even-odd, W_2 stands for odd-even, and W_3 stands for odd-odd. The usual yield algebra and the corresponding evaluation algebra as described in Subsection 2.3.1 gives the recursion on the sets of possible generations and counts them. Since the grammar is unambiguous, the number of possible generations is the number of possible sequences.

Exercise 10. Let a sequence from the alphabet $\{0, 1\}$ represent a subset of indexes of locations, such that the sequence $X = x_1 x_2 \ldots x_k$ represents

$$
\{i \mid x_i = 1\}.
$$

Let such a sequence be called legal if the corresponding set of locations is a legal opening plan. Define the following yield algebra. A contains the legal 0-1-sequences. Each legal sequence has a parameter (k, i) where k is the length of the string, and i is the largest position where there is a 1 in the string. On this parameter set, $(k_1, i_1) \leq (k_2, i_2)$ if $k_1 \leq k_2$. Obviously, $i \leq k$ for any parameter (k, i). The unary operator $\circ_1$ concatenates a 1 to the end of a sequence, the unary operator $\circ_0$ concatenates a 0 to the end of a sequence. The recursions are

$$
\begin{aligned}
S((k, i)) &= \circ_0 \left(S((k - 1, i)) \right) \quad \forall i < k & (2.140) \\
S((k, i)) &= \bigsqcup_{j \mid d_k - d_j \geq d} \circ_1 \left(S(k - 1, j) \right) \quad \text{if } k = i & (2.141)
\end{aligned}
$$

with initial condition $S((0, 0)) = \{\epsilon\}$, where ϵ denotes the empty sequence. Define d_0 to be $-\infty$, thus $d_k - d_0 > d$ for any k.

If the task is to calculate the maximum expected profit, then the evaluation algebra can be constructed in the following way. K is the dual tropical semiring, that is, the tropical addition is taking the maximum instead of the minimum, and the additive unit is $-\infty$ instead of ∞. The f function is defined as

$$
f(x_1 x_2 \ldots x_k) := \bigodot_{i \mid x_i = 1} p_i.
$$

The T_1 function for the operator $\circ_1$ depends on parameter (k, k); it is the tropical multiplication (that is, the usual addition) with p_k. The T_0 function for operator $\circ_0$ is the identity function. The maximum expected profit is

$$\oplus_{i=0}^{n} F(S((n, i))).$$

If the task is to calculate the number of legal opening plans, then $R = \mathbb{Z}$, and f is the constant 1 function. Both T_1 and T_0 are the identity functions. The number of legal opening plans is

$$\sum_{i=0}^{n} F(S((n, i))).$$

If the task is to calculate the variance of the expected profit of the uniform distribution of legal opening plans, then R is set to $(\mathbb{R}^3, \oplus, \odot)$, where $\oplus$ is the coordinate-wise addition, and

$$(a_0, a_1, a_2) \odot (b_0, b_1, b_2) = (a_0 b_0, a_1 b_0 + a_0 b_1, a_2 b_0 + 2a_1 b_1 + a_0 b_2).$$

The f function is defined in the following way

$$f(x_1 x_2 \ldots x_k) := \left(1, \sum_{i | x_i = 1} p_i, \left(\sum_{i | x_i = 1} p_i \right)^2 \right).$$

The function T_1 for the operator $\circ_1$ depends on the parameter (k, k); it is the multiplication with $(1, p_k, p_k^2)$ in K. T_0 for the operation $\circ_0$ is the identity function.

Define (N, M, Z) in the following way:

$$(N, M, Z) := \oplus_{i=0}^{n} F(S((n, i))).$$

Here N is the number of legal opening plans, M is the sum of the expected profits in all possible legal opening plans, and Z is the sum of the expected profits squared in all possible legal opening plans. The variance of the expected profits in the uniform distribution of possible legal opening plans is

$$\frac{Z}{N} - \left(\frac{M}{N} \right)^2$$

since for any distribution

$$V[x] = E[x^2] - E^2[x].$$

That is, the variance is the expectation of the squared values minus the expectation squared.

Exercise 11. First, observe that only the following 6 patterns are possible in a column:

$$a) \begin{bmatrix} 1 \\ 0 \\ 0 \\ 0 \end{bmatrix} \quad b) \begin{bmatrix} 0 \\ 1 \\ 0 \\ 0 \end{bmatrix} \quad c) \begin{bmatrix} 0 \\ 0 \\ 1 \\ 0 \end{bmatrix} \quad d) \begin{bmatrix} 0 \\ 0 \\ 0 \\ 1 \end{bmatrix} \quad e) \begin{bmatrix} 1 \\ 0 \\ 1 \\ 0 \end{bmatrix} \quad f) \begin{bmatrix} 0 \\ 1 \\ 0 \\ 1 \end{bmatrix}$$

Define the following regular grammar:

$$
\begin{aligned}
S &\rightarrow aA \mid bB \mid cC \mid dD \mid eE \mid fF \mid a \mid b \mid c \mid d \mid e \mid f \\
A &\rightarrow bB \mid cC \mid dD \mid fF \mid b \mid c \mid d \mid f \\
B &\rightarrow aA \mid cC \mid dD \mid eE \mid a \mid c \mid d \mid e \\
C &\rightarrow aA \mid bB \mid dD \mid fF \mid a \mid b \mid d \mid f \\
D &\rightarrow aA \mid bB \mid cC \mid eE \mid a \mid b \mid c \mid e \\
E &\rightarrow bB \mid dD \mid fF \mid b \mid d \mid f \\
F &\rightarrow aA \mid cC \mid eE \mid a \mid c \mid e
\end{aligned}
$$

It is easy to see that this grammar defines the possible sequences of columns such that no horizontal or vertical neighbors have pebbles. Define the score of a generation as the sum of the numbers on the squares that have pebbles. The yield algebra is the standard one for generations in a regular grammar, in the evaluation algebra. The semiring in the evaluation algebra should be appropriately chosen according to the given computational task.

Exercise 13. The index i is the number of columns in the alignment without gap symbols, that is, the alignment columns with a match or a mismatch. Their number might vary between 0 and $\min\{n, m\}$. If there are i matches or mismatches in an alignment, then there are $n-i$ deletions and $m-i$ insertions. The total length of such an alignment is $n + m - i$. The number of alignments with these properties is indeed

$$\binom{n+m-i}{n-i, m-i, i} = \frac{(n+m-i)!}{(n-i)!(m-i)!i!}.$$

Exercise 17. Define the following yield algebra. Let A be the possible parenthesizations of the possible substrings. Recall that a substring is a consecutive part of a sequence. The possible parenthesizations of a sequence X are sequences from the alphabet $\{(,), a, b, c\}$ satisfying the following rules.

- The subsequence obtained by removing the "(" and ")" symbols is X.

- The subsequence obtained by removing characters a, b and c is a Dyck word.

- There are at least two characters between an opening bracket and its corresponding closing bracket.

- If there are two consecutive opening brackets, their corresponding closing brackets are not neighbors. A similar statement is true for consecutive closing brackets.

The parameters are triplets in the form (i, j, x) where i and j are the first and last indexes of the substring, and x is the result of the evaluation of the parenthesization. The partial ordering of the parameters is defined in the following way: $(i_1, j_1, x_1) \leq (i_2, j_2, x_2)$ if $i_2 \leq i_1$ and $j_1 \leq j_2$. The binary operation $\circ$ is defined as

$$X_1 \circ X_2 := (X_1 X_2).$$

The recursions are

$$
\begin{aligned}
S((i, j, a)) &= \left(\sqcup_{k=i}^{j-1} S((i, k, a)) \circ S((k+1, j, c)) \right) \sqcup \\
&\quad \left(\sqcup_{k=i}^{j-1} S((i, k, b)) \circ S((k+1, j, c)) \right) \sqcup \\
&\quad \left(\sqcup_{k=i}^{j-1} S((i, k, c)) \circ S((k+1, j, a)) \right) \\
S((i, j, b)) &= \left(\sqcup_{k=i}^{j-1} S((i, k, a)) \circ S((k+1, j, a)) \right) \sqcup \\
&\quad \left(\sqcup_{k=i}^{j-1} S((i, k, a)) \circ S((k+1, j, b)) \right) \sqcup \\
&\quad \left(\sqcup_{k=i}^{j-1} S((i, k, b)) \circ S((k+1, j, b)) \right) \\
S((i, j, c)) &= \left(\sqcup_{k=i}^{j-1} S((i, k, b)) \circ S((k+1, j, a)) \right) \sqcup \\
&\quad \left(\sqcup_{k=i}^{j-1} S((i, k, c)) \circ S((k+1, j, b)) \right) \sqcup \\
&\quad \left(\sqcup_{k=i}^{j-1} S((i, k, c)) \circ S((k+1, j, c)) \right)
\end{aligned}
$$

with the initial values

$$S((i, i, x)) = \begin{cases} \{x\} & \text{if } x_i = x \\ \emptyset & \text{if } x_i \neq x \end{cases}.$$

In the evaluation algebra, $R = \mathbb{Z}$, f is the constant 1 function and the T function for the operation $\circ$ is the multiplication. The number of parenthesizations resulting a is $F(S((1, n, a)))$.

Exercise 20. First, observe the following. If sequence X is aligned to sequence Y resulting i insertions and d deletions, then $|X| + i - d = |Y|$. Therefore the two sequences and the number of deletions already determine the number of insertions. Having said this, define the following yield algebra. Let A be the possible alignments of prefixes of X and Y. The parameter set contains quadruples (i, j, s, d), where i is the length of the prefix of X, j is the length of the prefix of Y, s is the number of substitutions, and d is the number of deletions in the alignment. The partial ordering is defined such that a parameter (i_1, j_1, s_1, d_1) is lower than or equal to the parameter (i_2, j_2, s_2, d_2) if $i_1 \leq i_2$, $j_1 \leq j_2$, $s_1 \leq s_2$ and $d_1 \leq d_2$. The operation $\circ_{i,j}$ concatenates the alignment column $\begin{smallmatrix} x_i \\ y_j \end{smallmatrix}$ to the end of an alignment. Similarly, $\circ_{-,j}$ concatenates the alignment column $\begin{smallmatrix} - \\ y_j \end{smallmatrix}$ to the end of an alignment and $\circ_{i,-}$ concatenates $\begin{smallmatrix} x_i \\ - \end{smallmatrix}$ to the end of an alignment. The recursions are

$$
\begin{aligned}
S((i,j,s,d)) \;=\; & \circ_{i,j}\left(S((i-1,j-1,s-1+\delta_{x_i,y_j},d))\right) \sqcup \\
& \circ_{-,j}\left(S((i,j-1,s,d))\right) \sqcup \\
& \circ_{i,-}\left(S((i-1,j,s,d-1))\right)
\end{aligned}
$$

where d_{x_i,y_j} is the Kronecker delta function. The initial condition is $S((0,0,0,0)) - \left\{ \begin{smallmatrix} \epsilon \\ \epsilon \end{smallmatrix} \right\}$, where $\begin{smallmatrix} \epsilon \\ \epsilon \end{smallmatrix}$ denotes the empty alignment. The evaluation algebra simply counts the size of the sets, that is, R is the integer ring, f is the identity function, and all T functions corresponding to operators $\circ_{i,j}$, $\circ_{-,j}$ and $\circ_{i,-}$ are the identity functions.

An alternative solution is also possible. In this solution, the yield algebra is the following. A is the possible alignments of the possible prefixes. The parameters are pairs (i,j) denoting the length of the prefixes, $(i_1, j_1) \leq (i_2, j_2)$ if $i_1 \leq i_2$ and $j_1 \leq j_2$. The operators $\circ_{i,j}$, $\circ_{-,j}$ and $\circ_{i,-}$ are the same as defined above. The recursions are

$$
S((i,j)) = \circ_{i,j}\left(S((i-1,j-1))\right) \sqcup \circ_{-,j}\left(S((i,j-1))\right) \sqcup \circ_{i,-}\left(S((i-1,j))\right)
$$

with initial condition $S((0,0)) = \left\{ \begin{smallmatrix} \epsilon \\ \epsilon \end{smallmatrix} \right\}$. In the evaluation algebra, $R = \mathbb{Z}[z_1, z_2]$, the two variable polynomial ring over $\mathbb{Z}$. The function f maps $z_1^s z_2^d$ to an alignment with s number of substitutions and d number of insertions. The $T_{i,j}$ function for operator $\circ_{i,j}$ is the multiplication with $z_1^{1-\delta_{x_i,y_j}}$, $T_{-,j}$ is the identity function and $T_{i,-}$ is the multiplication with z_2. The number of alignments with s number of substitutions and d number of deletions is the coefficient of $z_1^s z_2^d$ of the polynomial $F(S((n,m)))$, where n and m are the length of X and Y, respectively.

Exercise 21. To solve the first subexercise, consider a sequence X whose characters are the $m+1$ segments of the original sequence. Each character in x has a weight $w(x)$ defined as the number of characters in the corresponding segment. A breaking scenario is a series of breaks, $b_{i_1} b_{i_2} \dots b_{i_k}$, where b_i is the break at the border of segments x_i and x_{i+1}. A breaking scenario $b_{i_1} b_{i_2} \dots b_{i_k}$ is canonical if for any $j < l$ either $i_j < i_l$ or the break b_{i_j} acts on a substring that includes x_{i_l} (or both conditions hold).

In the yield algebra, A is the possible canonical breaking scenarios of the possible substrings of X. The parameter (i,j) denotes the first and the last index of the substring. In the partial ordering of the parameters, $(i_1, j_1) \leq (i_2, j_2)$ if $i_2 \leq i_1$ and $j_1 \leq j_2$. The operation $\circ$ operates on breaking scenarios of two consecutive substrings. If Q and R are breaking scenarios with parameters $p(Q) = (i, k)$ and $p(R) = (k + 1, j)$, then

$$Q \circ R = b_k, Q, R.$$

The recursions are

$$S((i, j)) = \sqcup_{k=1}^{j-1} S((i, k)) \circ S((k + 1, j))$$

with the initial conditions $S(i, i)) = \{\epsilon\}$, where ϵ is the empty sequence (of breaking steps).

In the evaluation algebra, $R = \mathbb{Z}[z]$, the polynomial ring over the integers. The function f maps z^W to a breaking scenario, where W is the total running time of the scenario. The function T for the operator $\circ$ depends on the parameters; if $p(Q) = (i, k)$ and $p(R) = (k + 1, j)$, then

$$T(Q, R) = z^{\sum_{l=i}^{j} w(x_l)} f(Q) f(R)$$

and similarly, for two values $k_1, k_2 \in K$,

$$T(k_1, k_2, (i, k), (k + 1, j)) = z^{\sum_{l=i}^{j} w(x_l)} k_1 k_2.$$

The number of breaking scenarios with total running time w is the coefficient of z^w in $F(S((1, m + 1)))$.

In the second subexercise, A, the base set of the yield algebra contains the possible canonical breaking scenarios of the possible substrings of the original string into smaller substrings (not necessarily into single characters). The canonical breaking scenarios are defined similarly as above. The parameter (i, j, l) denotes the beginning of the substring, the end of the substring and the number of breaks, respectively. In the partial ordering of the parameters, $(i_1, j_1, l_1) \leq (i_2, j_2, l_2)$ if $i_2 \leq i_1$, $j_1 \leq j_2$ and $l_1 \leq l_2$. The operation $\circ$ operates on a pair of breaking scenarios operating on consecutive substrings; if $p(Q) = (i, k, l_1)$ and $p(R) = (k + 1, j, l_2)$, then

$$Q \circ R = b_k, Q, R.$$

The recursions are

$$S((i, j, l)) = \sqcup_{k=1}^{j-1} \sqcup_{l_1=0}^{l-1} S((i, k, l_1)) \circ S((k + 1, j, l - l_1 - 1))$$

with initial conditions $S((i, j, 0)) = \{\epsilon\}$, where ϵ denotes the empty sequence (of breaking steps).

In the evaluation algebra, $R = \mathbb{Z}[z]$, the polynomial ring over the integers. The function f maps z^W to a breaking scenario, where W is the total running time of the scenario. The function T for the operator $\circ$ depends on the parameters; if $p(Q) = (i, k, l_1)$ and $p(W) = (k + 1, j, l_2)$ then

$$T(Q, W) = z^{j-i+1} f(Q) f(W)$$

and similarly, for any two values $r_1, r_2 \in R$,

$$T(r_1, r_2, (i, k, l_1), (k+1, j, l_2)) = z^{j-i+1} r_1 r_2.$$

The number of breaking scenarios with total running time w in m number of steps is the coefficient z^w in $F(S((1, n, m)))$.

There is an alternative solution of the second subexercise. In the yield algebra, A and $\circ$ is the same. The parameter (i, j) denotes the beginning and the end of the substring. In the partial ordering, $(i_1, j_1) \leq (i_2, j_2)$ if $i_1 \leq i_2$ and $j_2 \leq j_1$. The recursions are

$$S((i, j)) = \sqcup_{k=i}^{j-1} S((i, k)) \circ S((k+1, j))$$

with initial conditions $S((i, i,)) = \{\epsilon\}$.

In the evaluation algebra, $R = \mathbb{Z}[z_1, z_2]$, the two variable polynomial ring over the integers. The function f maps $z_1^m z_2^w$ to a breaking scenario with m number of breaks and w total running time. The function T depends on the parameters; if $p(Q) = (i, k)$ and $p(W) = (k+1, j)$ then

$$T(Q, W) = z_1 z_2^{j-i+1} f(Q) f(W),$$

and similarly, for any $r_1, r_2 \in R$,

$$T(r_1, r_2, (i, k), (k+1, j)) = z_1 z_2^{j-i+1} r_1 r_2.$$

The number of breaking scenarios with total running time w in m number of steps is the coefficient of $z_1^m z_2^w$ in $F(S((1, n)))$.

Exercise 24. Notice that any edge e of a polygon is part of a triangle in a triangulation, therefore, at least one diagonal incident to e is in the collection of the diagonals describing the triangulation. If $e = (v_i, v_{i+1})$ is an edge, consider the triangle in which e is participating. If it contains a diagonal incident to v_1 it is called the left neighbor of e. Similarly, if it contains a diagonal incident to v_2, it is called the right neighbor. For a set of diagonals of a triangulation, define the following canonical ordering of the diagonals. Let $v_1, v_2, \ldots, v_n$ be the vertices, let $(v_n, v_1, v_2 \ldots v_{n-1}, v_n)$ denote the polygon, and for any diagonal (v_i, v_j), let $(v_i, v_j, v_{j+1}, v_{j+2}, \ldots, v_{i-1}, v_i)$ and $(v_j, v_i, v_{i+1}, v_{i+2}, \ldots, v_{j-1}, v_j)$ denote the two sub-polygons emerging by cutting the polygon at diagonal (v_i, v_j). The canonical ordering first contains the left neighbor of (v_n, v_1), denoted by (v_n, v_i), if it exists. Then the canonical ordering is continued with the canonical ordering of the triangulation of the sub-polygon $(v_n, v_i, v_{i+1}, \ldots, v_{n-1}, v_n)$, if the left neighbor of the edge (v_n, v_1) exists. Then the canonical ordering is continue with the right neighbor of (v_n, v_1), denoted by (v_i, v_1), if it exists. Then the canonical ordering is continued with the canonical ordering of the sub-polygon $(v_i, v_1, v_2, \ldots v_i)$, if the right neighbor of (v_n, v_1) exists. Note that in the sub-polygon $(v_n, v_i, v_{i+1}, \ldots, v_{n-1}, v_n)$, the first edge is the former diagonal (v_n, v_i), which defines the canonical ordering of the diagonals in this sub-polygon. Similarly, (v_i, v_1) is the first edge in $(v_i, v_1, v_2, \ldots v_i)$.

Define the following yield algebra. The base set A contains the triangulations of sub-polygons $(v_i, v_j, v_{j+1}, \ldots v_i)$ described as the canonical ordering of the diagonals participating in the triangulations, where $i > j$. The parameters are the pairs (i, j) describing the former diagonal (v_i, v_j) of the sub-polygon. In the partial ordering of the parameters, $(i_1, j_1) \leq (i_2, j_2)$ if $i_1 \leq i_2$ and $j_2 \leq j_1$ (observe that $i > j$). If $p(Q) = (i, k)$ and $p(R) = (k, j)$ then the operator $\circ$ acts on them as

$$Q \circ R = (v_i, v_k), Q, (v_k, v_j), R.$$

The unary operator $\circ_l$ adds the former diagonal (v_i, v_j) to a triangulation to get a triangulation of a larger sub-polygon. If $p(Q) = (i, j)$, then

$$\circ_l(Q) = (v_i, v_j), Q.$$

The recursions are

$$S((i, j)) = \circ_l\left(S((i-1, j))\right) \sqcup \left(\sqcup_{k=j+2}^{i-2} S((i, k)) \circ S((k, j))\right) \sqcup \circ_l\left(S((i, j+1))\right)$$
$$(2.142)$$

with the initial condition $S((i+2, i)) = \{\epsilon\}$, where ϵ is the empty sequence (of diagonals). If the score of the triangulation is additive, and the task is to find the triangulation with minimum score, then the evaluation algebra is the following. R is the tropical semiring. The function f assigns the sum of the lengths of the diagonals to a triangulation. The function T for operator $\circ$ depends on the parameters; if $p(Q) = (i, k)$ and $p(W) = (k, j)$, then

$$T(Q, W) = |(v_i, v_k)| \odot f(Q) \odot |(v_k, v_j)| \odot f(W),$$

and similarly, for any two values $r_1, r_2 \in R$,

$$T(r_1, r_2, (i, k), (k, j)) = |(v_i, v_k)| \odot r_1 \odot |(v_k, v_j)| \odot r_2.$$

Here $\odot$ is the tropical multiplication, that is, the usual addition, and $|(v_i, v_j)|$ denotes the length of the edge (v_i, v_j). The function T_l for the operator $\circ_l$ also depends on the parameter; if $p(Q) = (i, j)$, then

$$T_l(Q) = |(v_i, v_j)| \odot f(Q),$$

and similarly, for any $r \in R$,

$$T_l(r, (i, j)) = |(v_i, v_j)| \odot r.$$

The minimum score of the triangulations is $F(S((n, 1)))$.

If the score of the triangulation is additive, and the task is to find the average score, then the evaluation algebra is the following. $R = (\mathbb{R}^2, \oplus, \odot)$, where $\oplus$ is the coordinatewise addition and the multiplication is defined as

$$(x_1, y_1) \odot (x_2, y_2) := (x_1 x_2, x_1 y_2 + y_1 x_2).$$

The function f is defined as

$$f((e_1, e_2, e_k)) := \left(1, \sum_{i=1}^{k} |e_i|\right).$$

The function T for operator $\circ$ depends on the parameters. If $p(Q) = (i, k)$ and $p(W) = (k, j)$, then

$$T(Q, W) = (1, |(v_i, v_k)|) \odot f(Q) \odot (1, |(v_k, v_j)|) \odot f(W),$$

and similarly, for any $r_1, r_2 \in R$,

$$T(r_1, r_2, (i, k), (k, j)) = (1, |(v_i, v_k)|) \odot r_1 \odot (1, |(v_k, v_j)|) \odot r_2.$$

The function T_l for the operator $\circ_l$ depends on the parameter, if $p(Q) = (i, j)$, then

$$T_L(Q) = (1, |(v_i, v_j)|) \odot f(Q)$$

and similarly, for any $r \in R$,

$$T(r, (i, j)) = (1, |(v_i, v_j)|) \odot r$$

If $F(S((n, 1))) = (x, y)$, then x is the number of possible triangulations (it is easy to check that x is the $n - 2^{\text{nd}}$ Catalan number) and y is the total sum of the scores of the triangulations. The average score is simply y/x.

When the score is multiplicative and the task is to calculate the minimum score of the triangulations, then R is the exponentiated tropical semiring, that is, $R = (\mathbb{R}^+ \cup \infty, \oplus, \odot)$, where $\oplus$ is the minimum, and $\odot$ is the usual multiplication. The function f is defined as the product of the diagonal lengths in the triangulation. The function T for operator $\circ$ depends on the parameters; if $p(Q) = (i, k)$ and $p(W) = (k, j)$, then

$$T(Q, W) = |(v_i, v_k)| \odot f(Q) \odot |(v_k, v_j)| \odot f(W),$$

and similarly, for any $r_1, r_2 \in R$

$$T(r_1, r_2, (i, k), (k, j)) = |(v_i, v_k)| \odot r_1 \odot |(v_k, v_j)| \odot r_2.$$

The function T_l for operator $\circ_l$ depends on the parameters; if $p(Q) = (i, j)$, then

$$T_l(Q) = |(v_i, v_j)| \odot f(Q),$$

and similarly, for any $r \in R$

$$T_l(r, (i, j)) = |(v_i, v_j)| \odot r.$$

The minimum score is $F(S((n, 1)))$.

Finally, if the score is multiplicative and the task is to calculate the average score of the triangulations, then $R = (\mathbb{R}^2, \oplus, \odot)$, where $\oplus$ is the coordinatewise addition and $\odot$ is the coordinatewise multiplication. The function f is defined as

$$f((e_1, e_2, \ldots, e_k)) := \left(1, \prod_{i=1}^{k} |e_i|\right).$$

The function T for operator $\circ$ depends on the parameters; if $p(Q) = (i, k)$ and $p(W) = (k, j)$, then

$$T(Q, W) = (1, |(v_i, v_k)|) \odot f(Q) \odot (1, |(v_k, v_j)|) \odot f(W),$$

and similarly, for any $r_1, r_2 \in R$,

$$T(r_1, r_2, (i, k), (k, j)) = (1, |(v_i, v_k)|) \odot r_1 \odot (1, |(v_k, v_j)|) \odot r_2.$$

The function T_l for operator $\circ_l$ depends on the parameters; if $p(Q) = (i, j)$, then

$$T_l(Q) = (1, |(v_i, v_j)|) \odot f(Q),$$

and similarly, for any $r \in R$,

$$T_l(r, (i, j)) = (1, |(v_i, v_j)|) \odot r.$$

If $F(S((n, 1))) = (x, y)$, then x is the number of triangulations, and y is the sum of the scores. The average score is y/x.

Remark. If P is a convex polygon, then x, the number of triangulations, is the $n - 2^{\text{nd}}$ Catalan number, and thus, in the last subexercise, R could be chosen as the real number field, with the usual addition and multiplication. However, if P is concave then all the calculations can be modified such that in the yield algebra, the recursion in Equation (2.142) can be modified in such a way that only those operations are considered for which the emerging diagonals are inside the polygon. Then the appropriate evaluation algebra still counts the number of triangulations, which will no longer be the $n - 2^{\text{nd}}$ Catalan numbers.

Exercise 25. The two sub-exercises need different yield algebras and evaluation algebras. If the order of the breaking counts, then the yield algebra is simple, however, the evaluation algebra is tricky. When the order of the breakings does not count, the yield algebra is complicated and the evaluation algebra is simple.

When the order of the breaks does not count, a canonical ordering of the breakings must be defined. For any chocolate bar larger than 1×1, there is either at least one vertical break running through the entire chocolate bar or there is at least one horizontal break running through the entire chocolate bar, however it is impossible to have both horizontal and vertical breaks running through the entire chocolate bar. Call these breaks long breaks. If there are one or more vertical long breaks, the canonical order starts with the leftmost break breaking the bar into pieces B_1 and B_2 (left and right pieces, respectively), then followed by the canonical order of breaks of B_1 which must start with a horizontal break, then followed by the canonical order of the breaks of B_2, which might start with both a vertical and a horizontal break. If there are one or more horizontal long breaks, the canonical order starts with the top break breaking the bar into pieces B_1 and B_2 (top and bottom pieces, respectively), then followed by the canonical ordering of the breaks of B_1 which must start with a vertical break, then followed by the canonical order of the breaks of B_2, which might start with both a vertical and a horizontal break.

Construct the following yield algebra. The base set A contains the canonical ordering of chocolate bars with dimensions $i \times j$. The parameters are triplets (i, j, x), where i and j are the dimensions of the chocolate bar, and $x \in \{h, v\}$ denotes if the first break is horizontal or vertical. The partial ordering on the parameters is such that $(i_1, j_1, x_1) \leq (i_2, j_2, x_2)$ if $i_1 \leq i_2$ and $j_1 \leq j_2$, with the following exception. When both $i_1 = i_2$ and $j_1 = j_2$, then the two parameters are not comparable unless $x_1 = x_2$, when naturally the two parameters are the same.

If $p(Q) = (i_1, j, v)$ and $p(R) = (i_2, j, x)$, then the $\circ_h$ operator is defined as

$$Q \circ_h R := b_{i_1,h}, Q, R$$

where $b_{i_1,h}$ denotes the horizontal break breaking the chocolate bar at the horizontal line after the first i_1 rows. If $p(Q) = (i, j_1, h)$ and $p(R) = (i, j_2, x)$, then the $\circ_v$ operator is defined as

$$Q \circ_v R := b_{j_1,v}, Q, R$$

where $b_{j_1,v}$ denotes the vertical break breaking the chocolate bar at the vertical line after the first j_1 columns. The recursions are

$$S((i, j, h)) = \bigsqcup_{k=1}^{i-1} \bigsqcup_{x \in \{h,v\}} S((k, j, v)) \circ_h S((i - k, j, x))$$
$$S((i, j, v)) = \bigsqcup_{k=1}^{j-1} \bigsqcup_{x \in \{h,v\}} S((i, k, h)) \circ_v S((i, j - k, x))$$

with the initial condition $S((1,1)) = \{\epsilon\}$, where ϵ is the empty string (of breaking steps). The evaluation algebra is the standard one for computing the size of the sets, that is, R is the integer ring, f is the constant 1 function, and both functions for the operators $\circ_h$ and $\circ_v$ are the multiplication.

When the order does count, the possible breaking scenarios are clustered based on a canonical ordering defined below. The first breaking of the canonical ordering of a breaking scenario is its first break b breaking the bar into pieces B_1 and B_2 (the left and right or the top and bottom pieces, respectively), then followed by the canonical ordering of the breaks in B_1, $b_{1,1}, b_{1,2}, \ldots, b_{1,k_1}$ and then the canonical ordering of the breaks in B_2, $b_{2,1}, b_{2,2}, \ldots, b_{2,k_2}$. If there are g_1 number of breaking scenarios of B_1 whose canonical ordering is $b_{1,1}, b_{1,2}, \ldots, b_{1,k_1}$ and there are g_2 number of breaking scenarios of B_2 whose canonical ordering is $b_{2,1}, b_{2,2}, \ldots, b_{2,k_2}$, then there are

$$\binom{k_1 + k_2}{k_1} g_1 g_2$$

breaking scenarios whose canonical ordering is

$$b, b_{1,1}, b_{1,2}, \ldots, b_{1,k_1}, b_{2,1}, b_{2,2}, \ldots, b_{2,k_2}.$$

The idea is to define a yield algebra building the canonical ordering of the breaking scenarios, then cz^k will be assigned to each canonical ordering, where c is the number of breaking scenarios that the canonical ordering represents, and k is the number of breaks in it.

Having said this, define the following yield algebra. The base set A contains the canonical ordering of the possible breaking scenarios of the chocolate bars with dimensions $i \times j$. The parameters (i, j) denote these dimensions, $(i_1, j_1) \leq (i_2, j_2)$ if $i_1 \leq i_2$ and $j_1 \leq j_2$. If $p(Q) = (i_1, j)$ and $p(R) = (i_2, j)$, then the operator $\circ_h$ is defined as

$$Q \circ_h R := b_{i_1, h}, Q, R$$

where $b_{i_1, h}$ denotes the horizontal break breaking the chocolate bar at the horizontal line after the first i_1 rows. If $p(Q) = (i, j_1)$ and $p(R) = (i, j_2)$, then the $\circ_v$ operator is defined as

$$Q \circ_v R := b_{j_1, v}, Q, R$$

where $b_{j_1, v}$ denotes the vertical break breaking the chocolate bar at the vertical line after the first j_1 columns. The recursions are

$$
S((i, j)) = \left(\sqcup_{k=1}^{i-1} S((k, j)) \circ_h S((i - k, j)) \right) \sqcup \\
\left(\sqcup_{k=1}^{j-1} S((i, k)) \circ_v S((i, j - k)) \right)
$$

In the evaluation algebra, $R = \mathbb{Z}[z]$, the polynomial ring over the integers. The function f maps cz^k to a canonical ordering, where c is the number of breaking scenarios that the canonical ordering represent, and k is the number of breaks in it. Both functions T_h and T_v for the operators $\circ_h$ and $\circ_v$ are the following convolution of polynomials.

$$\left(\sum_{i=0}^{k_1} a_i z^i \right) \diamond \left(\sum_{j=0}^{k_2} b_j z^j \right) = \sum_{l=1}^{k_1+k_2+1} c_l z_l$$

where

$$c_{k+1} = \sum_{i=0}^{k} \binom{k}{i} a_i b_{k-i}.$$

The total number of breaking scenarios is the sum of the coefficients in $F(S((n, m)))$.

The answer to the question for fun is that any break takes one piece of chocolate bar and results two pieces, therefore, any breaking scenario of an $n \times m$ chocolate bar needs $nm - 1$ breakings (the number of pieces should be increased from 1 to nm, and each break increases the number of pieces by 1). Thus, $F(S((n, m)))$ is only one monoid, cz^{nm-1}. Therefore, the evaluation algebra can be simplified in the following way. R is the integer ring. The function f maps the number of breaking scenarios whose canonical representation is the one in question. Both T_h and T_v depend on the parameters; if $p(Q) = (i_1, j)$ and $p(W) = (i_2, j)$, then

$$T_h(Q, W) = \binom{i_1 j + i_2 j - 2}{i_1 j - 1} f(Q) f(W),$$

and similarly, for any two integers, r_1 and r_2,

$$T_h(r_1, r_2, (i_1, j), (i_2, j)) = \binom{i_1 j + i_2 j - 2}{i_1 j - 1} r_1 r_2.$$

T_v is defined in a similar way.

Exercise 27. Assume that the labels $l_1, l_2, \ldots l_n$ are ordered. Observe that in any subtree, the indexes of the labels of the vertices cover an interval $[i, j]$. Construct the following yield algebra. The base set A is the possible binary search trees of the substrings. The parameters (i, j) define the first and last indexes of the substring. In the partial ordering of the parameters, $(i_1, j_1) \leq (i_2, j_2)$ if $i_2 \leq i_1$ and $j_1 \leq j_2$. The operator $\circ$ might have two binary search trees with parameters $p(Q) = (i, k - 1)$ and $p(W) = (k + 1, j)$ and connects the roots of Q and W via the new root of the larger subtree. The new root is labeled by l_k. The unary operator $\circ_l$ takes a sub-tree with parameter (i, j), and connects its root to the new root of the larger sub-tree labeled by l_{j+1}. The child of the new root is a left child. Finally, the unary operator $\circ_r$ takes a sub-tree with parameter (i, j) and connects its root to the new root labeled by l_{i-1}. The child of the new root is a right child. The recursions are

$$S((i, j)) = \circ_r (S((i + 1, j))) \sqcup \left(\sqcup_{k=i+1}^{j-1} S((i, k - 1)) \circ S((k + 1, j)) \right) \sqcup$$
$$\sqcup \circ_l (S((i, j - 1))).$$

For each i, the initial value $S((i, i))$ is the set containing the tree with a single vertex labeled by l_i.

If the task is to find the minimum possible cost, then the algebraic structure R in the evaluation algebra is the tropical semiring. The f function assigns the score of the tree for any binary search tree. The function T for operator $\circ$ depends on the parameters; if $p(Q) = (i, k - 1)$ and $p(W) = (k + 1, j)$, then

$$T(Q, W) = f(Q) \odot f(W) \odot \left(\odot_{m=i}^{j} \pi(l_m) \right)$$

where $\odot$ is the tropical multiplication, that is, the usual addition. The explanation for this definition is that each distance in Q and W is increased by

1, furthermore, the new root is labeled with l_k, and this new vertex also has distance 1. Similarly, for any $r_1, r_2 \in R$,

$$T(r_1, r_2, (i, k-1), (k+1, j)) = r_1 \odot r_2 \odot \left(\odot_{m=i}^{j} \pi(l_m) \right).$$

The function T_l and T_r also depend on the parameter. If $p(Q) = (i, j)$, then

$$T_r(Q) = f(Q) \odot \left(\odot_{m=i-1}^{j} \pi(l_m) \right),$$

and for any $r_1 \in R$,

$$T_r(r_1, (i, j)) = r_1 \odot \left(\odot_{m=i-1}^{j} \pi(l_m) \right).$$

Similarly, for $p(Q) = (i, j)$,

$$T_l(Q) = f(Q) \odot \left(\odot_{m=i}^{j+1} \pi(l_m) \right),$$

and for any $r_1 \in R$,

$$T_l(r_1, (i, j)) = r_1 \odot \left(\odot_{m=i}^{j+1} \pi(l_m) \right).$$

If the task is to calculate the average score of the possible binary search trees, then $R = (\mathbb{R}^2, \oplus, \odot)$, where the addition is coordinatewise and the multiplication is defined as

$$(x_1, y_1) \odot (x_2, y_2) := (x_1 x_2, x_1 y_2 + y_1 x_2).$$

For any binary search tree with score x, the f function assigns the value $(1, x)$. The function T for operator $\circ$ depends on the parameters; if $p(Q) = (i, k-1)$ and $p(W) = (k+1, j)$, then

$$T(Q, W) = f(Q) \odot f(W) \odot \left(1, \sum_{m=i}^{j} \pi(l_m) \right),$$

and similarly, for any $r_1, r_2 \in K$,

$$T(r_1, r_2, (i, k-1), (k+1, j)) = r_1 \odot r_2 \odot \left(1, \sum_{m=i}^{j} \pi(l_m) \right).$$

The functions T_r and T_l for operators $\circ_r$ and $\circ_l$ also depend on the parameter. If $p(Q) = (i, j)$, then

$$T_r(Q) = f(Q) \odot \left(1, \sum_{m=i-1}^{j} \pi(l_m) \right)$$

$$T_l(Q) = f(Q) \odot \left(1, \sum_{m=i}^{j+1} \pi(l_m)\right)$$

and for any $r_1 \in R$,

$$T_r(r_1, (i,j)) = r_1 \odot \left(1, \sum_{m=i-1}^{j} \pi(l_m)\right),$$

and

$$T_l(r_1, (i,j)) = r_1 \odot \left(1, \sum_{m=i}^{j+1} \pi(l_m)\right).$$

If $F(S((1,n))) = (x,y)$, then the average score is y/x.

Exercise 28. Let v be an arbitrary leaf of $T = (V, E)$ and define a partial ordering $\leq$ on V such that $v_1 \leq v_2$ if the unique path from v_1 to v contains v_2. Let T_{v_i} denote the subtree which contains the vertices which are smaller than or equal to v_i. A matching of a subtree $T_{v_i} = (E_{v_i}, V_{v_i})$ is a mapping $M : E_{v_i} \to \{0,1\}$ such that for any two edges $e_1, e_2 \in E_{v_i}$, if $M(e_1) = M(e_2) = 1$, then e_1 and e_2 are disjoint. The size of a matching is $|\{e | M(e) = 1\}|$.

Define the following yield algebra. The base set A contains all matchings of all subtrees T_{v_i}. The parameter set θ is $V \times \{0,1\}$. On this parameter set, $(v_i, x_i) \leq (v_j, x_j)$ if $v_i < v_j$ in the above defined partial ordering of the vertices of T or $v_i = v_j$ and $x_i = x_j$. If $M \in K$ is a matching on T_{v_i} such that one of the edges covers v_i, then $p(M) = (v_i, 1)$, otherwise $p(M) = (v_i, 0)$.

The following operators are defined in the yield algebra. Let u be an internal node, let w_1 and w_2 be its children (the vertices that are smaller than u in the partial ordering). Let $e_1 = (u, w_1)$ and $e_2 = (u, w_2)$. If M_1 is a matching of T_{w_1} and M_2 is a matching of T_{w_2}, then $M_1 \circ M_2$ is a matching M of T_u such that $M(e_1) = M(e_2) = 0$ and other assignments follow the mappings in M_1 and M_2.

If M_1 is a matching with parameter $(w_1, 0)$ and M_2 is a matching of T_{w_2}, then $M_1 \circ_l M_2$ is a matching M of T_u such that $M(e_1) = 1$, $M(e_2) = 0$ and other assignments follow the mappings in M_1 and M_2.

If M_1 is a matching of T_{w_1} and M_2 is a matching with parameter $(w_2, 0)$, then $M_1 \circ_r M_2$ is a matching M of T_u such that $M(e_1) = 0$, $M(e_2) = 1$ and other assignments follow the mappings in M_1 and M_2.

If the child of v is w, $e = (v, w)$, and M is a matching with parameter $(w, 0)$, then $M_1 = \circ_1(M)$ is a matching such that $M_1(e) = 1$ and all other mappings follow M. Finally, if M is a matching of M_w, then $M_0 = \circ_0(M)$ is a matching such that $M_0(e) = 0$ and all other assignments follow M.

The recursions are

$$
\begin{aligned}
S((u,0)) &= \bigsqcup_{\substack{x_1 \in \{0,1\} \\ x_2 \in \{0,1\}}} S((w_1, x_1)) \circ S((w_2, x_2)) &\text{(2.143)} \\
S((u,1)) &= \left(\bigsqcup_{x_2 \in \{0,1\}} S((w_1, 0)) \circ_l S((w_2, x_2))\right) \sqcup \\
&\quad \left(\bigsqcup_{x_1 \in \{0,1\}} S((w_1, x_1)) \circ_r S((w_2, 0))\right) &\text{(2.144)}
\end{aligned}
$$

for any internal node u and its children w_1 and w_2 and

$$S((v,0)) = \circ_0 (S((w,0))) \sqcup \circ_0 (S((w,1))) \tag{2.145}$$
$$S((v,1)) = \circ_1 (S((w,0))) \tag{2.146}$$

for v and its child w.

If the task is to calculate the number of matchings, then the evaluation algebra can be set in the following way. R is the integer ring, f is the constant 1 function, the T functions for $\circ$, $\circ_l$, $\circ_r$ are multiplications, the T functions for $\circ_0$ and $\circ_1$ is the identity. The total number of matchings is $F(S((v,0))) + F(S((v,1)))$.

If the task is to calculate the size of the maximum matching, then R is the dual tropical semiring (where the addition is the maximum and not the minimum), f is the size of the matching. The T function for $\circ$ is the tropical multiplication (that is, the usual addition of integers); for $\circ_l$ and $\circ_r$, it is the tropical multiplication further tropically multiplied with 1 (that is, the usual addition of integers and further adding 1), the T function for $\circ_0$ is the identity, and the T function for $\circ_1$ is the tropical multiplication with 1, namely, adding 1 in the usual arithmetic. The size of a maximal matching is $F(S((v,0))) \oplus F(S((v,1)))$, that is $\max\{F(S((v,0))), F(S((v,1)))\}$ in the usual arithmetic.

If the task is to count the number of maximum matchings, then $R = \mathbb{Z}[x]$, $f(M) = x^{|M|}$. The T function for $\circ$ is the multiplication. For $\circ_l$ and $\circ_r$, it is the multiplication, such that the product is further multiplied with x. It is the identity function for $\circ_0$, and finally, it is the multiplication with x for $\circ_1$. The number of maximal matchings is the coefficient of the largest monoid in $F(S((v,0))) + F(S((v,1)))$.

Finally, if the task is to calculate the average edges in a matching, then the evaluation algebra might be set similar to the one in Subsection 2.1.3.4. Alternatively, since the number of edges in a matching is a non-negative integer and upper bounded by half the number of vertices, the same evaluation algebra can be used as the one used to count the number of maximum matchings. The number of matchings is the value of the polynomial $F(S((v,0))) + F(S((v,1)))$ at $x = 1$ and the total number of edges in all matchings is the value of the polynomial $\frac{d}{dx}[F(S((v,0))) + F(S((v,1)))]$ at $x = 1$. The average number of edges in a matching is the total number of edges in the matchings divided by the number of matchings.

Exercise 32. Let the input text be $w_1 \ w_2 \ \dots \ w_n$, where each w_i is a word. For any $i \le j$, define

$$g(i,j) := \left(t - \sum_{k=i}^{j} |w_k| + j - i \right)^2.$$

Define the following yield algebra. The base set A contains the possible wrappings of the prefix text $w_1 \ w_2 \ \dots \ w_i$ into lines. The parameter i denotes the number of words in the prefix, and the parameters are naturally ordered. The operator $\circ_{i,j}$ adds a new line to the wrapped text containing the words from w_i till w_j. The recursions are

$$S(j) = \bigsqcup_{i \mid \sum_{k=i}^{j} |w_k| + j - i \leq t} \circ_{i,j} \left(S(i-1) \right)$$

with the initial condition $S(0)$ containing an empty text.

In the evaluation algebra, $R = \mathbb{Z}[z_1, z_2]$, the two variable polynomial ring. The function f maps $z_1^k z_2^x$ to each wrapping, where k is the number of lines, and x is the total score. The function $T_{i,j}$ for operator $\circ_{i,j}$ is the multiplication with $z_1 z_2^{g(i,j)}$. The number of possible wrappings of the text into k lines with score x is the coefficient of $z_1^k z_2^x$ in $F(S(n))$.

Exercise 34. First a few notations are introduced. Recall that a sequence Y is a *supersequence* of sequence X if X is a subsequence of Y. Y is a *palindromic supersequence* of X if the following conditions hold.

- Y is a supersequence of X

- Y is palindromic

- X can be injected onto to Y such that for each pair (y_i, y_{m-i}), at least one of the characters is an image of a character from X, where $m = |Y|$, furthermore, if m is odd, then $y_{\frac{m+1}{2}}$ is an image of a character of X.

Note that for an X and one of its palindromic supersequences Y, there might be more than one injection that certifies that Y is indeed a supersequence of X. However, each such injection indicates a possible (and different!) way to make X palindromic. Any injection can be represented as an alignment of X and Y that does not contain mismatches (substitutions) and deletions, only matches and insertions. Therefore the algebraic dynamic approach counts these alignments.

Based on the above, define the following yield algebra. The base set A is the possible alignments of substrings of X to their possible palindromic supersequences. The parameters (i, j) indicate the first and the last indexes of the substrings. In the partial ordering of the parameters, $(i_1, j_1) \leq (i_2, j_2)$ if $i_2 \leq i_1$ and $j_1 \leq j_2$. The unary operator $\circ$ takes an alignment with parameters (i, j), and adds an alignment column $\begin{matrix} x_{i-1} \\ x_{i-1} \end{matrix}$ at the beginning of the alignment and adds an alignment column $\begin{matrix} x_{j+1} \\ x_{j+1} \end{matrix}$ at the end of the alignment. The unary operator $\circ_l$ takes an alignment with parameters (i, j), adds an alignment column $\begin{matrix} x_{i-1} \\ x_{i-1} \end{matrix}$ at the beginning of the alignment, and adds an alignment column $\begin{matrix} - \\ x_{i-1} \end{matrix}$ at the end of the alignment. The unary operator $\circ_r$ takes an alignment with parameters (i, j), adds an alignment column $\begin{matrix} - \\ x_{j+1} \end{matrix}$ at the beginning of the alignment, and adds an alignment column $\genfrac{}{}{0pt}{}{x_{j+1}}{x_{j+1}}$ at the end of the alignment. The recursions are

$$S((i,j)) = \circ_l\left(S((i+1,j))\right) \sqcup \circ_r\left(S((i,j-1))\right) \quad \text{if } x_i \neq x_j$$

and

$$S((i,j)) = \circ_l\left(S((i+1,j))\right) \sqcup \circ_r\left(S((i,j-1))\right) \sqcup \circ\left(S((i,j))\right) \quad \text{if } x_i = x_j$$

with initial conditions $S((i,i)) = \left\{ \genfrac{}{}{0pt}{}{x_i}{x_i} \right\}$ and $S((i+1,i)) = \left\{ \genfrac{}{}{0pt}{}{\epsilon}{\epsilon} \right\}$ where $\genfrac{}{}{0pt}{}{\epsilon}{\epsilon}$ is the empty alignment.

In the evaluation algebra, the algebraic structure R is the integer polynomial ring $\mathbb{Z}[z]$. The function f maps z^k to an alignment, where k is the number of insertions in it. The function T for the operator $\circ$ is the identity function, the function T_l and T_r for the operators $\circ_l$ and $\circ_r$ are both the multiplication with z. The number of ways to insert a minimum number of characters into X to make it palindromic is the coefficient of the smallest monomial in $F(S((1,n)))$, where n is the length of X.

Exercise 35. Construct a partially ordered set that contains all (a_i, b_i), (b_i, a_i), (a_i, c_i), (c_i, a_i), (b_i, c_i) and (c_i, b_i) pairs for each box with dimensions (a_i, b_i, c_i). Add a 0 and a 1 to this partially ordered set. The ordering is such that $(x_1, y_1) \leq (x_2, y_2)$ if $x_1 \leq x_2$ and $y_1 \leq y_2$. The possible stacks of boxes are the subsequences of the possible paths from 0 to 1. The yield algebra should build up stacks such that the last box is a given one with a given rotation. The evaluation algebras are the standard ones with the tropical semiring, the integer ring and the ring introduced in Subsection 2.1.3.4.

Exercise 36. Recall that the sub-trees of a rooted binary tree are the rooted binary trees that are rooted in a vertex of the tree. Define the following yield algebra. For each subexercise, the base set A contains the colored subtrees of the given tree $\mathcal{T}$ colored as follows:

(a) The vertices are colored with black and white, and no neighbor vertices are both colored with black. The black vertices form an independent set.

(b) The vertices are colored with three colors, and no neighbor vertices have the same color.

(c) The vertices are partially colored with $k \geq 2$ colors, and no neighbor vertices have the same color, however, they might both be uncolored.

The parameters of a coloring are pairs (v, c), where v denotes the root of the subtree and c is its color (no coloring can be considered as an additional color). The partial ordering of the parameters is such that $(v_1, c_1) \leq (v_2, c_2)$ if v_2 is on the path from v_1 to the root of $\mathcal{T}$. The operator $\circ_c$ takes two subtrees with sibling nodes, connects them via their common parent, and colors this parent with color c. In the following recursions, the children of v are u_1 and u_2, w denotes white, b denotes black, r denotes red, n denotes "no color". The recursions are

(a)

$$S((v,w)) = (S((u_1,b)) \sqcup S((u_1,w))) \circ_w (S((u_2,b)) \sqcup S((u_2,w)))$$
$$S((v,b)) = S((u_1,w)) \circ_b S((u_2,w))$$

(b)

$$S((v,w)) = (S((u_1,b)) \sqcup S((u_1,r))) \circ_w (S((u_2,b)) \sqcup S((u_2,r)))$$
$$S((v,b)) = (S((u_1,w)) \sqcup S((u_1,r))) \circ_w (S((u_2,w)) \sqcup S((u_2,r)))$$
$$S((v,r)) = (S((u_1,w)) \sqcup S((u_1,b))) \circ_w (S((u_2,w)) \sqcup S((u_2,b)))$$

(c)

$$S((v,c_i)) = \left(\left(\bigsqcup_{c_j \neq c_i} S((u_1,c_j))\right) \sqcup S((u_1,n))\right) \circ_{c_i}$$
$$\left(\left(\bigsqcup_{c_j \neq c_i} S((u_2,c_j))\right) \sqcup S((u_2,n))\right)$$
$$S((v,n)) = \left(\left(\bigsqcup_{c_i} S((u_1,c_i))\right) \sqcup S((u_1,n))\right) \circ_n$$
$$\left(\left(\bigsqcup_{c_i} S((u_2,c_i))\right) \sqcup S((u_2,n))\right).$$

$$(2.147)$$

The evaluation algebra is the standard one counting the size of sets, that is, R is the integer ring, f is the constant 1 function, and each T operator is the multiplication.

Exercise 38. Note that inverting the path from the bottom right corner to the top left makes it also a path from the top left corner to the bottom right. This inverted path will step to a shared square in the same number of steps as the top-down path. Having said this, construct the following yield algebra. The base set A contains the pair of prefixes of the top-down path and the inverted down-top path with the same length. The parameter quadruple (i_1, j_1, i_2, j_2) gives the indexes of the last squares of the prefixes, (i_1, j_1) and (i_2, j_2). In the partial ordering of the parameters, $(i_1, j_1, i_2, j_2) \leq (l_1, m_1, l_2, m_2)$ if $i_k \leq l_k$ and $(j_k \leq m_k)$ for both $k = 1, 2$. The four unary operators, $\circ_{hh}$, $\circ_{hv}$, $\circ_{vh}$ and $\circ_{vv}$ extend the prefixes with horizontal or vertical steps. The recursions are

$$S((i_1, j_1, i_2, j_2)) = \circ_{hh} (S((i_1 - 1, j_1, i_2 - 1 j_2))) \sqcup$$
$$\sqcup \circ_{hv} (S((i_1 - 1, j_1, i_2, j_2 - 1))) \sqcup$$
$$\sqcup \circ_{vh} (S((i_1, j_1 - 1, i_2 - 1 j_2))) \sqcup$$
$$\sqcup \circ_{vv} (S((i_1, j_1 - 1, i_2, j_2 - 1))),$$

with the initial condition $S((1,1,1,1))$ containing a pair of paths both containing the top left square.

If the task is to count the paths, then the evaluation algebra is the usual one with $R = \mathbb{Z}$, the f function is the constant 1, and each function T_{xy}, $x, y \in \{h, v\}$ is the identity. The number of paths is $F(S((n, m)))$.

If the task is to calculate the tour with the maximum score, then the semiring in the evaluation algebra is the dual tropical semiring (with maximum instead of minimum). The f function is the sum (tropical product) of the number of coins along the paths, without multiplicity. The T functions depend on the parameter. If the parameter of the resulting couple of paths is (i_1, j_1, i_2, j_2) and $i_1 = i_2$ (and then $j_1 = j_2!$), then the function is the tropical multiplication (usual addition) with the number of coins on square m_{i_1, j_1}. Otherwise, the function is the tropical multiplication (usual addition) of the sum (tropical product) of the coins on the two different squares. The maximum possible sum of coin values is $F(S((n, m)))$.

Finally, if the task is to calculate the average score of the tours, then $R = (\mathbb{R}^2, \oplus, \odot)$, where $\oplus$ is the coordinatewise addition and

$$(x_1, y_1) \odot (x_2, y_2) = (x_1 x_2, x_1 y_2 + y_1 x_2).$$

The f function assigns $(1, w)$ of a pair of paths, where w is the sum of the coins on a pair of paths without multiplicity. The T functions depend on the parameters. If the resulting pair of paths have parameters (i_1, j_1, i_2, j_2) and $i_1 = i_2$, then it is the multiplication with $(1, w_{i_1, j_1})$, where w_{i_1, j_1} is the number of coins on square m_{i_1, j_1}. Otherwise, it is the multiplication with $(1, w_{i_1, j_1} + w_{i_2, j_2})$. If $F(S((n, m))) = (x, y)$, then the average score is y/x.

Exercise 39. The trick of inverting the down-top path still works. However, the lengths of the paths might differ, therefore the prefix of the top-down path and the prefix of the inverted path might step to the same square after a different number of steps. Therefore the pair of prefixes must be in the same column to be able to check which squares are shared. An extension of both prefixes contains a diagonal or horizontal step and possibly a few down steps.

Exercise 40. Construct the following yield algebra. The base set A contains the possible tripartitions $X \sqcup Y \sqcup Z$ of the possible prefixes of $a_1, a_2, \ldots, a_n$. The parameter (l, d_1, d_2) describes the length of the prefix, l, and the differences in the sums of the sets, $d_1 := \sum_{a_i \in X} a_i - \sum_{a_j \in Y} a_j$, $d_2 := \sum_{a_i \in X} a_i - \sum_{a_k \in Z} a_k$. The partial ordering is based on the length of the prefixes, and a parameter with shorter prefix length is smaller. The unary operators $\circ_{a_i, X}$, $\circ_{a_i, Y}$, and $\circ_{a_i, Z}$ add a_i to the sets X, Y and Z, respectively. The recursions are

$$S(l, d_1, d_2) = \circ_{a_l, X} \left(S((l - 1, d_1 - a_l, d_2)) \right) \quad \sqcup$$
$$\circ_{a_l, Y} \left(S((l - 1, d_1 + a_l, d_2)) \right) \quad \sqcup \quad \circ_{a_l, Z} \left(S((l - 1, d_1, d_2 + a_l)) \right)$$

with the initial condition $S((0, 0, 0))$ containing an empty set for all X, Y and Z.

If the task is to count the tripartitions with equal sums, then the evaluation algebra contains $R = \mathbb{Z}$, f is the constant 1 function, and all T functions are the identity functions. The number of tripartitions with equal sums is $S((n, 0, 0))$.

The average of the sum of squared differences of the subset sums can be calculated with the same evaluation algebra. The solution is

$$\frac{\sum_{d_1} \sum_{d_2} \left(F(S((n, d_1, d_2)))(d_1^2 + d_2^2 + (d_1 - d_2)^2) \right)}{\sum_{d_1} \sum_{d_2} F(S((n, d_1, d_2)))}.$$

Exercise 42. First, observe that

$$(c^*)^* = c$$

since the $*$ operation is the mirroring to the $y = x$ axis when c is drawn as a column diagram. This means that there is a bijection between the possible c sequences and c^* sequences. Furthermore,

$$\sum_{i=1}^{m} c_i \geq \sum_{j=1}^{n} c_j^*$$

with equality if $c_1 \leq n$. Therefore the pair of conditions

$$\sum_{i=1}^{n} b_i = \sum_{j=1}^{m} c_i$$

and

$$\sum_{i=1}^{n} b_i \leq \sum_{j=1}^{n} c_j^*$$

is equivalent with

$$\sum_{i=1}^{n} b_i = \sum_{j=1}^{n} fc_j^*$$

and $c_1 \leq n$ (and thus $c_{n+1}^* = 0$). Therefore the number of graphical bidegree sequences $\{b, c\}$ with length $|b| = n$ and $|c| = m$ is the number of decreasing non-negative sequences $\{b, c^*\}$ with length $|b| = n$ and $|c^*| = n$ such that they satisfy for all $k = 1, \ldots, n$

$$\sum_{i=1}^{k} b_i \leq \sum_{j=1}^{k} c_j^* \tag{2.148}$$

with equality when $k = n$, and $b_1^* \leq m$.

Based on the above, define the following yield algebra. The base set A contains the equal long (b, c^*) pair of sequences satisfying condition in Equation (2.148) and $c_1^* \leq m$. The parameter set contains quadruples (s, t, δ, k) describing the last number in b, the last number in c^*, the difference $\sum_{i=1}^{k} c_i^* - \sum_{j=1}^{k} b_j$, and the length of the sequences, respectively. In the partial ordering of the parameters, $(s_1, t_2, \delta_1, k_1) \leq (s_2, t_2, \delta_2, k_2)$ is $k_1 \leq k_2$. The unary operator $\circ_{x,y}$ extends the sequence b with number x and the sequence c^* with number y. The recursions are

$$S((s, t, d, k)) = \sqcup_{s_1 \geq s} \sqcup_{t_1 \geq t} \circ_{s,t} \left(S((s_1, t_1, d + t - s, k - 1)) \right)$$

with initial conditions $S((s, t, t - s, 1)) = \{\{s, t\}\}$ and $S((s, t, \delta, 1)) = \emptyset$ for all $\delta \neq t - s$.

The evaluation algebra is the standard one counting the size of the sets, that is, $R = \mathbb{Z}$, f is the identity function and each $T_{x,y}$ function is the identity function. The number of graphical bidegree sequences with lengths $|b| = n$ and $|c| = m$ is

$$\sum_{s=0}^{m} \sum_{t=0}^{m} F(S((s, t, 0, n))).$$

Chapter 3

Linear algebraic algorithms. The power of subtracting

In this chapter, we are going to introduce algorithms that count discrete mathematical objects using three operations: addition, subtraction and multiplication. The determinant and the Pfaffian of certain matrices will be the center of interest. The minors of Laplacians of graphs are the number of spanning trees or in case of directed matrices, the number of in-trees. The number of in-trees appears in counting the Eulerian circuits in directed Eulerian graphs. The Pfaffians of appropriately oriented adjacency matrices of planar graphs are the number of perfect matchings in those graphs.

The above-mentioned discrete mathematical objects might have weights, and these weights might come from arbitrary commutative rings. We might also want to calculate the sum of these weights. Divisions are not available in commutative rings, therefore it is necessary to give division-free polynomial running time algorithms for calculating the determinant and Pfaffian of matrices in an arbitrary commutative ring. Surprisingly, such algorithms are available, and actually, they are dynamic programming algorithms. These dynamic programming algorithms are on some combinatorial objects called *clow sequences*. Clow sequences are generalizations of permutations. The sum of signed weights of clow sequences coincide with the determinants and Pfaffians, which, by definition, are the sum of signed weights of permutations. The coincidence is due to cancellation of terms in the summation, which might not be available without subtractions. We are going to show that subtractions are inevitable in efficient algorithms: theorems exist that there are no algorithms that can count the above-mentioned discrete mathematical objects in polynomial time without subtraction.

3.1   Division-free algorithms for calculating the determinant and Pfaffian

It is well known that the determinant of a matrix can be calculated in polynomial time using Gaussian elimination. However, Gaussian elimination contains divisions that we might want to avoid for several reasons. One possible reason is that division might cause numerical instability [77]. Another reason is that we might want to calculate the determinant of a matrix over a commutative ring where division is not possible. This is the case, for example, when the characteristic polynomial of a matrix is to be calculated, which is formally a determinant of a matrix over the polynomial ring $\mathbb{R}[\lambda]$. The first such method was published in the textbook by Faddayev and Faddayeva [66], referring to a work of Samulelson [149]. Berkowitz published an algorithm [18] to calculate the determinant with parallel processors and he called it Samuelson's method referring to Faddayev and Faddyeva's textbook. Valiant gave a combinatorial explanation of the method and called it the Samuelson-Berkowitz algorithm. In fact, Samuelson's method was not division-free yet, so the reason for the erroneous attribution is that the first division-free approach is in the text book by Faddayev and Faddayeva in the section about Samuelson's method. The method was further developed by Mahayan and Vinay [124]. Here we introduce the combinatorial description of the method.

The combinatorial approach is also a dynamic programming algorithm. It calculates the determinant also in polynomial time, although in $O(n^4)$ running time instead of $O(n^3)$. On the other hand, it can be applied for a matrix over an arbitrary commutative ring. By definition, the determinant of a matrix $M = \{m_{i,j}\}$ is

$$det(M) = \sum_{\sigma \in \mathcal{S}_n} sign(\sigma) \prod_{i=1}^{n} m_{i,\sigma(i)}. \tag{3.1}$$

The determinant can be considered as the sum of the weights of permutations, where the weight of a permutation depends on the matrix M, and it is a multiplicative function. Since there are efficient dynamic programming algorithms to calculate the sum of weights of combinatorial objects, a naïve idea is to build up permutations in a yield algebra, and apply a standard evaluation algebra that calculates the sum of the weights. However, this is impossible, since there is no yield algebra that can build all the permutations in S_n and there are only $O(poly(n))$ number of parameters below that θ for which $S(\theta) = \mathcal{S}_n$, see Section 3.5. Instead, clow sequences have to be used, as defined below.

Definition 40. *Let $\vec{G} = (V, E)$ be a directed graph with an arbitrary total ordering of its vertices. A clow (closed ordered walk) is a walk $C = v_{i_0}, e_{j_1}, v_{i_1}, \ldots e_{j_k}, v_{i_k}$ such that for all $l = 1, \ldots, k-1$, $v_{i_l} > v_{i_0}$ and $v_{i_k} = v_{i_0}$. For sake of simplicity, the edges and the last vertex might be omitted in the* description of walks. When the last vertex is omitted, the series of vertices are put into parenthesis, indicating that the walk is closed. Vertex v_{i_0} is called the head of the walk, and is denoted by $head(C)$. The length of a walk is the number of edges in it, and is denoted by $l(C)$.

A clow sequence is a series of clows, $\mathcal{C} = C_1, C_2, \ldots C_m$ such that for all $i = 1, \ldots, m-1$, $head(C_i) < head(C_{i+1})$. The length of a clow sequence, $l(\mathcal{C})$, is the sum of the length of the clows in it.

If $\vec{G}$ is edge-weighted, where $w : E \to \mathbb{R}$ denotes the weight function, the score of a clow sequence is defined. The score of a clow $C = v_{i_0}, e_{j_1}, v_{i_1}, \ldots e_{j_k}, v_{i_k}$ is defined as

$$W(C) := \prod_{i=1}^{k} w(e_i). \tag{3.2}$$

and the score of a clow sequence $\mathcal{C} = C_1, C_2 \ldots C_m$ is defined as

$$W(\mathcal{C}) := (-1)^{l(\mathcal{C})+m} \prod_{i=1}^{m} W(C_i) \tag{3.3}$$

When $\vec{G}$ is the complete directed graph with loops, all permutations appear amongst clow sequences. Indeed, define the canonical cycle representation of a permutation as

$$(\sigma_{1,1}\sigma_{1,2} \ldots \sigma_{1,l_1})(\sigma_{2,1} \ldots) \ldots (\sigma_{m,1}, \ldots \sigma_{m,l_m})$$

such that for all i, j, $\sigma_{i,1} < \sigma_{i,j}$ and for all $i = 1, \ldots, m-1$, $\sigma_{i,1} < \sigma_{i+1,1}$. This canonical ordering represents a clow sequence containing m clows, and the ith clow is

$$(v_{\sigma_{i,1}}, v_{\sigma_{i,2}}, \ldots, v_{\sigma_{i,l_i}}).$$

On the other hand, there are clow sequences that are not permutations. Indeed a clow might visit a vertex several times and two different clows might visit the same vertex if both heads of the clows are smaller than the visited vertex. However, these are the only exceptions when a clow sequence is not a permutation, as stated in the lemma below.

Lemma 3. Let $\vec{K}_n$ be the complete directed graph on n vertices, in which loops are allowed. Let $\mathfrak{P}_n$ denote the set of clow sequences on $\vec{K}_n$ for which each clow is a cycle and each vertex is in at most one cycle. Then the mapping φ that maps each permutation

$$(\sigma_{1,1}\sigma_{1,2} \ldots \sigma_{1,l_1})(\sigma_{2,1} \ldots) \ldots (\sigma_{m,1}, \ldots \sigma_{m,l_m})$$

to the clow sequence

$$(v_{\sigma_{1,1}}, v_{\sigma_{1,2}}, \ldots, v_{\sigma_{1,l_1}} v_{\sigma_{1,1}}) \ldots (v_{\sigma_{m,1}}, \ldots, v_{\sigma_{m,l_m}})$$

is a bijection between $\mathcal{S}_n$ and $\mathfrak{P}_n$.

Proof. The bijection is based on the canonical representation of the permutation given above. Each permutation has one canonical representation. Let $\varphi(\sigma)$ denote the clow sequence obtained by the canonical representation. It is clear that φ is an injection. It is also clear that φ is a surjection, since any clow sequence $\mathcal{C} \in \mathfrak{P}_n$ appears as an image of a permutation. Indeed, if $\mathcal{C} = C_1, C_2, \ldots, C_m$ is a clow sequence in $\mathfrak{P}$, then it is image of the permutation that contains m cycles and the ith cycle contains the indexes of the vertices in C_i. That is indeed a cycle representation of a permutation, since each index appears at most once, however, there are n possible indexes and exactly n indexes appear, thus each index appears exactly once. □

Above this bijection, permutations and clow sequences in $\mathfrak{P}$ have the same score when a directed graph is weighted based on a matrix. This is precisely stated in the following theorem

Theorem 28. *Let M be an $n \times n$ matrix. Assign weights to the directed complete graph $\vec{K}_n$ such that $w((v_i, v_j)) := m_{i,j}$. Then*

$$det(M) = \sum_{\mathcal{C} \in \mathfrak{P}} W(\mathcal{C}). \tag{3.4}$$

Proof. Let σ be a permutation in $\mathcal{S}_n$, and define

$$W(\sigma) := sign(\sigma) \prod_{i=1}^{n} m_{i,\sigma(i)}.$$

It is sufficient to show that for any permutation σ,

$$W(\sigma) = W(\varphi(\sigma))$$

for the bijection φ given in Lemma 3. It is true, since $sign(\sigma) = (-1)^{n+m}$, where m is the number of cycles in σ, and that is indeed $(-1)^{l(\varphi(\sigma))+m}$. The weights in the products in $W(\sigma)$ and $W(\varphi(\sigma))$ are also the same due to the definition of φ and the definition of the edge weights in $\vec{K}_n$. □

Although there are clow sequences which are not permutations, the following theorem also holds.

Theorem 29. *Let M be an $n \times n$ matrix. Assign weights to the directed complete graph $\vec{K}_n$ such that $w((v_i, v_j)) := m_{i,j}$. Let $\mathfrak{C}_n$ denote the set of clow sequences of length n. Then*

$$det(M) = \sum_{\mathcal{C} \in \mathfrak{C}_n} W(\mathcal{C}). \tag{3.5}$$

Proof. According to Theorem 28, it is sufficient to show that

$$\sum_{\mathcal{C} \in \mathfrak{C}_n \setminus \mathfrak{P}_n} W(\mathcal{C}) = 0. \tag{3.6}$$

To show this, an involution g is given on $\mathfrak{C}_n \setminus \mathfrak{P}_n$ such that for any $\mathcal{C} \in \mathfrak{C}_n \setminus \mathfrak{P}_n$, it holds that

$$W(\mathcal{C}) = -W(g(\mathcal{C})).$$

Let $\mathcal{C}$ be a clow sequence in $\mathfrak{C}_n \setminus \mathfrak{P}_n$ containing m clows. Let i be the smallest index such that $C_{i+1}, C_{i+2}, \ldots C_m$ are vertex disjoint cycles. Since $\mathcal{C}$ is not in $\mathfrak{P}$, i cannot be 0. $C_i = (v_{i,1}, v_{i.2}, \ldots, v_{i,l+i})$ is not a cycle or C_i intersects with some C_k, $k > i$ or both. Let $v_{i,p}$ be called a witness if $v_{i,p} \in C_k$ for some $k > i$ or there exists $v_{i,q} \in C_i$ such that $q < p$ and $v_{i,p} = v_{i,q}$. Take the smallest index j such that $v_{i,j}$ is a witness. Observe that either $v_{i,j} \in C_k$ or there is a $j' < j$ such that $v_{i,j} = v_{i,j'}$, but the two cases cannot hold in the same time. Indeed, otherwise $v_{i,j'}$ was a witness since it is in C_k, contradicting the minimality of j. If $v_{i,j} \in C_k$, and it has an index pair k, j' in C_k, then define

$$C_i' := (v_{i,1}, v_{i,2}, \ldots v_{i,j}, v_{k,j'+1}, v_{k,j+2}, \ldots v_{,k,l+k}, v_{k,1}, \ldots, v_{k,j'}, v_{i,j+1}, \ldots, v_{i,l_i}).$$

namely, glue C_k into C_i. Define

$$g(\mathcal{C}) := C_1, C_2, \ldots, C_i', C_{i+1} \ldots, C_{k-1}, C_{k+1}, \ldots, C_m$$

Observe that the smallest index i' such that $C_{i'+1}, C_{i'+2}, \ldots$ are disjoint cycles in $g(\mathcal{C})$ is still i, and the smallest index witness in C_i is still vertex $v_{i,j}$ (just now it has a larger index, $j + l(C_k)$). Indeed, the cylces $C_{i+1} \ldots, C_{k-1}, C_{k+1}, \ldots, C_m$ are still disjoint. The set of vertices $v_{i,1}, \ldots, v_{i,j-1}$ were not witnesses in $\mathcal{C}$, so they cannot be witnesses in $g(\mathcal{C})$. Furthermore, the new vertices in C_i' coming from C_k cannot be witnesses since C_k was a cycle disjoint from cycles $C_{i+1} \ldots, C_{k-1}, C_{k+1}, \ldots, C_m$, and there were no witnesses in C_i with a smaller index than j, so C_k does not contain any vertex from $v_{i,1}, \ldots, v_{i,j-1}$.

Now consider a clow sequence $\mathcal{C}'$ such that its smallest index witness $v_{i,j}$ is a witness since a $v_{i,j'}$ exists such that $j' < j$ and $v_{i,j} = v_{i,j'}$. Then define

$$C_i' := (v_{i,1}, v_{i,2}, \ldots, v_{i,j'}, v_{i,j+1}, \ldots, v_{i,l_i}),$$

and define a new cycle

$$C' := (v_{i,j'}, v_{i,j'+1}, \ldots, v_{i,j-1}).$$

C' might have to be rotated to start with the smallest vertex, and thus to be a clow sequence. Define

$$g(\mathcal{C}') := C_1, \ldots, C_i', C_{i+1}, \ldots, C', \ldots, C_m.$$

Observe that the head of C' is larger than $v_{i,1}$, so it will appear after C_i in the clow sequence. Furthermore, C' is vertex disjoint from cycles $C_{i+1}, \ldots, C_m$, since its vertices were not witnesses in C_i, except $v_{i,j'} = v_{i,j}$. Then the smallest index i' such that $C_{i'+1}, C_{i'+2}, \ldots$ are disjoint cycles is still i, and the smallest index witness in C_i is still vertex $v_{i,j}$, since it is in C'.

Based on the above, observe that whatever is the reason that $v_{i,j}$ is the smallest index witness, the g function is defined such that

$$g(g(\mathcal{C})) = \mathcal{C},$$

therefore g is indeed an involution. Furthermore the lengths of $\mathcal{C}$ and $g(\mathcal{C})$ are the same, they contain exactly the same edges, however, their number of cycles differs by 1. Thus

$$W(\mathcal{C}) = -W(g(\mathcal{C})),$$

and thus

$$\sum_{\mathcal{C} \in \mathfrak{C} \backslash \mathfrak{P}} W(\mathcal{C}) = 0.$$

But in that case,

$$\sum_{\mathcal{C} \in \mathfrak{C}} W(\mathcal{C}) = \sum_{\mathcal{C} \in \mathfrak{P}} W(\mathcal{C}) = det(M).$$

$\square$

It is possible to build a yield algebra for the clow sequences. More precisely, the base set A contains partial clow sequences on a given directed graph, $\vec{G} = (V, E)$. A partial clow sequence contains some clows and an ordered walk that might not be closed yet. The parameters are (v_h, v, k), where v_h is the head of the current clow, v is the actual vertex, and k is the length of the partial clow sequence, that is, the number of edges in the clow sequence. In the partial ordering of the parameters, $(v_h, v, k) \le (v'_h, v', k')$ if $v_h \le v'_h$ and $k \le k'$. Obviously, only those parameters are valid, in which $v_h < v$. The unary operator $\circ_{(u,w)}$ adds the new edge (u, w) to the partial clow sequence. If (v_h, v) is an edge in $\vec{G}$, then u might be v_h and w might be v, and the edge (u, w) starts a new partial clow. Otherwise u must be v, and the current (partial) clow is extended with an edge. Therefore, the recursions are

$$
\begin{aligned}
S((v_h, v, k)) &= \textstyle\bigsqcup_{u > v_h \wedge (u,v) \in E} \circ_{(u,v)} \left(S((v_h, u, k-1))\right) \\
&\qquad\qquad \text{if } (v_h, v) \notin E \qquad\qquad\qquad (3.7) \\
S((v_h, v, k)) &= \left(\textstyle\bigsqcup_{u > v_h \wedge (u,v) \in E} \circ_{(u,v)} \left(S((v_h, u, k-1))\right)\right) \sqcup \\
&\quad \left(\textstyle\bigsqcup_{v'_h < v_h} \circ_{(v_h, v)} \left(S((v'_h, v'_h, k-1))\right)\right) \\
&\qquad\qquad \text{if } (v_h, v) \in E. \qquad\qquad\qquad (3.8)
\end{aligned}
$$

The initial conditions are the following. $S((v_h, v, 1))$ is the empty set if $(v_h, v) \notin E$ and contains the partial clow sequence having only the edge (v_h, v) in its first clow.

In the evaluation algebra, the algebraic structure R is the real number field. The function f is defined as

$$f(\mathcal{C}) := (-1)^{l(\mathcal{C})+m} \prod_{e \in \mathcal{C}} w(e) \qquad\qquad\qquad (3.9)$$

where $l(\mathcal{C})$ is the number of edges in the partial clow sequence $\mathcal{C}$ with multiplicity, m is the number of ordered walks in $\mathcal{C}$ including the last, possibly yet not closed ordered walk, and the product also considers the multiplicity of the edges in $\mathcal{C}$. It is easy to see that

$$f(\mathcal{C}) = W(\mathcal{C})$$

for any clow sequence $\mathcal{C}$. The $T_{(u,v)}$ function for the operator $\circ_{(u,v)}$ depends on the parameter. If in the parameter (v_h, v, k), $v_h = v$, then $T_{(u,v)}$ is the multiplication with $w((u,v))$, otherwise it is the multiplication with $-w((u,v))$. The determinant of M can be calculated as

$$det(M) = \sum_{v_h} F(S((v_h, v_h, n))).$$

Observe that the real numbers can be replaced with any commutative ring in the evaluation algebra. Therefore the following theorem holds.

Theorem 30. *Let M be an $n \times n$ matrix over an arbitrary commutative ring R. Then the determinant of M can be calculated in $O(n^4)$ time.*

Proof. The algebraic dynamic programming approach provides this solution. With the classical notations, the dynamic programming algorithm fills in a dynamic programming table $w(i, j, k)$ for each $i \leq j$, $i, j, k = 1, \ldots, n$. The initial conditions are

$$w(i, j, 1) = m_{i,j},$$

and the recursions are

$$w(i, j, k) \quad = \quad \sum_{j' > i} -m_{j',j} w(i, j', k-1) + \sum_{i' < i} m_{i,j} w(i', i', k-1). \quad (3.10)$$

Finally,

$$det(M) = \sum_{i=1}^{n} w(i, i, n).$$

Since the number of entries is $O(n^3)$ and each entry can be computed in $O(n)$ time, the overall running time is $O(n^4)$. □

Observe that this running time is an order larger than the standard Gaussian elimination, which runs in $O(n^3)$ running time. On the other hand, this algorithm does not need any division, thus it can be used for matrices over arbitrary commutative rings. Specially, if the matrix is over the integer ring, then the calculations in the recursion remain in the integer ring.

Also observe that the yield algebra builds the clow sequences and not the permutations. Therefore, the evaluation algebra will calculate the sum of scores of clow sequences with some given parameters. For some special scores, this might coincide with the determinant of a matrix. However, if $P \neq NP$, no similar algorithm exists for calculating the permanent of a matrix. This is explained in detail in Chapter 4.

Another important value assigned to skew-symmetric, even-dimensional matrices is the Pfaffian, which appears in counting perfect matchings of planar graphs.

Definition 41. *A matrix A is* skew-symmetric *if for all* i, j, $a_{i,j} = -a_{i,j}$. *The* Pfaffian *of a* $2n \times 2n$ *matrix A is defined as*

$$pf(A) := \frac{1}{2^n n!} \sum_{\sigma \in S_{2n}} sign(\sigma) \prod_{i=1}^{n} a_{\sigma(2i-1),\sigma(2i)}. \qquad (3.11)$$

It is easy to see that the normalizing constant $\frac{1}{2^n n!}$ in the definition of the Pfaffian is for cancelling the $2^n n!$ cases of $\sigma \in S_{2n}$ providing the same score,

$$sign(\sigma) \prod_{i=1}^{n} a_{\sigma(2i-1),\sigma(2i)}. \qquad (3.12)$$

Indeed, there are $2^n n!$ ways to form n pairs on $2n$ indexes, corresponding to $2^n n!$ permutations. The permutations have different signs, however, the signs of the permutations and the skew-symmetry of the matrix cancel each other. Swapping two consecutive elements in a permutation changes its sign, however, it is also true that $a_{\sigma(2i-1),\sigma(2i)} = -a_{\sigma(2i),\sigma(2i-1)}$. Swapping two consecutive pairs does not change the sign of the permutation, and it also does not change the product in Equation (3.12). Any two permutations representing the same n pairs of the $2n$ indexes can be transformed into each other with these elementary transformations (swapping the order of two indexes in the same pair and swapping two consecutive pairs). Since these elementary transformations do not change the score of the permutation, we get that for all the $2^n n!$ permutations, the score of the permutation in Equation (3.12) is the same.

If we want to extend the definition of the Pfaffian for matrices over an arbitrary commutative ring, we have to modify the definition since dividing by $2^n n!$ might not be available. Define the equivalence relation $\sim$ such that $\sigma \sim \sigma'$ if σ and σ' represent the same n pairs on $2n$ indexes, and let $S_{2n}/_{\sim}$ denote the set containing one permutation for each equivalence class. Then the Pfaffian of the matrix is

$$pf(A) := \sum_{\sigma \in S_{2n}/_{\sim}} sign(\sigma) \prod_{i=1}^{n} a_{\sigma(2i-1),\sigma(2i)}. \qquad (3.13)$$

Since the signed product is the same for any σ coming from the same equivalence class, this definition is well defined.

Similar to the determinant, the Pfaffian can be calculated in polynomial time in an arbitrary commutative ring using the so-called alternating clow sequences. First, we define them.

Definition 42. *A clow sequence is an* alternating clow sequence *if each even indexed edge in each clow is either* (v_{2i-1}, v_{2i}) *or* (v_{2i}, v_{2i-1}) *for some i.*

We are ready to state and prove the following theorem.

Theorem 31. *The Pfaffian of a $2n \times 2n$ skew-symmetric matrix A over any commutative ring R can be calculated using only $O(n^4)$ operations (additions, subtractions and multiplications) in the ring R.*

Proof. We define an injective mapping from $S_{2n}/\sim$ to alternating clow sequences of the complete directed graph $\vec{K}_{2n}$. For a permutation $\sigma(1), \sigma(2), \ldots, \sigma(2n)$, first consider the set M_σ containing the unoriented edges

$$(v_{\sigma(1)}, v_{\sigma(2)}), (v_{\sigma(3)}, v_{\sigma(4)}), \ldots, (v_{\sigma(2n-1)}, v_{\sigma(2n)}),$$

and the set M containing the unoriented edges

$$(v_1, v_2), (v_3, v_4), \ldots, (v_{2n-1}, v_{2n}).$$

Note that both M_σ and M are a perfect matching of the complete graph on $2n$ vertices, therefore the multiset union of $M_\sigma \uplus M$ contains even long alternating cycles, where along a walk on each cycle, one edge comes from M_σ and the other from M. Edges in $M_\sigma \cap M$ are represented twice in $M_\sigma \uplus M$, and they are considered as cycles with two edges. The clow sequence assigned to σ consists of the oriented versions of these cycles, such that each clow starts with the smallest index vertex, and the first edge in each clow comes from M_σ. For example, if σ is

$$1, 5, 2, 12, 3, 4, 6, 11, 7, 9, 8, 10,$$

then the clow sequence is

$$(v_1, v_5, v_6, v_{11}, v_{12}, v_2), (v_3, v_4), (v_7, v_9, v_{10}, v_8).$$

As we can see, each clow has even length, and every second edge is either (v_{2i-1}, v_{2i}) or (v_{2i}, v_{2i-1}) for some i, namely, the images are alternating clow sequences.

We have to define the weights of the edges in the clow sequence to define the score of the clow sequence. If an edge (v_i, v_j) comes from M_σ, then its weight is $a_{i,j}$, and if an edge (v_i, v_j) comes from M, its weight is $j - i$ (recall that it is either 1 or -1).

We claim if a clow sequence $\mathcal{C}$ is the image of σ, then the score of $\mathcal{C}$ is the score of σ as defined in Equation (3.12) if n is even, and the score of $\mathcal{C}$ is the additive inverse of the score of σ if n is odd. It is clear that the score of the clow sequence contains the product of the appropriate $a_{i,j}$ terms; we only have to check the signs. Since the score of σ is well defined on $S_{2n}/\sim$, without loss of generality, we might assume that σ is

$$i_{1,1}, i_{1,2}, \ldots, i_{1,k_1}, i_{2,1}, \ldots, i_{2,k_2}, \ldots, i_{m,1}, \ldots, i_{m,k_m}$$

when the clow sequence is

$$(v_{i_{1,1}}, v_{i_{1,2}}, \ldots, v_{i_{1,k_1}}), (v_{i_{2,1}}, \ldots, v_{i_{2,k_2}}), \ldots, (v_{i_{m,1}}, \ldots, v_{i_{m,k_m}}).$$

Observe that the obtained clow sequence corresponds to the permutation

$$(i_{1,1}, i_{1,2}, \ldots, i_{1,k_1}), (i_{2,1}, \ldots, i_{2,k_2}), \ldots, (i_{m,1}, \ldots, i_{m,k_m}),$$

which is the product of permutation σ and the permutation

$$2, 1, 4, 3, \ldots, 2n - 1, 2n.$$

Since this latter permutation contains n cycles and its length is $2n$, its sign is $(-1)^n$. Therefore we get that

$$sign(\sigma) \prod_{i=1}^{n} a_{\sigma(2i-1), \sigma(2i)} = (-1)^n W(\mathcal{C}), \tag{3.14}$$

since the sign of a clow sequence is the same as the sign of its corresponding permutation.

The mapping is clearly an injection, since two permutations σ and σ' from different equivalence classes contain different edges in M_σ and $M_{\sigma'}$, and therefore their images will be different, too. We claim that the mapping is actually a bijection between $S_{2n}/\sim$ and those alternating clow sequences that are permutations and contain only even long cycles. Indeed, if there is a permutation π corresponding to an alternating clow sequence $\mathcal{C}$, then the permutation

$$\sigma = \pi * (1,2)(3,4) \ldots (2n-1, 2n)$$

is the permutation whose image is $\mathcal{C}$, where $*$ denotes the multiplication in the group S_{2n}.

Next, we are going to show that there is an involution g on those alternating clow sequences which are not permutations and which are permutations but contain an odd cycle, and if the alternating clow sequence $\mathcal{C}$ is any of these, then

$$W(g(\mathcal{C})) = -W(\mathcal{C}). \tag{3.15}$$

The involution can be decomposed into two involutions on two disjoint sets. The involution g_1 is on alternating clow sequences that contain a clow in which a vertex is revisited after an odd number of steps. Let $\mathcal{C}$ be such an alternating clow sequence, and let C be the clow in it which contains the smallest revisited vertex v (if there are more than one such clows or revisited vertices). Then $g_1(\mathcal{C})$ contains the same clows as $\mathcal{C}$, except in C, the odd length closed path from v to its next visit is inverted. Due to the skew-symmetry of A and the definition of edge weights of edges (v_{2i-1}, v_{2i}) and (v_{2i}, v_{2i-1}),

$$W(g_1(\mathcal{C})) = -W(\mathcal{C}). \tag{3.16}$$

The involution g_2 is on clow sequences that are not permutations and each revisited vertex is revisited after an even number of steps. The involution given in the proof of Theorem 29 suffices for g_2. We already showed that it is an involution satisfying Equation (3.15); we only have to show that the image of an alternating clow sequence with the given property also has this property. The involution operates with cutting and merging cycles. These cycles must have even length, since each vertex is revisited after an even number of steps, therefore their images also satisfy the prescribed properties.

We get that

$$pf(A) = (-1)^n \sum_{\mathcal{C} \in \mathfrak{AC}} W(\mathcal{C}) \tag{3.17}$$

where $\mathfrak{AC}$ is the set of alternating clow sequences. It is easy to see that the dynamic programming algorithm given in the proof of Theorem 30 can calculate this sum if a restriction is added that in each clow, in each even step, the edge must be (v_{2i-1}, v_{2i}) or (v_{2i}, v_{2i-1}). $\qquad\square$

In the following sections, such counting problems are introduced that can be solved via calculating the determinant or the Pfaffian of a matrix.

3.2   Kirchhoff's matrix-tree theorem

Kirchoff's matrix-tree theorem is about the number of spanning trees in a graph [113]. It shows that this number can be calculated in polynomial time via computing the determinant of a matrix derived from the graph. The theorem uses several lemmas. First, we are going to introduce them below.

Lemma 4. *Let $G = (V, E)$ be an arbitrary graph. Orient its edges in an arbitrary way, and consider the vertex-edge adjacency matrix of this orientation, that is, if the edge e is oriented from vertex u to vertex v, then the entry in the matrix for the $\{u, e\}$ pair is -1, and the entry for the $\{v, e\}$ entry is 1. All other entries in the column of e are 0. Let C denote this incidence matrix, and let C^{-v} denote the submatrix obtained by deleting the row corresponding to vertex v. Furthermore for any $F \subseteq E$, let $C^{-r}[F]$ denote the submatrix which contains the columns corresponding to the edges in F in matrix C^{-r}. Let F be such that $|F| = |V| - 1$. Then*

1. *$|det(C^{-v}[F])| = 1$ if and only if F are edges of a spanning tree of G, and*

2. *$det(C^{-v}[F]) = 0$ if and only if F are not the edges of a spanning tree of G.*

Proof. Observe that the two options for F (it is a set of edges of a spanning tree or it is not) are complements, so the "only if" parts of both statements follow from the "if" part of the other statement, and thus it is sufficient to prove the "if" parts.

If the edges in F form the edges of a spanning tree T, we first set up a partial ordering of the vertices such that $u \leq w$ if w is on the way from u to v. Extend this partial ordering to an arbitrary total ordering and rearrange the rows of the matrix based on this total ordering. There is a bijection φ between the vertices in $V \setminus \{v\}$ and the edges in F: u is mapped to e if e connects u to its parent in the spanning tree rooted in v. The bijection indicates a total ordering of the edges in F, $e_1 < e_2$ if $\varphi^{-1}(e_1) < \varphi^{-1}(e_2)$. Let the columns of $C^{-v}[F]$ be ordered according to this total ordering.

Observe that the rearranged matrix is a lower diagonal matrix with all 1 and -1 in the diagonals. Therefore its determinant is either 1 or -1. However, the determinant of this matrix and the original $C^{-v}[F]$ are the same in absolute value.

If the edges in F do not form a spanning tree, then it contains a cycle. Indeed, $|F| = |V| - 1$, and any cycle free graph with $|V| - 1$ edges is a tree. Since F is not the set of edges of a tree, it must contain a cycle. Let F' denote the edges in this cycle. The corresponding columns in $C^{-v}[F]$ are linearly dependent since an appropriate linear combinations of them is the 0 vector. Indeed, fix an arbitrary walk on the edges around the cycle, $v_0, e_1, v_1, e_2, \ldots, v_{|F'|-1}, e_{|F'|}, v_0$. Let the linear coefficient of the column representing e_i be 1 if the matrix entry for the pair v_{i-1}, e_i in C is 1, otherwise let the coefficient be -1. Then this linear combination of column vectors indeed the 0 vector, since for each row representing v_i, there is a 1 and a -1 in the appropriately weighted column vectors. Since the columns of $C^{-v}[F]$ are not linearly independent, $det(C^{-v}[F]) = 0$.

$\square$

The product of C^{-v} and its transpose is a $(|V| - 1) \times (|V| - 1)$ matrix. The Cauchy-Binet theorem [20, 39] is about the determinant of the product of two matrices, and thus, can tell $det(C^{-v}C^{-v^T})$.

Theorem 32. *Let $A, B \in \mathbb{R}^{n \times m}$ be matrices. Then*

$$det(AB^T) = \sum_{F \subset \{1,2,\ldots,m\} \wedge |F|=n} det(A[F])det(B[F]) \qquad (3.18)$$

where $A[F]$ denotes the $n \times n$ submatrix of A whose column's indexes are the indexes in F.

Proof. Consider the determinant of the following matrix

$$D = \left[\begin{array}{c|c} 0 & A \\ \hline B^T & I \end{array} \right]$$

where 0 denotes the all-zero matrix and I denotes the identity matrix. We are going to calculate the determinant in two different ways.

The first way is based on the Laplace expansion on the first n rows of D. Define the set of indexes $\mathcal{C} = \{n+1, \ldots, n+m\}$, and index the column of A by these indexes. It is sufficient to consider subsets of these columns since in other subsets of indexes F, the matrix $(0|A)[F]$ contains an all-zero column, and thus, its determinant is 0. Therefore

$$det(D) = \sum_{F \in \mathcal{C} \wedge |F| = n} (-1)^{\sum_{i=1}^{n} i + \sum_{f \in F} f} det(A[F]) det((B^T|I)[\bar{F}]). \quad (3.19)$$

We also calculate $det((B|I)[F])$ by Laplace expansion, but now based on the first n columns. Then a submatrix containing some rows should be defined. If F' is a subset of indexes, let $[F']B$ denote the submatrix of B whose rows indexes are in F'. Then

$$det((B^T|I)[F]) =$$
$$\sum_{F' \in \mathcal{C} \wedge |F'| = n} (-1)^{\sum_{i=1}^{n} i + \sum_{f' \in F'} (f'-n)} det([F']B^T) det([\bar{F}']I[\bar{F}]). (3.20)$$

It is easy to see that all terms in this sum are 0 except the one in which $F' = F$. Indeed, if $F' \neq F$, then $\bar{F}' \neq \bar{F}$. In that case $[\bar{F}']I[\bar{F}]$ contains an all-0 row (and also, an all-0 column), and thus, $det([\bar{F}']I[\bar{F}]) = 0$. If $F' = F$, then $det([\bar{F}']I[\bar{F}]) = 1$. Therefore,

$$det((B^T|I)[F]) = (-1)^{\sum_{i=1}^{n} i + \sum_{f \in F} (f-n)} det([F]B^T) =$$
$$(-1)^{\sum_{i=1}^{n} i + \sum_{f \in F} (f-n)} det(B[F]). \quad (3.21)$$

We get that

$$det(D) = \sum_{F \subset \{1,2,\ldots,m\} \wedge |F| = n} (-1)^{2 \sum_{i=1}^{n} i + 2 \sum_{f \in F} f - n^2} det(A[F]) det(B[F])$$

$$= \sum_{F \subset \{1,2,\ldots,m\} \wedge |F| = n} (-1)^n det(A[F]) det(B[F]), \quad (3.22)$$

since

$$-n^2 \equiv n \pmod{2}. \quad (3.23)$$

The second way to calculate $det(D)$ is based on Gaussian elimination. For each $i = 1, \ldots, n$, we add to row i the following linear combination: $-a_{i,1}$ times the $n + 1^{st}$ row plus $-a_{i,2}$ times the $n + 2^{nd}$ row, etc., plus $-a_{i,j}$ times the $n + j^{th}$ row. We get a matrix

$$D' = \left[\begin{array}{c|c} C & 0 \\ \hline B^T & I \end{array} \right]$$

where

$$c_{i,j} = \sum_{k=1}^{m} -a_{i,k} b_{k,j} \quad (3.24)$$

namely, $C = -AB^T$. Note that

$$det(D) = det(D') = det(C) = (-1)^n det(AB^T), \qquad (3.25)$$

therefore

$$det(AB^T) = \sum_{F \subset \{1,2,\ldots,m\} \wedge |F|=n} det(A[F])det(B[F]). \qquad (3.26)$$

$\square$

Now we are ready to prove Kirchhoff's theorem on the number of spanning trees.

Theorem 33. *Let* $G = (V, E)$ *be an arbitrary graph, and let* C^{-v} *be the matrix constructed from* G *in Lemma 4. Then the number of spanning trees of* G *is*

$$det(C^{-v}C^{-v^T}). \qquad (3.27)$$

Proof. From the Cauchy-Binet theorem we get that

$$det(C^{-v}C^{-v^T}) = \sum_{F \subset \{1,2,\ldots,m\} \wedge |F|=n-1} det(C[F])^2 \qquad (3.28)$$

where m is the number of edges, and n is the number of vertices in G. However, from Lemma 4, we know that $det(C[F])^2$ is 1 if and only if F contains the edge indexes of a spanning tree. $\square$

By calculating the matrix $C^{-v}C^{-v^T}$, we can give another form of Kirchhoff's matrix-tree theorem.

Theorem 34. *Let* $G = (V, E)$ *be an arbitrary graph. Let* D *denote the diagonal matrix containing the degrees of the graph, and let* A *denote the adjacency matrix of* G. *Let* D^{-v} *and* A^{-v} *denote the matrices obtained from* D *and* A *by deleting the row and column for vertex* v. *The number of spanning trees of* G *is*

$$det(D^{-v} - A^{-v}). \qquad (3.29)$$

Proof. It is sufficient to show that

$$D^{-v} - A^{-v} = C^{-v}C^{-v^T}. \qquad (3.30)$$

The element in position (i, j) in the matrix $C^{-v}C^{-v^T}$ is the scalar product of the ith and j^{th} row of matrix C^{-v}. If $i \neq j$, then it is -1 if there is an edge between vertices v_i and v_j, and 0 otherwise. Indeed, $c_{i,k}c_{j,k} = -1$ if e_k is the edge between v_i and v_j and 0 otherwise. If $i = j$, then the scalar product is the degree of vertex v_i ($= v_j$), since for each e_k incident to v_i, $c_{i,k}^2 = 1$ and for all k' such that $e_{k'}$ is not incident to v_i, $c_{i,k'}^2 = 0$. Therefore the equality in Equation (3.30), and thus the theorem, hold. $\square$

The Kirchhoff theorem says that the number of spanning trees can be calculated in polynomial time for any graph. We can use Theorem 34 to count the leaf-labeled trees.

Example 7. *Count the leaf-labeled trees on n vertices.*

Solution. The number of leaf-labeled trees on n vertices are the spanning trees of the complete graph K_n. By Theorem 34, it is the determinant of the matrix

$$\begin{pmatrix} n-1 & -1 & \cdots & -1 & -1 \\ -1 & n-1 & \cdots & -1 & -1 \\ \vdots & \vdots & \ddots & \vdots & \vdots \\ -1 & -1 & \cdots & n-1 & -1 \\ -1 & -1 & \cdots & -1 & n-1 \end{pmatrix}$$

where the matrix has $n-1$ rows and $n-1$ columns. We can add linear combinations of lines to some lines of this matrix without changing the determinant. First, add all lines starting from the second line to the first line. We get the following matrix:

$$\begin{pmatrix} +1 & +1 & \cdots & +1 & +1 \\ -1 & n-1 & \cdots & -1 & -1 \\ \vdots & \vdots & \ddots & \vdots & \vdots \\ -1 & -1 & \cdots & n-1 & -1 \\ -1 & -1 & \cdots & -1 & n-1 \end{pmatrix}.$$

Adding the first line to all other lines, we get

$$\begin{pmatrix} +1 & +1 & \cdots & +1 & +1 \\ 0 & n & \cdots & 0 & 0 \\ \vdots & \vdots & \ddots & \vdots & \vdots \\ 0 & 0 & \cdots & n & 0 \\ 0 & 0 & \cdots & 0 & n \end{pmatrix}.$$

Since it is an upper triangle matrix, its determinant is the product of the elements in the diagonal. Therefore, the number of leaf-labeled trees on n vertices is n^{n-2}. ∎

Kirchhoff's theorem can be extended to weighted spanning trees.

Theorem 35. *Let $G = (V, E)$ be a graph, and let $w : E \to R$ be a weight function, where R is an arbitrary commutative ring. Let T be a spanning tree of G. Define the weight of T as*

$$W(T) := \prod_{e \in T} w(e). \tag{3.31}$$

Define the diagonal matrix D such that its diagonal entry

$$d_{i,i} := \sum_{e \in I(v_i)} w(e), \tag{3.32}$$

where $I(v_i)$ is the set of edges incident to v_i. Define the weighted adjacency matrix A such that $a_{i,j} = w(e)$ if e is the edge (v_i, v_j). Let v be an arbitrary vertex, and define A^{-v} and D^{-v} by deleting the row and column corresponding to vertex v from A and D. Then

$$\sum_{T \in \mathcal{T}} W(T) = det(D^{-v} - A^{-v}) \tag{3.33}$$

where $\mathcal{T}$ is the set of spanning trees of G.

Proof. Give an arbitrary orientation of the edges, and let C denote the usual oriented incidence matrix of G. Construct a weighted version of the oriented incidence matrix C_w whose rows are the vertices of G, columns are the edges of G, and if an edge e is oriented from vertex u to vertex v, then the entry for the pair u, e is $-w(e)$, and the entry for the pair v, e is $w(e)$. All other entries are 0.

First, we are going to show that

$$C^{-v} C_w^{-v^T} = D^{-v} - A^{-v}. \tag{3.34}$$

Indeed, the element in position (i, j) in the matrix $C^{-v} C_w^{-v^T}$ is the scalar product of the ith row in C^{-v} and the jth row in Cw^{-v}. If $i \neq j$, then it is $-w(e)$ if e is an edge between v_i and v_j and 0 otherwise. If $i = j$, then the scalar product is $\sum_{e \in I(v_i)} w(e)$. Therefore Equation (3.34) indeed holds. We have to use the Cauchy-Binet theorem:

$$det(C^{-v} C_w^{-v^T}) = \sum_{F \subset \{1,2,\ldots,m\} \wedge |F| = n-1} det(C[F]) det(C_w[F]). \tag{3.35}$$

Observe that $|det(C[F])| = 1$ and $det(C_w[F])$ is either $W(T)$ or $-W(T)$ if and only if F is the edge set of spanning tree T. Furthermore $det(C[F]) = 1$ exactly when $det(C_w[F]) = W(T)$. Therefore, their product is always $W(T)$ when F is the edge set of spanning tree T.   □

We can use Theorem 35, for example, to count spanning trees with given (small integer) weights.

Example 8. *Let $G = (V, E)$ be a graph, and let $w : E \to \mathbb{Z}$ be a weight function. Let m be the maximum of the absolute values of the edge weights, and let $n = |V|$. Define the weight of a spanning tree T as*

$$W(T) := \sum_{e \in T} w(e). \tag{3.36}$$

Count the spanning trees with a given weight k. The running time must be a polynomial function of both n and m.

Solution. Define a new weight function, $w'(e) := x^{w(e)}$. This new weight function contains monomials from the Laurent polynomial ring, $\mathbb{Z}[x, -x]$. Furthermore, for any spanning tree T it holds that

$$W'(T) = x^{W(T)}. \tag{3.37}$$

Since $\mathbb{Z}[x, -x]$ is a commutative ring, the summation over the spanning trees T of the weights $W'(T)$ can be calculated using only a polynomial number of ring operations. Furthermore, the degrees of the monomials can vary between $-mn$ and mn, where n is the number of vertices in G, therefore any ring operation can be done in polynomial time in mn. The coefficient of the monomial x^k in $det(D^{-v} - A^{-v})$ is the number of spanning trees of weight k. ∎

The number of minimum (or dually, the maximum) spanning trees can be counted in polynomial time with both the size of the graph and the logarithm of the weights, see Exercise 13. On the other hand, it is NP-complete to decide if there is a spanning tree with a given sum of weights, see Exercise 14.

3.3 The BEST (de Bruijn-Ehrenfest-Smith-Tutte) algorithm

Kirchoff's matrix-tree theorem can be extended to directed graphs, and the directed trees (also known as *arborescences*) in a directed graph can be counted. Such directed trees are related to Eulerian circuits summarized in the BEST theorem. BEST is the acronym of four mathematicians, de Bruijn, Ehrenfest, Smith and Tutte, who developed the theory we are going to introduce in this section [2, 170]. We first define the directed trees.

Definition 43. *A directed tree or out-tree or arborescence is a directed graph such that wiping out the direction of the edges makes it a tree. Furthermore, there is a vertex v called the* root *of the tree such that from any vertex u, there is a directed path from v to u.*

Flipping the direction of all the edges yields an in-tree *or anti-arborescence, so in an in-tree, there is a directed path to the root v from any other vertex u of the in-tree. Vertex v is still called the root in in-trees.*

We are going to count the in-trees having a fixed root. There is a theorem for counting in-trees similar to Kirchhoff's matrix-tree theorem. In fact, the following theorem is a generalization of Kirchhoff's matrix-tree theorem.

Theorem 36. *Let $\vec{G} = (V, E)$ be a directed graph. Let A denote its adjacency matrix, and let D_{out} denote the diagonal matrix containing the out-degrees for each vertex. Let A^{-v} and D_{out}^{-v} denote the matrices we get by deleting the row and column for vertex v. Then the number of in-trees rooted into v is*

$$det(D_{out}^{-v} - A^{-v}). \tag{3.38}$$

Proof. Let M^{-v} denote $D_{out}^{-v} - A^{-v}$. The proof is based on an induction on the number of edges in G. The base cases are the graphs having at most $n-1$ edges, where n is the number of vertices. If the number of edges is less than $n-1$, there are at least two components in the graph, and at least one of them does not contain v. The sum of the columns in M^{-v} corresponding to the vertices in such a component is the zero vector. Therefore the determinant is 0, which is indeed the number of trees in the graph.

If the number of edges is $n-1$, then there are three cases.

1. G is not connected in the weak sense, namely, its undirected version is not connected. Then still the columns of M^{-v} are linearly dependent, thus, the determinant is 0, which is indeed the number of trees in G.

2. The undirected version of G is a tree, however, it is not an in-tree due to the wrong direction of some of the edges. If the direction of the edges were correct, then the out-degree of v would be 0 and the out-degree of all other edges is 1. Since the sum of the out-degrees is $n-1$, it follows that there is a vertex $v' \neq v$ whose out-degree is 0. However, in that case the row of vertex v' contains all 0, therefore the determinant of M^{-v} is 0, which is the correct number of in-trees in G.

3. G is an in-tree. Then we first set up a partial ordering of the vertices such that $u \leq w$ if w is on the way from u to v. We extend this partial ordering to an arbitrary total ordering and apply a similarity transformation on M^{-v} such that the rows and columns follow the order of the vertices in the total ordering. Such similarity transformation does not change the value of the determinant. After the transformation, the matrix becomes an upper triangular matrix with all 1s in its diagonal. Therefore, the determinant is 1, which is the number of in-trees in G.

We proved the correctness of the base cases, now we do the induction. Assume that the number of edges in G is $m > n-1$. We can delete any outgoing edge from v without changing the number of in-trees in G and without changing M^{-v}. Therefore, we can assume that the number of outgoing edges from v is 0, and there are still $m > n-1$ edges. Therefore, there must be a vertex $v' \neq v$ such that the out-degree of v' is greater than 1. Consider any edge e going out from v'. Generate two graphs. G_1 contains all the edges of G except e, G_2 contains all the edges of G except it does not contain any edge going out from v' but e. Since in any in-tree, there is exactly one edge going out from v', any in-tree of G is either an in-tree of G_1 or an in-tree of G_2. Since both G_1 and G_2 have less edges than G, the determinants of the corresponding matrices M_1^{-v} and M_2^{-v} calculates correctly the number of in-trees in G_1 and G_2. The sum of these two is the number of in-trees in G. However, due to the linearity of the determinant,

$$det(M^{-v}) = det(M_1^{-v}) + det(M_2^{-v}), \tag{3.39}$$

therefore, $det(M^{-v})$ is indeed the number of in-trees in G. □

It is easy to see that Theorem 34 is a special case of Theorem 36. Indeed, let G be an arbitrary undirected graph. Create a directed graph $\vec{G}$, such that each edge e in G is replaced with a pair of directed edges going to both directions between the vertices incident to e. Then, on one hand, the matrix $D^{-v} - A^{-v}$ constructed from G is the same as the matrix $D_{out}^{-v} - A^{-v}$ constructed from G'. On the other hand, there is a bijection between the spanning trees in G and the in-trees rooted into v in G'. Indeed, take any spanning tree T in G, and for any edge $e \in T$, select the edge in G' corresponding to e with the appropriate direction.

Theorem 36 also holds for directed multigraphs in which parallel edges are allowed but loops are not allowed. Two in-trees are considered different if the same pair of vertices are connected in the same direction, however, for at least for one pair of vertices, the directed edges connecting them are different parallel edges in the multigraph. For such graphs, the out-degree is defined as the number of outgoing edges, and there is k in the adjacency matrix in position (i, j) if there are k parallel edges going from vertex v_i to vertex v_j. Indeed, it is easy to see that the base cases hold, and the induction also holds since the linearity of the determinant holds for any matrices.

The number of in-trees appears in the formula counting the directed Eulerian circuits in directed Eulerian graphs. First, we define them.

Definition 44. *A directed Eulerian graph is a directed, connected graph in which for each vertex v, its in-degree equals to its out-degree. A directed Eulerian circuit (or short, a Eulerian circuit) is a directed circuit that travels each edge exactly once.*

It is easy to see that each directed Eulerian graph has at least one directed Eulerian circuit. Their number can be calculated in polynomial time, stated by the following theorem.

Theorem 37. *Let $\vec{G} = (V, E)$ be a directed Eulerian graph, and let v^* be an arbitrary vertex in it. Then its number of directed Eulerian circuits is*

$$|\mathcal{T}_{v^*}^{in}| \prod_{v \in V} (d_{out}(v) - 1)! \tag{3.40}$$

where $\mathcal{T}_{v^*}^{in}$ is the set of intrees rooted into v^*.

Proof. The Eulerian circuit might start with an arbitrary edge, therefore, fix an outgoing edge e of v^*. Start a walk on it, and due to the pigeonhole rule, this walk can be continued until v^* is hit $d_{out}(v^*)$ times. The last hit closes the walk, thus obtaining a circuit. This circuit might not be a Eulerian circuit, since there is no guarantee that it used all the edges in $\vec{G}$. However, assume that the circuit is a Eulerian circuit. Let E' be the set of last outgoing edges along this circuit for each vertex in $V \setminus \{v^*\}$. We claim that these edges are the edges of an in-tree rooted into v^*.

Indeed, there are $n - 1$ edges. We prove that from each particular vertex

u_0, there is a directed path to v^*. Let e_1 be the last edge going out from u_0 in the Eulerian circuit. It goes to some u_1, and thus from u_1, the last outgoing edge is later in the Eulerian circuit than e_1. Let it be denoted by e_2. This edge goes into u_2, etc. We claim that the edges $e_1, e_2, \ldots$ eventually go to v^* due to the pigeonhole rule. Indeed, it is impossible that some e_i goes to u_j for some $j < i$ since e_i is later in the Eulerian circuit than e_j. However, if e_i goes to u_j then the Eulerian path is continued by going out from u_j, and thus, the last outgoing edge of u_j is after e_i, a contradiction.

Therefore the $n-1$ edges form a connected graph, in the weak sense, and thus, the undirected version of a graph is a tree. Furthermore, all edges are directed towards v^*, therefore the directed version of the graph is an in-tree rooted into v^*.

Furthermore, for each vertex v, give an arbitrary but fixed ordering of the outgoing edges. If $v \neq v^*$ and the last outgoing edge is the kth in this list, then decrease the indexes of each larger indexed edge by 1. Thus, the indexes of the outgoing edges except the last one form a permutation of length $d_{out}(v) - 1$. For v^*, decrease the indexes of each edge which have a larger index than the index of e. Therefore, the indexes of the last $d_{out}(v^*) - 1$ edges form a permutation of length $d_{out}(v^*) - 1$.

In this way, we can define a mapping of Eulerian circuits onto

$$\mathcal{T}_{v^*}^{in} \times \bigtimes_{v \in V} S_{d_{out}(v)-1} \tag{3.41}$$

where S_n denotes the set of permutations of length n in the following way. The direct product of the in-tree of the last outgoing edges and the aforementioned permutations for each vertex is the image of a Eulerian circuit.

It is clear that this mapping is an injection. Indeed, consider the first step where two Eulerian circuits, C_1 and C_2 deviate. If these edges go out from v^*, then their image will be different on the permutation in $S_{d_{out}(v^*)-1}$. Otherwise, they have different edges going out from some $v \neq v^*$. If the last outgoing edges from v are different in the two circuits, then the images of the two circuits have different in-trees. If the last outgoing edges are the same, then the permutations in $S_{d_{out}(v)-1}$ are different in the images.

This mapping is also a surjection. Indeed, take any in-tree $T \in \mathcal{T}_{v*}^{in}$ and for each $v \neq v*$, take a permutation $\pi_v \in S_{d_{out}(v)-1}$. We are going to construct a Eulerian circuit whose image is exactly

$$T \times \bigtimes_{v \in V \setminus \{v^*\}} \pi_v. \tag{3.42}$$

For each v, if the index of the edge which is in T and going out from v is k, then increase by 1 all the indexes in permutation π_v which are greater than or equal to k. This is now a list of indexes. Extend this list by k; this extended list L_v will be the order of the outgoing edges from v. That is, start a circuit with the edge e going out from v^*, and whenever the circuit arrives at v, go out on the next edge in the list L_v. Continue till all the outgoing edges from v^* are used. We claim that the so obtained circuit is a Eulerian circuit whose image is exactly the one in Equation (3.42).

Indeed, assume that some edges going out from v_0 is not used. Then particularly, the last edge e_1 going out from v_0 is not used. It goes out to some v_1, and since it is not used, then there are also outgoing edges in v_1 which is not used. Particularly, its last edge e_2 is not used, which goes to some v_2. However, these last edges $e_1, e_2, \ldots$ goes to v^*, namely, there are ingoing edges of v^* which are not used in the circuit. However, in that case not all outgoing edges from v^* is used, contradicting that the walk generating the circuit is finished.

Therefore, the circuit is a Eulerian circuit. Its image is indeed the one in Equation (3.42), due to construction.

Since the mapping is injective and surjective, it is a bijection. Thus, the number of Eulerian circuits is

$$\left| \mathcal{T}_{v^*}^{in} \times \mathop{\bigtimes}_{v \in V \setminus \{v^*\}} S_{d_{out}(v)-1} \right| = |\mathcal{T}_{v^*}^{in}| \prod_{v \in V \setminus \{v^*\}} (d_{out}(v) - 1)! \qquad (3.43)$$

$\square$

There is an interesting corollary of this theorem.

Corollary 5. *Let $\vec{G} = (V, E)$ be a directed Eulerian graph, and $v_1, v_2 \in V$. Then*

$$|\mathcal{T}_{v_1}^{in}| = |\mathcal{T}_{v_2}^{in}|. \qquad (3.44)$$

It is easy to see that the BEST theorem also holds for directed Eulerian multigraphs, even with loops. Indeed, if $\vec{G}$ is a multigraph, then add a vertex to the middle of each edge. Namely, each directed edge going from u to v is replaced by a directed edge going from u to w and an edge going from w to v, where w is a new vertex. This new graph $\vec{G}'$ is a directed Eulerian graph, and it is not a multigraph. There is a natural bijection between the Eulerian circuits in $\vec{G}$ and $\vec{G}'$, therefore they have the same number of Eulerian circuits. Furthermore, it is clear that any subpaths of length 2 $(u, w_1), (w_1, v)$ and $(u, w_2), (w_2, v)$ in a Eulerian circuit in $\vec{G}'$ are interchangeable if w_1 and w_2 are inserted vertices in $\vec{G}'$, therefore each Eulerian circuit differing only in the permutation of parallel edges appears

$$\prod_{u,v \in V} m(u, v)! \qquad (3.45)$$

times, where $m(u, v)$ is the number of parallel edges going from u to v. (Here $m(v, v)$ is the number of loops on v.) Having said this, the following theorem holds.

Theorem 38. *Let $\vec{G} = (V, E)$ be a directed multigraph, so loops (and even multiple loops) are possible. Let $m(u, v)$ denote the number of parallel edges from u to v. Let $v*$ be an arbitrary vertex in $\vec{G}$. The number of Eulerian circuits where two circuits are not distinguished if they differ only in the permutations of parallel edges is*

$$\frac{|\mathcal{T}_{v^*}^{in}| \prod_{v \in V} (d_{out}(v) - 1)!}{\prod_{u,v \in V} m(u, v)!} \tag{3.46}$$

where $\mathcal{T}_{v^*}^{in}$ is the set of in-trees rooted into v^*.

We are going to use Theorem 38 to count sequences with prescribed statistics of consecutive character pairs.

Example 9. *Let Σ be a finite alphabet, and let $s : \Sigma \times \Sigma \to \mathbb{N}$ statistics function be given. Count the sequences in which for each $(\sigma_1, \sigma_2) \in \Sigma \times \Sigma$, $\sigma_1 \sigma_2$ is a substring $s(\sigma_1, \sigma_2)$ times.*

Solution. Construct a directed multigraph $\vec{G}$ whose vertices are the characters in Σ, and there are $s(\sigma_1, \sigma_2)$ edges from the vertex representing σ_1 to the vertex representing σ_2, and there are $s(\sigma, \sigma)$ loops on the vertex representing σ. There are 3 cases.

1. There are more than 2 vertices v such that $d_{in}(v) \neq d_{out(v)}$ or there are 2 vertices v such that $|d_{in}(v) - d_{out(v)}| > 1$. Then no prescribed sequence exists. Indeed, assume that there is a sequence A with the prescribed statistics of consecutive character pairs. Then each character in the sequence is part of two substrings of length 2; in one of them it is the first character and in one of them it is the last character, except the first and the last character of A. Therefore, if a solution exists, for each vertex v, $d_{out}(v)$ must be $d_{in}(v)$ except for those 2 vertices that represent the first and last character of A. The vertex representing the first character has one more outgoing edge than incoming edge, and the vertex representing the last character has one more incoming edge than outgoing. However, if the first and the last character in A are the same, then the constructed graph must be Eulerian.

2. For each vertex v, $d_{in}(v) = d_{out}(v)$ except for two vertices v_b and v_e, for which $d_{out}(v_b) = d_{in}(v_b) + 1$ and $d_{in}(v_e) = d_{out}(v_e) + 1$. Extend $\vec{G}$ with a new vertex v^*, and add an edge going from $v*$ to v_b and also an edge from v_e to v^*. The so-modified $\vec{G}'$ is a directed Eulerian graph. We claim that the number of sequences is the number of Eulerian circuits in $\vec{G}'$ factorized by the permutations of parallel edges. Indeed, fix an arbitrary ordering on each set of parallel edges. Call a Eulerian circuit canonical if the outgoing edges are used in the prescribed order when the walk on the circuit starts in v^*. It is easy to see that there is a bijection between sequences $a_1 a_2 \ldots a_n$ with prescribed statistics and

the canonical circuits that start in v^*, go to the vertex representing a_1, then to the vertex representing a_2, etc., finally going from the vertex representing a_n to v^* using the outgoing edges in canonical order.

3. The constructed graph is Eulerian. Then take $\sigma \in \Sigma$, and consider the sequences which start and end with σ. Extend $\vec{G}$ with a vertex v^* and add two edges, one going from v^* to the vertex representing σ and one going from this vertex back to v^*. This modified graph is still Eulerian, and the number of canonical Eulerian circuits in it is the number of sequences starting and ending with σ and satisfying the prescribed character pair statistics. Doing this for all $\sigma \in \Sigma$ and summing the number of canonical Eulerian circuits answers the question.

∎

3.4 The FKT (Fisher-Kasteleyn-Temperley) algorithm

The FKT algorithm is for counting perfect matchings in planar graphs. The problem is originated from statistical physics where the question was: How many ways are there to arrange diatomic molecules on a surface? In the simplest case, the two atoms of each molecule occupy two vertices on a grid, thus the simplest case can be reduced to count domino tilings on an $n \times m$ grid. In 1961, Kasteleyn [112] and Temperley and Fisher [167] independently solved this problem. Later, Kasteleyn generalized this result to all planar graphs [110, 111].

In Chapter 4, we are going to prove that counting the perfect matchings in a graph is #P-complete. Furthermore, it is also #P-complete to count the (not necessarily perfect) matchings, even for planar graphs. However, counting the perfect matchings in planar graphs is in FP. This fact is the base of the holographic algorithms, which is an exciting, rapidly developing research topic, and which is briefly introduced in Chapter 5.

The main idea of the FKT algorithm is to find the square of the number of perfect matchings. As we are going to explain, this is the number of coverings of the graph with disjoint edges and oriented even cycles. If the edges of a graph are appropriately oriented, the number of such coverings is the determinant of the oriented adjacency matrix. Therefore we have to start with the definition of this appropriate orientation.

Definition 45. *Let* $G = (V, E)$ *be a planar embedding of a planar graph. An orientation of the edges is a* Pfaffian orientation *if an odd number of edges are oriented clockwise on each internal face.*

Below we provide a fast (polynomial running time) algorithm to construct a Pfaffian orientation of a planar graph, thus also proving that each planar graph has a Pfaffian orientation.

Given a planar embedding of a planar graph $G = (V, E)$, construct its dual graph $G^* = (V^*, E^*)$ in the following way. V^* is the set of faces of G, also one vertex for the external face. Two vertices in V^* are connected with an edge if the corresponding faces are neighbors. Take any spanning tree T^* of G^*, and root it into the vertex corresponding to the outer face of G. Let this vertex be denoted by v^*. The edges of T^* correspond to edges of G separating the neighbor faces. Let E' be the subset of these edges in G. Give an arbitrary orientation of the edges in $E \setminus E'$. We claim that this orientation can be extended to a Pfaffian orientation by visiting and removing the edges of T^* and giving an appropriate orientation of the corresponding edges in E'. While there is an edge in T^*, take any edge $e^* \in E^*$ connecting a leaf which is not v^* to the rest of T^*. The corresponding edge $e' \in E'$ is the last edge of a face F not having an orientation yet. Give e' an orientation such that F has an odd number of clockwise-oriented edges. The edge e' separates two faces, F and F'. Observe that due to construction, F' is either the outer face or a face which still has unoriented edges. If F' is not the outer face, the orientation of e' does not violate the property that F' will also have an odd number of clockwise-oriented edges. Remove e^* from T^*, and continue this procedure. In the last step, T^* contains one edge, which connects a leaf to v^*. Orient the corresponding edge in E' such that the last face in G also has an odd number of clockwise-oriented edges.

This procedure indeed generates a Pfaffian orientation, since T^* at the beginning contains a vertex for each face in G, and once the last edge of each face is oriented in a Pfaffian manner, the orientation of the edges of the face are not changed.

Pfaffian orientations have a very important property, stated in the following lemma.

Lemma 6. *Let $G = (V, E)$ be a planar embedding of a planar graph and let its edges be in a Pfaffian orientation. Let C be a cycle containing an even number of edges, surrounding an even number of vertices in the planar embedding. Then C contains an odd number of clockwise-oriented edges.*

Proof. Consider the subgraph G' that contains C and the vertices and edges surrounded by C. G' also has Pfaffian orientation since all of its internal faces are also internal faces of G. Let F' denote the number of internal faces in G', let E' denote the number of edges of G', and let V' denote the number of vertices in G'. From Euler's theorem, we know that

$$F' - E' + V' = 1. \qquad (3.47)$$

(Note that the outer face is not counted by F'.) We know that V' is even, since there are an even number of vertices in the cycle C and the cycle surrounds an even number of vertices. Therefore we get that

$$F' + E' \equiv 1 \qquad \text{mod } 2. \qquad (3.48)$$

Each internal edge separates two faces, and the orientation of an internal edge is clockwise in one of the faces and anticlockwise in the other face. Namely, if we put those edges into E_c which have clockwise orientation in some faces, then we put each internal edge into E_c and those edges in C which has clockwise orientation. Since there are an odd number of clockwise-oriented edges in each face, E_c has the same parity as F'. (Note that each internal edge is clockwise only in one of the faces!)

If F' is odd, then there are an even number of edges in G', due to Equation (3.48). Since there are an even number of edges in C, there are an even number of internal edges. The parity of E_c is odd, and since there are an even number of internal edges, the number of clockwise-oriented edges in C is odd.

On the other hand, if F' is even, there are an odd number of edges in G', due to Equation (3.48). Since there are an even number of edges in C, the number of internal edges is odd. The parity of E_c is even, so removing the odd number of internal edges from E_c, we get that the number of clockwise-oriented edges in C is still odd. □

Even long cycles surrounding an even number of vertices in a planar embedding are important since any cycle appearing in the union of two perfect matchings are such cycles. More specifically, there is a bijection of oriented even cycle coverings and ordered pairs of perfect matchings, stated and proved below. First, we have to define oriented even cycle coverings.

Definition 46. *Let $G = (V, E)$ be a graph. An* oriented even cycle covering *of G is a set of cycles with the following properties.*

1. *Each cycle is oriented, and removing the orientations, the edges are all in E.*

2. *Each cycle has even length. A cycle length of 2 is allowed; in that case, the unoriented versions of the edges are the same, however, they are still in E.*

3. *Each vertex in V is in exactly one cycle.*

Theorem 39. *Let $G = (V, E)$ be an arbitrary graph. The number of oriented even cycle coverings of G is the square of the number of perfect matchings of G.*

Proof. We give a bijection between the oriented even cycle coverings and ordered pair of perfect matchings. Since the number of ordered pairs of perfect matchings is the number of perfect matchings squared, it proves the theorem.

Fix an arbitrary total ordering of the vertices. We define two injective functions, one from the ordered even cycle coverings to the ordered pair of perfect matchings, and one in the other way, and prove that they are inverses of each other.

Let $\mathcal{C} = C_1, C_2, \ldots C_k$ be a cycle covering. For each cycle C_i, consider the smallest vertex v_i in it. Take a walk on C_i starting at v_i in the given orientation. Along the walk, put the edges into the sets M_1 and M_2 in an alternating manner, the first edge into M_1, the second into M_2, etc. Remove the orientations of all edges, both in M_1 and M_2. The so-constructed sets will be perfect matchings, since it is a matching and all vertices are covered. In this way, we constructed a mapping from the oriented even cycle coverings to the ordered pairs of perfect matchings.

We claim that this mapping is an injection. Indeed, if the unoriented versions of two coverings, $\mathcal{C}_1$ and $\mathcal{C}_2$, differ in some edges, then they have different images. If the edges are the same, and just some of the cycles have different orientation, then different edges of that cycle will go to the first and the second perfect matchings, and thus the images are still different.

The inverse mapping is the following. Let M_1 and M_2 be two perfect matchings. First, take the union of them. The union of them consists of disjoint even cycles and separated edges. Make an oriented cycle of length 2 from each separated edge. Orient each cycle C_i such that the edge which is incident to the smallest vertex v_i and comes from M_1 goes out from v_i and the edge which is from M_2 and incident to v_i comes into v_i. We constructed an oriented even cycle covering.

We claim that this mapping is an injection. Indeed, consider two ordered pairs of perfect matchings, (M_1, M_2) and (M_1', M_2'). If the set of edges of $M_1 \cup M_2$ is not the set of edges of $M_1' \cup M_2'$, then the images of the two pairs of perfect matchings are clearly different. If the two sets of edges in the unions are the same, but $M_1 \neq M_1'$, then consider an edge e which is in M_1 and not in M_1'. This edge is in a cycle C_i in the union of the two perfect matchings. We claim that C_i is oriented in a different way in the two matchings. Indeed, e is an edge from M_2', and since the edges are alternating in C_i, any edge in C_i which comes from M_1 is an edge from M_2'. Similarly, any edge from M_2 is an edge from M_1'. Especially, the two edges incident to the smallest vertex v_i in C_i comes from M_1 and M_1', therefore C_i is oriented differently in the two images, thus, the two images are different. Finally, it is easy to see if the two sets of edges in the unions are the same and $M_1 = M_1'$, then also $M_2 = M_2'$, thus the two ordered pairs of perfect matchings are the same.

It is also easy to see that the two injections are indeed the inverses of each other. $\qquad\square$

We are ready to prove the main theorem.

Theorem 40. *Let $G = (V, E)$ be a planar embedding of a planar graph, with edges having a Pfaffian orientation. Define the oriented adjacency matrix A in the following way. Let the entry $a_{i,j}$ be 1 if there is an edge going from v_i to v_j, and let $a_{i,j}$ be -1 if there is an edge going from v_j to v_i. All other entries are 0. Then the number of perfect matchings of G is*

$$\sqrt{det(A)}. \tag{3.49}$$

Proof. Based on Theorem 39, it is sufficient to prove that $det(A)$ is the number of oriented even cycle coverings in G. By definition, the determinant of the $n \times n$ matrix A is

$$det(A) := \sum_{\pi \in S_n} sign(\pi) \prod_{i=1}^{n} a_{i,\pi(i)}. \qquad (3.50)$$

Let O_n denote the set of those permutations that contain at least one odd cycle. We prove that

$$\sum_{\pi \in O_n} sign(\pi) \prod_{i=1}^{n} a_{i,\pi(i)} = 0. \qquad (3.51)$$

If a permutation π has a fixed point, then

$$\prod_{i=1}^{n} a_{i,\pi(i)} = 0, \qquad (3.52)$$

since if j is a fixed point then $a_{j,\pi(j)} = a_{j,j} = 0$. For those permutations π that contain an odd cycle of length at least 3, we set up an involution g and show that

$$sign(\pi) \prod_{i=1}^{n} a_{i,\pi(i)} = -sign(g(\pi)) \prod_{i=1}^{n} a_{i,g(\pi)(i)}. \qquad (3.53)$$

The involution is the following. Let π be a permutation containing at least one odd cycle of length at least 3. Amongst the odd cycles, let C_i be the cycle containing the smallest number. Then the image of π, $g(\pi)$ is the permutation that contains the same cycles as π except C_i is inverted. Since A is skew-symmetric, namely, for all i, j, $a_{i,j} = -a_{j,i}$, C_i has an odd length and $sign(\pi) = sign(g(\pi))$, so Equation (3.53) holds. Therefore Equation (3.51) also holds, since any permutation with odd cycles has a fixed point or an odd cycle with length at least 3 (or both, but then Equation (3.53) still holds, both sides are 0).

We get that

$$det(A) := \sum_{\pi \in S_n \setminus O_n} sign(\pi) \prod_{i=1}^{n} a_{i,\pi(i)}. \qquad (3.54)$$

Therefore it is sufficient to show that for any $\pi \in S_n \setminus O_n$,

$$sign(\pi) \prod_{i=1}^{n} a_{i,\pi(i)} = 1 \qquad (3.55)$$

if π is an oriented even cycle covering and 0 otherwise. If π is not an even cycle covering, then there exists a j such that $a_{j,\pi(j)} = 0$, thus the product is 0. If π is an oriented even cycle covering, observe the following. Since n is even, $sign(\pi)$ is 1 if it contains an even number of cycles and -1 if it contains an odd number of cycles. However, each even cycle appearing in an oriented even cycle covering must contain an odd number of clockwise and an odd number of anti-clockwise edges, since G has a Pfaffian orientation. Therefore, the contribution of each cycle in the product in Equation (3.55) is -1. Thus the product is -1 if the number of cycles is odd and it is 1 if the number of cycles is even. Since the same is true for $sign(\pi)$, Equation (3.55) holds if π is an oriented even cycle covering.

Therefore $det(A)$ is indeed the number of oriented even cycle coverings. Due to Theorem 39, $det(A)$ is the number of perfect matchings in G squared, so its square root is indeed the number of perfect matchings. □

Example 10. *Count the 2×1 domino tilings on a 3×4 square.*

Solution. Consider the planar graph whose vertices are the unit squares and two vertices are connected if the corresponding squares are neighbors. A possible Pfaffian orientation is

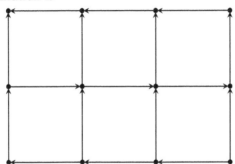

Number the vertices from top to bottom, and from left to right, row by row. Then the oriented adjacency matrix is

$$A = \begin{pmatrix} 0 & 1 & 0 & 0 & 1 & 0 & 0 & 0 & 0 & 0 & 0 & 0 \\ -1 & 0 & 1 & 0 & 0 & 1 & 0 & 0 & 0 & 0 & 0 & 0 \\ 0 & -1 & 0 & 1 & 0 & 0 & 1 & 0 & 0 & 0 & 0 & 0 \\ 0 & 0 & -1 & 0 & 0 & 0 & 0 & 1 & 0 & 0 & 0 & 0 \\ -1 & 0 & 0 & 0 & 0 & -1 & 0 & 0 & 1 & 0 & 0 & 0 \\ 0 & -1 & 0 & 0 & 1 & 0 & -1 & 0 & 0 & 1 & 0 & 0 \\ 0 & 0 & -1 & 0 & 0 & 1 & 0 & -1 & 0 & 0 & 1 & 0 \\ 0 & 0 & 0 & -1 & 0 & 0 & 1 & 0 & 0 & 0 & 0 & 1 \\ 0 & 0 & 0 & 0 & -1 & 0 & 0 & 0 & 0 & 1 & 0 & 0 \\ 0 & 0 & 0 & 0 & 0 & -1 & 0 & 0 & -1 & 0 & 1 & 0 \\ 0 & 0 & 0 & 0 & 0 & 0 & -1 & 0 & 0 & -1 & 0 & 1 \\ 0 & 0 & 0 & 0 & 0 & 0 & 0 & -1 & 0 & 0 & -1 & 0 \end{pmatrix}$$

Using standard methods to compute the determinant, we get that $det(A) = 121$. Therefore, the number of possible domino tilings on a 3×4 rectangle is 11. ■

Note that there were two places where we used the planarity of G. First, G can have a Pfaffian orientation, and second, any even cycle appearing in an oriented even cycle covering has an odd number of clockwise edges. Indeed, Theorem 39 holds for any graph. This suggests a weaker definition of Pfaffian orientation.

Definition 47. *Let $G = (V, E)$ be a graph. A Pfaffian orientation of the edges of G in the weak sense is an orientation that determines, for the corresponding oriented adjacency matrix A of G, that*

$$\sqrt{det(A)} \tag{3.56}$$

is the number of perfect matchings in A.

Recall that the Pfaffian of the matrix is

$$pf(A) := \sum_{\sigma \in S_{2n}/\sim} sign(\sigma) \prod_{i=1}^{n} a_{\sigma(2i-1),\sigma(2i)}. \tag{3.57}$$

If A is the oriented adjacency matrix of a graph G, then the score of a permutation from $S_{2n}/\sim$ in Equation (3.57) is 1 or -1 if and only if for all i, $(v_{\sigma(2i-1)}, v_{\sigma(2i)})$ is an edge in G. In such a case, these edges form a perfect matching in G. Therefore, for any orientation of G, we get that

$$|pf(A)| \leq PM(G) \tag{3.58}$$

where A is the oriented adjacency matrix of G, and $PM(G)$ is the number of perfect matchings in G. Furthermore, equality holds if and only if for any perfect matching of G, the score of the corresponding permutations in Equation (3.57) are all 1 or all -1. Since for any skew-symmetric matrix, the square of the Pfaffian is the determinant (see also Exercise 19), we get that an orientation of the edges of graph G is a Pfaffian orientation (by Definition 47) if and only if the absolute value of the Pfaffian of the corresponding oriented adjacency matrix is the number of perfect matchings in G.

We can assign weights to the edges of a planar graph from an arbitrary commutative ring, and then we might ask the sum of the weights of perfect matchings in a planar graph, where the weight of a perfect matching is the product of the weights of the edges. It is easy to see that the square of this sum is the determinant of the weighted, oriented adjacency matrix, where the orientation is Pfaffian.

The determinant can be calculated in polynomial time in an arbitrary commutative ring, thus it is easy to see that the square of the sum of perfect matching weights can be calculated in polynomial time. However, the Pfaffian can be directly calculated in polynomial time in an arbitrary commutative ring. This yields to the following theorem.

Theorem 41. *Let $G = (V, E)$ be a planar graph, let R be an arbitrary commutative ring, and let a weight function $w : E \to R$ be given. Define the partition function of G as*

$$Z(G) := \sum_{M \in PM(G)} \prod_{e \in M} w(e) \tag{3.59}$$

where $PM(G)$ is the set of perfect matchings of G.

Take a Pfaffian orientation of G in the string sense, namely by Definition 45, and generate two matrices. A is the usual oriented adjacency matrix, and A_w is the weighted adjacency matrix, that is, for an edge $e = (v_i, v_j)$, the corresponding matrix entry $a_{i,j}$ is $w(e)$ if edge e is oriented from v_i to v_j, and it is $-w(e)$ if edge e is oriented from v_j to v_i. All other entries are 0 in A_w.
Then $Z(G) = pf(A_w)$ if $pf(A) \geq 0$ and $Z(G) = -pf(A_w)$ if $pf(A) < 0$.

Proof. Define the signed weight of a perfect matching M as

$$sign(\sigma) \prod_{i=1}^{n} a_{\sigma(2i-1),\sigma(2i)} \tag{3.60}$$

where σ is a permutation satisfying that for all $i = 1, \ldots, n$, $(v_{\sigma(2i-1)}, v_{\sigma(2i)})$ is an edge in M, and $a_{k,j}$ is an entry of the oriented weighted adjacency matrix A_w. We showed that the weight is well-defined. Indeed, any swapping of indexes $\sigma(2i - 1)$ and $\sigma(2i)$ changes the sign of the permutation, and also changes the sign of one term in the product of the weights. Swapping a pair of indexes changes neither the sign of the permutation nor the product of the weights. We define the sign of the signed weight of a perfect matching as positive, if it is the product of the edge weights in the perfect matching, and this sign as negative, if it is the additive inverse of the product of the edge weights in the perfect matching. Clearly, this sign depends only on the sign of σ and on how many times $a_{\sigma(2i-1),\sigma(2i)}$ is the weight of the corresponding edge, and how many times it is the additive inverse of the weight of the corresponding edge.

The next observation is that the sign of the signed weight of each perfect matching is the same for a fixed Pfaffian orientation. We are going to prove it via the following steps. Let M be a perfect matching of G. An alternating cycle C is a cycle in G such that along that cycle, its edges are alternatingly presented and not presented in M. It is easy to see that $M \Delta C$ is also a perfect matching, where Δ denotes the symmetric difference. We claim that C surrounds an even number of vertices in a planar embedding of G. Indeed, M is a perfect matching, and the vertex set of C and $M \cap C$ is the same. Therefore each vertex inside C must be paired in M with another vertex inside C. Thus, the number of vertices inside C is even. Further, we claim that the signs of the signed weights of M and $M \Delta C$ are the same. To see this, first fix two permutations σ and σ', with the following properties. The permutation σ contains the edges of M while σ' contains the edges of $M \Delta C$.

Outside of C, σ and σ' are the same, and on C, both permutations contain the edges in anticlockwise orientation. That is, if $(v_{\sigma(2i-1)}, v_{\sigma(2i)})$ is an edge in C, then $v_{\sigma(2i-1)}$ is before $v_{\sigma(2i)}$ in anticlockwise direction. The same is true for σ'. For sake of simplicity and without loss of generality, we can assume that σ starts with $i_1, i_2, \ldots, i_{2k}$, where $v_{i_1}, v_{i_2}, \ldots, v_{i_{2k}}$ are the vertices of C in anticlockwise direction, and σ' starts with $i_{2k}, i_1, i_2, \ldots, i_{2k-1}$. We show that σ and σ' have different signs. Indeed, to obtain σ' from σ, we have to bubble down i_{2k} to the first position. This is an odd number of transpositions, which changes the sign of the permutation.

Now C contains an odd number of clockwise edges, since G is Pfaffian oriented. Then M and $M\Delta C$ contain different parity of clockwise-oriented edges along C. What follows is that in the signed weight of M and $M\Delta C$, different parity of entries of A_w are the additive inverses of the weights of the corresponding edges when the signed weight is written using σ and σ'. However, σ and σ' have different signs, therefore the sign of the signed weight of M and $M\Delta C$ is the same.

To see that the signs of the signed weights of all perfect matchings are the same, consider two perfect matchings of G, M and M', and take their symmetric difference. $M\Delta M'$ is the disjoint union of cycles, take them in an arbitrary order, $C_1, C_2, \ldots C_k$. Define $M_1 := M\Delta C_1$ and $M_i = M_{i-1}\Delta C_i$, for all $i = 2, \ldots k$. C_1 is an alternating cycle with respect to M, and each C_i is an alternating cycle with respect to M_{i-1}. Then the sign of $M\Delta C_1$ is the sign of M, and the sign of M_i is the sign of M_{i-1}. Especially, the sign of M is the sign of M_k. Observe that $M_k = M'$, therefore the sign of M is the sign of M'.

We get that

$$|pf(A_w)| = \sum_{M \in PM(G)} \prod_{e \in M} w(e), \tag{3.61}$$

since there is a bijection between the non-vanishing products in the definition of the Pfaffian in Equation (3.13) and the perfect matchings in G. Furthermore, each such product in absolute value is the weight of the corresponding perfect matching M (that is, $\prod_{e \in M} w(e)$), and either all of these products are the weight of the perfect matchings or all of them are the additive inverses. The sign of $pf(A)$ can tell which case holds, as stated in the theorem.   □

Example 11. *Count how many ways there are to tile a 2×4 rectangle with k horizontal and $4 - k$ vertical dominos.*

Solution. Consider the planar graph whose vertices are the unit squares, and two vertices are connected if their corresponding squares are neighbors. A possible Pfaffian orientation is

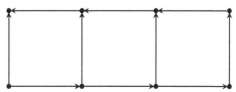

Using $\mathbb{Z}[x]$, assign a weight x to each horizontal edge, and weight 1 to each vertical edge. Number the vertices from top to bottom, and from left to right, row by row. Then the weighted oriented adjacency matrix is

$$A = \begin{pmatrix}
0 & x & 0 & 0 & 1 & 0 & 0 & 0 \\
-x & 0 & x & 0 & 0 & 1 & 0 & 0 \\
0 & -x & 0 & x & 0 & 0 & 1 & 0 \\
0 & 0 & -x & 0 & 0 & 0 & 0 & 1 \\
-1 & 0 & 0 & 0 & 0 & -x & 0 & 0 \\
0 & -1 & 0 & 0 & x & 0 & -x & 0 \\
0 & 0 & -1 & 0 & 0 & x & 0 & -x \\
0 & 0 & 0 & -1 & 0 & 0 & x & 0
\end{pmatrix}$$

Using the introduced algorithm, we get that

$$|pf(A)| = x^4 + 3x^2 + 1. \tag{3.62}$$

Namely, there is 1 tiling with 4 horizontal dominos, 3 tilings with 2 horizontal dominos, and 1 tiling with no horizontal (all vertical) dominos. The careful reader might observe that these coefficients come from the shallow diagonal of Pascal's triangle, and their sum is a Fibonacci number. ∎

3.5 The power of subtraction

We already mentioned in Chapter 2 that Jerrum and Snir proved that the spanning tree polynomial needs an exponential number of additions and multiplications if subtractions are not allowed. On the other hand, its value can be computed in polynomial time at any point using additions, multiplications and subtractions, see Theorem 35. It is easy to see that the formal computations in Theorem 35 build up the spanning tree polynomial, and use only a polynomial number of operations (additions, multiplications and subtractions). This does not mean that the spanning tree polynomial could be calculated in polynomial time. Indeed, the size of the spanning tree polynomial might be exponential. To resolve these facts looking paradoxically, observe that the computation needed to perform an operation in a formal many-variable polynomial ring might take exponential time. The operations might be easy to perform in some of the homomorph images, and in those cases, the overall running time will be a polynomial function of the input graph, not only the number of operations. However, these algorithms use subtractions. We are going to prove that there is no polynomial-sized yield algebra building the set of spanning trees of a graph satisfying mild conditions.

Theorem 42. *Define* $K_{[n]}$ *as the complete graph on the first n positive integer* number. There is no yield algebra $(A, (\Theta, \leq), p, \mathcal{O}, \mathcal{R})$ satisfying the following properties.

1. For each n, there is a parameter θ_n such that $S(\theta_n)$ contains the spanning trees of the complete graph $K_{[n]}$.

2. There is a function $g : A \to 2^{\binom{\mathbb{Z}^+}{2}}$ with the following properties:

   (a) For any $\circ_i \in \mathcal{O}$, and for any of its operands,

   $$g(\circ_i \left((a_j)_{j=1}^{m_i} \right)) = (\sqcup_{j=1}^{m_i} g(a_j)) \sqcup h_{\circ_i}(p(a_1), \dots, p(a_{m_i}))$$

   where $h_{\circ_i}$ is a function mapping from Θ^{m_i} to $2^{\binom{\mathbb{Z}^+}{2}}$.

   (b) If $a \in A$ is a spanning tree of $K_{[n]}$, then $g(a)$ is the set of edges in it.

3. $|\theta_{n\downarrow}| = O(poly(n))$.

Proof. The proof is by contradiction. We show that if such a yield algebra exists, then there is a corresponding evaluation algebra that could build up the spanning tree polynomial using a polynomial number of additions and multiplications.

So assume that a yield algebra $Y = (A, (\Theta, \leq), p, \mathcal{O}, \mathcal{R})$ with the above described properties exists. Then construct the following evaluation algebra $(Y, R, f, \mathcal{T})$. Let R be the multivariate polynomial ring over $\mathbb{Z}$ that contains a variable $x_{i,j}$ for each unordered pair of positive integers (i, j). Define the f function as

$$f(a) := \prod_{(i,j) \in g(a)} x_{i,j}. \tag{3.63}$$

Then the T_i function for operator $\circ_i$ is defined as

$$T_i(r_1, \dots, r_{m_i}, \theta_1, \dots, \theta_{m_i}) := \prod_{(k,l) \in h_{\circ_i}(\theta_1, \dots, \theta_{m_i})} x_{k,l} \prod_{j=1}^{m_i} r_i. \tag{3.64}$$

This is indeed an evaluation algebra, since each T_i function satisfies Equation (2.52).

Since for each spanning tree $a \in A$, $f(a)$ is the monomial containing the variables corresponding to the edges in the spanning tree, $F(S(\theta_n))$ is indeed the spanning tree polynomial. Since $|\theta_{n\downarrow}| = O(poly(n))$, the spanning tree polynomial could be calculated using only a polynomial number of additions and multiplications. This contradicts the theorem of Jerrum and Snir stating that computing the spanning tree polynomial needs an exponential number of additions and multiplications. $\square$

Jerrum and Snir also proved an exponential lower bound on the number of additions and multiplications to calculate the permanent polynomial defined as

$$per(M) := \sum_{\sigma \in S_n} \prod_{i=1}^{n} x_{i,\sigma(i)} \tag{3.65}$$

where M is an $n \times n$ matrix containing indeterminants $x_{i,j}$ and S_n is the set of permutations of the first n positive integers. This result provides a theorem on the absence of certain yield algebras on permutations.

Theorem 43. *Let $\vec{K}_{[n]}$ denote the complete direct graph containing loops on the first n positive integers. There is no yield algebra $(A, (\Theta, \leq), p, \mathcal{O}, \mathcal{R})$ with the following properties.*

1. *For each n, there is a parameter θ_n such that $S(\theta_n)$ is S_n.*

2. *There is a function $g : A \to 2^{\mathbb{Z}^+ \times \mathbb{Z}^+}$ with the following properties:*

   (a) *For any $\circ_i$ and for any of its operands*

   $$g(\circ_i \left((a_j)_{j=1}^{m_i} \right)) = (\sqcup_{i=1}^{m_i} g(a_j)) \sqcup h_{\circ_i}(p(a_1), \dots, p(a_{m_i}))$$

   where $h_{\circ_i}$ is a function mapping from Θ^{m_i} to $2^{\mathbb{Z}^+ \times \mathbb{Z}^+}$.

   (b) *If $a \in A$ is a permutation of length n, then $g(a)$ is the set of edges in the cycle cover of $\vec{K}_{[n]}$ that the permutation indicates.*

3. *It holds that $|\theta_{n\downarrow}| = O(poly(n))$.*

Proof. The proof is similar to the proof of Theorem 42. If such yield algebra existed, then we could build up a corresponding evaluation algebra that could compute the permanent polynomial using only a polynomial number of additions and subtractions. □

Contrary to the spanning tree polynomial, no algorithm is known to compute the permanent polynomial using only a polynomial number of arithmetic operations. In fact, computing the permanent is in #P-hard, so no polynomial algorithm exists to compute the permanent of a matrix, assuming that P is not NP. From this angle, it looks really accidental that the determinant can be calculated with a dynamic programming algorithm due to cancellation of terms. What that dynamic programming algorithm really calculates is the sum of the weights of clow sequences which coincides with the determinant of a matrix. A large set of similar "accidental" cases are known where we can compute the number of combinatorial objects in polynomial time due to cancellations. These cases are briefly described in Chapter 5.

Leslie Valiant proved that computing the perfect matching polynomial needs exponentially many arithmetic operations if subtractions are forbidden [177]. His theorem holds also if the problem is restricted to planar graphs.

On the other hand, the perfect matching polynomial can be computed with a polynomial number of arithmetic operations if subtractions are allowed. Indeed, it is easy to see that the formal computations in Theorem 41 build up the perfect matching polynomial using only a polynomial number of arithmetic operations in the multivariate polynomial ring. This again shows the computational power of subtractions: subtractions might have exponential computational power.

3.6 Further reading

- Anna Urbańska gave an $\tilde{O}(n^{3.03})$ running time algorithm both for calculating the determinant and the Pfaffian using only additions, subtractions and multiplications [172]. Here $\tilde{O}$ is for hiding sub-power (logarithmic) terms.

- Kasteleyn [112], and independently Temperley and Fisher [167], gave a formula for the number of domino tilings of a rectangle. The number of ways that an $n \times m$ rectangle can be covered with $\frac{nm}{2}$ dominos is

$$
\prod_{i=1}^{\left\lceil \frac{n}{2} \right\rceil} \prod_{j=1}^{\left\lceil \frac{m}{2} \right\rceil} \left(4\cos^2 \left(\frac{i\pi}{n+1} \right) + 4\cos^2 \left(\frac{j\pi}{m+1} \right) \right). \tag{3.66}
$$

The number of domino tilings of a square with edge length $0, 2, 4, \ldots$ is

$$
1, 2, 36, 6728, 12988816, 258584046368, 53060477521960000, \ldots
$$

This is the integer sequence $A004003$ in OEIS.

- Charles H.C. Little extended Kasteleyn's algorithm to the case when a graph does not contain any subgraph homeomorphic to $K_{3,3}$ [120]. Graphs G_1 and G_2 are homeomorphic if one can be transformed into another with a series of transformations of the following two types.

  - Subdivision. A vertex w added to the middle of an edge (u, v), so the transformed graph contains edges (u, w) and (w, v).

  - Smoothing. It is the reverse operation of subdivision: the degree 2 vertex w is deleted together with its edges (u, w) and (w, v), and its neighbors u and v are connected with an edge.

Little's theorem is that any graph not containing a subgraph homeomorphic to $K_{3,3}$ can be Pfaffian oriented in the weak sense (Definition 47). Unfortunately, this theorem does not directly provide a polynomial running time algorithm that decides if a graph can be Pfaffian oriented.

- György Pólya asked the following question [142]. Let A be a square 0-1 matrix. Is it possible to transform it to a matrix B by changing some of its 1s to -1, such that the permanent of A is the determinant B? Neil Robertson, Paul Seymour and Robin Thomas solved this question [146]. Roughly speaking, their theorem says that a matrix A has the above-mentioned property if and only if it is the adjacency matrix of a bipartite graph that can be obtained by piecing together planar graphs and one sporadic non-planar bipartite graph. Their theorem provides a polynomial running time algorithm to decide if A has such a property and if so, it also generates a corresponding matrix B.

- Simon Straub, Thomas Thierauf and Fabian Wagner gave a polynomial time algorithm to count the perfect matchings in K_5-free graphs [164]. It is interesting to mention that some of the K_5-free graphs cannot be Pfaffian orieneted in the weak sense; an example of that is $K_{3,3}$, see also Exercise 23. Their algorithm decomposes a K_5-free graph into components and applies some matchgate techniques to get a polynomial time algorithm. Matchgates are introduced in this book in Chapter 5.

3.7   Exercises

1. List all clow sequences of length 3.

2. ∘ Let Cl_n denote the number of clow sequences of length n on the numbers $\{1, 2, \ldots, n\}$. Give a dynamic programming recursion that finds Cl_n.

3. * Show that there are exponentially many more clow sequences of a given length than permutations.

4. Prove that the rank of the adjacency matrix of a directed graph is the number of vertices minus the number of components of the graph.

5. Prove that any cycle-free graph $G = (V, E)$ for which $|E| = |V| - 1$ is a tree.

6. A graph G contains exactly two cycles, the length of the cycles are l_1 and l_2. Find the number of spanning trees of G.

7. * How many spanning trees are in an octahedron?

8. How many spanning trees are in a cube?

9. Let $G = (V, E)$ be a graph, and let $w : E \to \mathbb{Z}$ be a weight function assigning a polynomial of an order of at most t to each edge of a graph. Show that

$$\sum_{T \in \mathcal{T}_G} \prod_{e \in T} w(e)$$

can be calculated in $O(poly(|V|, t))$ time, where $\mathcal{T}_G$ denotes the set of spanning trees of G.

10. ○ Let $G = (V, E)$ be a graph, and let $w : E \to \mathbb{R}$ be a weight function assigning a weight to each edge of a graph. Show that

$$\sum_{T \in \mathcal{T}_G} \sum_{e \in T} w(e)$$

can be calculated in polynomial time.

11. * Let $G = (V, E)$ be a graph, whose edges are colored with red and blue and green. Find a polynomial running time algorithm that counts the spanning trees of G with k_1 blue edges, k_2 red edges, and $|V| - 1 - k_1 - k_2$ green edges.

12. Count the spanning trees of the complete bipartite graph $K_{n,m}$.

13. * Count the number of minimum spanning trees in an edge-weighted graph. The weights might be large integer numbers, that is, the size of the input is measured by the number of digits necessary to write the numbers.

14. * Prove that it is NP-complete to decide if there is a spanning tree in a graph with a given sum of weights. Hint: reduce the subset sum problem (Theorem 5) to this problem.

15. * Replace each edge of a tetrahedron with a pair of antiparallel directed edges. How many Eulerian circuits are there in the so-obtained directed graph?

16. Orient the edges of the octahedron such that each meridian and the equator are oriented. How many Eulerian circuits are in this directed graph?

17. * How many sequences are there that contain the *ab* substring 6 times, the *ba* substring 4 times, the *bc* substring 4 times, the *cb* substring 3 times, the *ca* substring once, and there are no other substrings of length 2?

18. How many sequences are there that contain the *ab* substring 12 times, the *ba* substring also 12 times, the *bc* substring 5 times, the *cb* substring 2 times, the *ac* substring 3 times, the *ca* substring 6 times, and there are no other substrings of length 2?

19. ***** Prove that for any skew-symmetric $2n \times 2n$ matrix A,

$$pf^2(A) = det(A).$$

20. ***** How many perfect matchings are in an octahedron?

21. How many perfect matchings are in a cube?

22. Prove that for any matrix A,

$$det(A) = pf \begin{pmatrix} 0 & A \\ -A^T & 0 \end{pmatrix}.$$

23. ***** The complete bipartite graph $K_{3,3}$ is not planar. Prove that it does not have a Pfaffian orientation in a weak sense (Definition 47).

24. ∘ Let $G = (V, E)$ be a planar graph, and let $w : E \to \mathbb{R}$ be a weight function. Show that

$$Z(G) := \sum_{M \in PM(G)} \sum_{e \in M} w(e)$$

can be calculated in polynomial time, where $PM(G)$ is the set of perfect matchings in G.

3.8 Solutions

Exercise 2. Apply the algebraic dynamic programming approach on Equations (3.7) and (3.8).

Exercise 3. Those clow sequences that contain only one clow with head 1 are already exponentially more than the permutations of the same length. Indeed, there are $(n-1)^{n-1}$ clows of length n with head n over the indexes $1, 2, \ldots, n$. Therefore the ratio of the number of clow sequences and the number of permutations is at least

$$\frac{(n-1)^{n-1}}{n!}.$$

Applying the Stirling formula, this fraction is

$$\frac{(n-1)^{n-1}}{\sqrt{2\pi n} \left(\frac{n}{e}\right)^n} = \frac{1}{n\sqrt{2\pi n}} \left(\frac{n-1}{n}\right)^n e^n$$

which clearly tends to infinity exponentially quickly.

Exercise 7. Index the vertices of the octahedron such that the north pole is the first vertex, the south pole is the last one, and the vertices on the equator are ordered along the walk on the equator. Let v be the north pole. Then

$$D^{-v} - A^{-v} = \begin{pmatrix} 4 & -1 & 0 & -1 & -1 \\ -1 & 4 & -1 & 0 & -1 \\ 0 & -1 & 4 & -1 & -1 \\ -1 & 0 & -1 & 4 & -1 \\ -1 & -1 & -1 & -1 & 4 \end{pmatrix}.$$

Using standard calculations, we get that $det(D^{-v} - A^{-v} = 384$, therefore, the octahedron has 384 spanning trees.

Exercise 10. Let G^{-e} be the graph obtained from G by removing edge e. Observe that the difference in the number of spanning trees of G and G^{-e} is the number of spanning trees containing e. If T^e is the number of spanning trees containing e, then the value to be calculated is

$$\sum_{e \in E} T^e w(e).$$

Alternatively, we can take the commutative ring R defined in Subsection 2.1.3.4, and assign values $(1, w(e))$ to each weight. According to Theorem 35, the partition function of the spanning trees can be calculated in polynomial time, from which the total sum of weights can be read out.

Exercise 11. Using the bivariate polynomial ring, $\mathbb{Z}[x_1, x_2]$, assign value x_1 to each blue edge, x_2 to each red edge and 1 to each green edge. With these weights, calculate

$$det(D^{-v} - A^{-v}).$$

The coefficient of the monomial $x_1^{k_1} x_2^{k_2}$ tells the number of spanning trees with k_1 blue edges and k_2 red edges.

Exercise 13. We have to make two observations. The first is that every minimum spanning tree can be obtained by Kruskal's algorithm if we put the edges in all possible total orderings that satisfy $e_i < e_j$ if $w(e_i) \leq w(e_j)$. The second observation is that any time during Kruskal's algorithm, we set up a spanning forest on the components defined by the edges $E' = \{e | w(e) \leq w'\}$, where the remaining edges in Kruskal's algorithm all have weights larger than w'. Therefore, how to finish the minimum spanning tree does not depend on the spanning forest so far built by Kruskal's algorithm. Therefore, the number of minimum spanning trees can be computed in the following way.

1. Set H to G.

2. While H is not the simple vertex graph, do the following:

   (a) Let H' be the graph spanned by the minimum weight vertices in H. Let the minimum weight in H be w, and let $f(w)$ be the number of spanning forests of H'. The number of spanning forests is the product of the number of spanning trees on each component, this can be calculated in polynomial time.

(b) Redefine H as contracting the vertices of each component in H'. Note that multiple edges might appear when a vertex is connected to different vertices of a component in H', however, the number of spanning trees can be easily calculated in the presence of parallel edges, too.

3. The number of minimum spanning trees in G is $\prod_{w \in W} f(w)$, where W is the set of weights appearing during the iteration in the previous step.

Exercise 14. Let $S = \{w_1, w_2, \ldots, w_n\}$ be a set of weights. Create an edge-weighted graph G in the following way. Define vertices $v_1, v_2, \ldots, v_{n+1}$ and $u_1, u_2, \ldots, u_{n+1}$. For each i, connect v_i to u_i. Assign a weight w_i to the edge (v_i, u_i) for each $i = 1, \ldots, n$, and assign weight 0 to the edge (v_{n+1}, u_{n+1}). Finally, we create a complete graph K_{n+1} on both the v_i and u_i vertices. Each edge in both K_{n+1} components gets a weight 0.

It is easy to see that for each subset A of S, G contains a spanning tree whose sum of weights is the sum of weights in A. Therefore if we could answer in polynomial time if there is a spanning tree in G whose weight is m, we could also tell in polynomial time if there is a subset A of S such that its sum of weights is m. Since this latter is in NP-complete (see Theorem 5), it is also in NP-complete to decide if there is a spanning tree of a graph with a weight m.

Exercise 15. The obtained graph is the directed complete graph on 4 vertices. For an arbitrary v, we get that

$$D_{out}^{-v} - A^{-v} = \begin{pmatrix} 3 & -1 & -1 \\ -1 & 3 & -1 \\ -1 & -1 & 3 \end{pmatrix}.$$

The determinant of this matrix is 16. Since for each vertex, the out-degree is 3, the number of Eulerian circuits is

$$16 \times (3-1)!^4 = 256.$$

Exercise 17. Since character a is 6 times in the first position of a substring and only 5 times in the second position of a substring, the sequence must start with a. Similarly, character b is 9 times in the second position in a substring, and only 8 times in the first position, the sequence must end with character b. Therefore, the sequences are the Eulerian circuits in the directed multigraph with vertices v_s, v_a, v_b and v_c. Vertex v_s sends one edge to v_a. Vertex v_a sends 6 edges to v_b. Vertex v_b sends 4 edges to v_a, 4 edges to v_c and 1 edge to v_s. Vertex v_c sends 1 edge to v_a and 3 edges to v_b. Therefore

$$D_{out}^{-v_s} - A^{-v_s} = \begin{pmatrix} 6 & -6 & 0 \\ -4 & 9 & -4 \\ -1 & -3 & 4 \end{pmatrix}$$

$det(D_{out}^{-v_s} - A^{-v_s}) = 24$. Therefore the number of sequences with the prescribed statistics is

$$24\frac{0!5!8!3!}{1!6!4!4!3!1!1!} = 280.$$

Exercise 19. Observe that for any $2n \times 2n$ skew-symmetric matrix, Equation (3.54) holds. Also observe that the permutations containing only even long cycles are exactly the oriented even cycle coverings. There is a bijection between oriented even cycle coverings and ordered pairs of perfect matchings. Using this bijection, for each permutation σ containing only even cycles, we can assign two permutations π_1 and π_2 representing the two perfect matchings. If

$$\sigma = (x_{1,1}, x_{1,2}, \ldots, x_{1,k_1})(x_{2,1}, \ldots, x_{2,k_2}) \ldots (x_{m,1}, \ldots, x_{m,k_m})$$

then π_1 is

$$\begin{pmatrix} 1 & 2 & \cdots & k_1 & k_1+1 & \cdots & k_1+k_2 & \cdots & 2n-k_m & \cdots & 2n \\ x_{1,1} & x_{1,2} & \cdots & x_{1,k_1} & x_{2,1} & \cdots & x_{2,k_2} & \cdots & x_{m,1} & \cdots & x_{m,k_m} \end{pmatrix}$$

and π_2 is

$$\begin{pmatrix} 1 & 2 & \cdots & k_1 & k_1+1 & \cdots & k_1+k_2 & \cdots & 2n-k_m & \cdots & 2n \\ x_{1,2} & x_{1,3} & \cdots & x_{1,1} & x_{2,2} & \cdots & x_{2,1} & \cdots & x_{m,2} & \cdots & x_{m,1} \end{pmatrix}.$$

Observe that

$$sign(\pi_1)sign(\pi_2) = sign(\sigma)$$

since

$$\pi_1^{-1}\pi_2 = \sigma,$$

and the sign of any permutation is the sign of its inverse. Therefore we get that for any so-constructed σ, π_1 and π_2,

$$\left(sign(\pi_1) \prod_{i=1}^{n} a_{\pi_1(2i-1),\pi_1(2i)} \right) \left(sign(\pi_2) \prod_{i=1}^{n} a_{\pi_2(2i-1),\pi_2(2i)} \right) =$$

$$= sign(\sigma) \prod_{i=1}^{2n} a_{i,\sigma(i)}. \tag{3.67}$$

Summing this for all permutation σ which are even cycle coverings, we get that

$$\left(\sum_{\pi_1 \in S_{2n}/\sim} sign(\pi_1) \prod_{i=1}^{n} a_{\pi_1(2i-1),\pi_1(2i)} \right) \times$$

$$\left(\sum_{\pi_2 \in S_{2n}/\sim} sign(\pi_2) \prod_{i=1}^{n} a_{\pi_2(2i-1),\pi_2(2i)} \right) =$$

$$= \sum_{\sigma \in S_{2n} \backslash O_{2n}} sign(\sigma) \prod_{i=1}^{2n} a_{i,\sigma(i)}.$$

Observe that on the left-hand side we have $pf^2(A)$ by definition, and on the right-hand side, we have $det(A)$ due to Equation (3.54).

Exercise 20. Index the vertices of the octahedron such that the north pole is the first vertex, the south pole is the last one, and the vertices on the equator are ordered along the walk on the equator. With an appropriate Pfaffian orientation, the oriented adjacency matrix is

$$A = \begin{pmatrix} 0 & 1 & -1 & -1 & -1 & 0 \\ -1 & 0 & 1 & 0 & -1 & -1 \\ 1 & -1 & 0 & -1 & 0 & -1 \\ 1 & 0 & 1 & 0 & -1 & 1 \\ 1 & 1 & 0 & 1 & 0 & -1 \\ 0 & 1 & 1 & -1 & 1 & 0 \end{pmatrix}.$$

Using standard matrix calculations, we get that $det(A) = 64$. Therefore, the number of perfect matchings in an octahedron is 8.

Exercise 23. From Exercise 22, we get that the Pfaffian of the oriented adjacency graph of the complete bipartite graph $K_{3,3}$ is the determinant of the matrix

$$A = \begin{pmatrix} x_1 & x_2 & x_3 \\ x_4 & x_5 & x_6 \\ x_7 & x_8 & x_9 \end{pmatrix}$$

where each x_i is either 1 or -1. The number of perfect matchings in $K_{3,3}$ is 6. However the determinant of matrix A has only 6 terms, so the determinant can be 6 only if each term is 1. This means that the $det(A)$ could be 6 if all of the following sets contain an even number of -1s:

$$\{x_1, x_5, x_9\}, \quad \{x_2, x_6, x_7\}, \quad \{x_3, x_4, x_8\}$$

and the following sets all contain an odd number of -1s:

$$\{x_1, x_6, x_8\}, \quad \{x_2, x_4, x_9\}, \quad \{x_3, x_5, x_7\}.$$

From the first 3 sets, we get that there must be an even number of -1s in A. However, from the second 3 sets, we get that there must be an odd number of -1s in A. It is impossible.

Exercise 24. Similar to Exercise 10, for each edge (v_i, v_j), we can remove vertices v_i and v_j to count the perfect matchings containing edge (v_i, v_j).

Chapter 4

#P-complete counting problems

We already learned in Chapter 1 that #SAT is #P-complete. This chapter provides a comprehensive list of #P-complete counting problems and introduces the main proving techniques. We have to distinguish two different ways to prove #P-completeness. The reason for the distinction is that one of the proving techniques for #P-completeness actually proves more: it also proves that the counting problem does not have an FPRAS unless RP = NP. These proving techniques apply polynomial reductions reducing #SAT to some counting problem #A that keeps the relative error. That is, if the computation of problem instances from #A was only approximate, then still separating the zero and non-zero number of solutions of a problem instance from #SAT could be done with high probability. That is, it would provide a BPP algorithm for SAT, which, as we already learned, would imply that RP = NP. Other proving techniques apply polynomial reduction in which at least one operation does not keep the relative error. This might be a subtraction or modulo prime number calculation. Many (but definitely not all) of those computational problems whose #P-completeness is proved in this way are in FPRAS and in FPAUS. Below we detail these proving techniques.

1. Polynomial reductions that keep the relative error. These reductions are also called *approximation-preserving* reductions.

    (a) A polynomial reduction that for a CNF Φ, it constructs a problem instance x in #A such that the number of solutions of x is exactly the number of satisfying assignments of Φ. Such reduction actually also proves the NP-completeness of the corresponding decision problem A. Thus, #A does not have an FPRAS unless RP = NP. An example in this chapter is the proof of the #P-completeness of the #3SAT. This type of reduction is also called a *parsimonious* [155] reduction.

    (b) A polynomial reduction that for a CNF Φ, it constructs a problem instance x in some function problem A such that the answer for x is a number which is k times the number of satisfying assignments of Φ, where k can be computed in polynomial time. Typically, the computational problem A is to compute the sum of weights of some discrete mathematical objects, for example, the weights of cycle covers in a directed graph. This happens to coincide with the permanent of the corresponding weighted adjacency graph, as shown in this chapter. Such proof proves neither that deciding if the set of the discrete mathematical objects is not empty is NP-complete nor that finding the minimum or maximum weighted object is NP-hard. However, an FPRAS algorithm for #A would prove that RP = NP.

    (c) A polynomial reduction that for a CNF Φ, it constructs a problem instance x in #A such that the number of solutions of x is $a + by$, where $0 \le a \ll b$, y is the number of satisfying assignments of Φ, and b is easy to compute. Such a reduction proves the #P-completeness of #A. Indeed, $\left\lfloor \frac{a+by}{b} \right\rfloor$ is the number of satisfying assignments of Φ. It is easy to see that an FPRAS for #A would imply that RP = NP. Examples in this chapter are the proof of #P-completeness of counting the most parsimonious substitution histories on a binary tree and the proof of #P-completeness of counting independent sets in a graph.

2. Polynomial reductions that do not keep the relative error. Such reductions do not exclude the possibility for an FPRAS approximation; on the other hand, they do not guarantee the existence of efficient approximations.

   (a) Reductions using only one subtraction. For example, this is the way to reduce #SAT to #DNF.

   (b) Modulo prime number calculations. This is applied in reducing the permanent computation of matrices of $-1, 0, 1, 2$ and 3 to matrices containing only small non-negative integers.

   (c) Polynomial interpolation. This reduction is based on the following fact. If the value of a polynomial of order n is computed in $n + 1$ different points, then the coefficients of the polynomial can be determined in polynomial time by solving a linear equation system. In the polynomial reduction, for a problem instance a in the #P-complete problem A, $m + 1$ problem instances b_j from problem B are constructed such that the number of solutions for b_j is the evaluation of some polynomial $\sum_{i=0}^{m} c_i x^i$ at distinct points $x = x_j$. For some k, c_k is the number of solutions of problem instance a. This is a very powerful approach that is used in many #P-completeness proofs of counting problems including counting the not necessarily perfect matchings in planar graphs and counting the subtrees of a graph.

   (d) Reductions using other linear equation systems. An example in this chapter is the reduction of the number of perfect matchings to the number of Eulerian orientations of a graph.

Finally, a class of #P-complete computational problems has become a center of interest recently. These problems are known to be equally hard, and it is conjectured that they do not have an FPRAS approximation. We will discuss them in Subsection 4.3.2.

4.1 Approximation-preserving #P-complete proofs

4.1.1 #3SAT

Definition 48. *A 3CNF is a conjunctive normal form in which each clause is a disjunction of exactly 3 literals. The 3SAT is the decision problem if a 3CNF can be satisfied. Accordingly, the #3SAT asks the number of satisfying assignments of a 3CNF.*

It is well known that 3SAT is NP-complete, and the usual reduction for proving NP-completeness is to break large clauses into smaller ones introducing auxiliary variables. For example, the CNF

$$(x_1 \vee \overline{x}_2 \vee x_3 \vee x_4) \wedge (x_2 \vee x_3 \vee \overline{x}_4) \tag{4.1}$$

can be rewritten into the 3CNF

$$(x_1 \vee \overline{x}_2 \vee y) \wedge (x_3 \vee x_4 \vee \overline{y}) \wedge (x_2 \vee x_3 \vee \overline{x}_4). \tag{4.2}$$

It is easy to see that for any satisfying assignment of the CNF in Equation (4.1), there is a satisfying assignment of the 3CNF in Equation (4.2) and vice versa. Indeed, if any of the literals satisfies the first clause in the CNF, then y can be set such that the other clause in the 3CNF will also be satisfied. Similarly, y on its own cannot satisfy the first two clauses, therefore another literal must be TRUE, and then it satisfies the first clause in the CNF, too. However, this reduction cannot keep the number of satisfying assignments. For example, if both x_1 and x_3 are TRUE, then y might be arbitrary. On the other hand, if x_1 is FALSE and x_2 is TRUE, then y must be TRUE to satisfy the 3CNF in Equation (4.2). Therefore there is neither one-to-one nor one-to-many correspondence between the two conjunctive normal forms. It is easy to see that the CNF in Equation (4.1) has 13 satisfying assignments, and on the other hand, the 3CNF in Equation (4.2) has 20 satisfying assignments.

Therefore, another reduction is needed if we also want to keep the number of satisfying assignments. Any CNF Φ can be described with a directed acyclic graph, $\vec{G} = (V, E)$, see also Figure 4.1. In $\vec{G}$, the input nodes are the logical variables and there is one output node, O, representing the result of the computation. Each internal node has two incoming edges and has exactly one outgoing edge. The number of outgoing edges of the input node for the logical variable x_i equals the number of clauses in which x_i or $\overline{x}_i$ is a literal. An edge crossed with a tilde ($\sim$) means negation. Each internal node is evaluated according to its label ($\vee$ or $\wedge$), and the result of the computation is propagated on the outgoing edge.

From $\vec{G}$, we construct a 3CNF Φ' such that Φ and Φ' have the same number of satisfying assignments. First, we describe the computation on $\vec{G}$ with a 3CNF, then we amalgamate it to get Φ' such that only those assignments satisfy Φ' that represent evaluations of Φ on $\vec{G}$ yielding a TRUE value propagated to its output node O.

For each internal node, we can write a 3CNF of 3 logical variables. The logical variables represent the two incoming and one outgoing edges, and those 4 assignments satisfiy the 3CNF that represen the computation performed by the node. If the two incoming edges of a node performing a logical OR operation are represented by x_1 and x_2, and the outgoing edge is represented by y, then the 3CNF is

$$(x_1 \vee x_2 \vee \overline{y}) \wedge (\overline{x}_1 \vee x_2 \vee y) \wedge (x_1 \vee \overline{x}_2 \vee y) \wedge (\overline{x}_1 \vee \overline{x}_2 \vee y). \tag{4.3}$$

Similarly, the 3CNF for the node performing a logical AND operation is

$$(x_1 \vee x_2 \vee \overline{y}) \wedge (\overline{x}_1 \vee x_2 \vee \overline{y}) \wedge (x_1 \vee \overline{x}_2 \vee \overline{y}) \wedge (\overline{x}_1 \vee \overline{x}_2 \vee y). \tag{4.4}$$

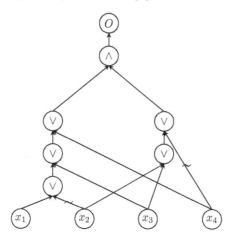

FIGURE 4.1: The directed acyclic graph representation of the CNF $(x_1 \vee \overline{x}_2 \vee x_3 \vee x_4) \wedge (x_2 \vee x_3 \vee \overline{x}_4)$. Logical values are propagated on the edges, an edge crossed with a tilde ($\sim$) means negation. Each internal node has two incoming edges and one outgoing edge. The operation performed by a node might be a logical OR ($\vee$) or a logical AND ($\wedge$). The outcome of the operation is the income at the other end of the outgoing edge.

Similar 3CNFs can be constructed when one or both incoming values are negated. The conjunction of the 3CNFs obtained for the internal nodes describes the computation on the directed acyclic graph, namely, the satisfying assignments represent the possible computations on the directed acyclic graph. To get the satisfying assignments of the initial CNF Φ, we require that the value propagated to O must be a logical TRUE value. We can represent it with a 3CNF adding two auxiliary logical variables, O' and O":

$$(O \vee O' \vee O") \wedge (O \vee O' \vee \overline{O"}) \wedge (O \vee \overline{O'} \vee O") \wedge (\overline{O} \vee O' \vee O") \wedge$$
$$(\overline{O} \vee \overline{O'} \vee O") \wedge (\overline{O} \vee O' \vee \overline{O"}) \wedge (O \vee \overline{O'} \vee \overline{O"}). \qquad (4.5)$$

It is easy to check that only $O = O' = O" = TRUE$ satisfies the 3CNF in Equation (4.5).

In this way, the CNF in Equation (4.1) can be rewritten as

$$(x_1 \vee \overline{x}_2 \vee \overline{y_1}) \wedge (\overline{x}_1 \vee \overline{x}_2 \vee y_1) \wedge (x_1 \vee x_2 \vee y_1) \wedge (\overline{x}_1 \vee x_2 \vee y_1) \wedge$$
$$(y_1 \vee x_3 \vee \overline{y}_2) \wedge (\overline{y}_1 \vee x_3 \vee y_2) \wedge (y_1 \vee \overline{x}_3 \vee y_2) \wedge (\overline{y}_1 \vee \overline{x}_3 \vee y_2) \wedge$$
$$(y_2 \vee x_4 \vee \overline{y}_3) \wedge (\overline{y}_2 \vee x_4 \vee y_3) \wedge (y_2 \vee \overline{x}_4 \vee y_3) \wedge (\overline{y}_2 \vee \overline{x}_4 \vee y_3) \wedge$$
$$(x_2 \vee x_3 \vee \overline{y}_4) \wedge (\overline{x}_2 \vee x_3 \vee y_4) \wedge (x_2 \vee \overline{x}_3 \vee y_4) \wedge (\overline{x}_2 \vee \overline{x}_3 \vee y_4) \wedge$$
$$(y_4 \vee \overline{x}_4 \vee \overline{y}_5) \wedge (\overline{y}_4 \vee \overline{x}_4 \vee y_5) \wedge (y_4 \vee x_4 \vee y_5) \wedge (\overline{y}_4 \vee x_4 \vee y_5) \wedge$$
$$(y_3 \vee y_5 \vee \overline{O}) \wedge (\overline{y}_3 \vee y_5 \vee O) \wedge (y_3 \vee \overline{y}_5 \vee \overline{O}) \wedge (\overline{y}_3 \vee \overline{y}_5 \vee O) \wedge$$
$$(O \vee O' \vee O") \wedge (O \vee O' \vee \overline{O"}) \wedge (O \vee \overline{O'} \vee O") \wedge (\overline{O} \vee O' \vee O") \wedge$$
$$(\overline{O} \vee \overline{O'} \vee O") \wedge (\overline{O} \vee O' \vee \overline{O"}) \wedge (O \vee \overline{O'} \vee \overline{O"}). \tag{4.6}$$

If the CNF Φ has k variables and m logical operations $\vee$ and $\wedge$, then Φ' has $n + m + 2$ logical variables, and has $4m + 7$ clauses. Since clearly Φ' can be constructed in polynomial time, we get the following theorem:

Theorem 44. *The counting problem #3SAT is in #P-complete.*

4.1.2 Calculating the permanent of an arbitrary matrix

Leslie Valiant proved in 1979 that calculating the permanent is #P-hard [175]. Recall that the permanent of an $n \times n$ square matrix A is defined as

$$per(A) := \sum_{\sigma \in S_n} \prod_{i=1}^{n} a_{i,\sigma(i)} \tag{4.7}$$

where S_n is the set of permutations of length n. The permanent is similar to the determinant, just the products are not weighted with the sign of the permutation. While the determinant can be calculated in polynomial time, there is no known polynomial running time algorithm to find the permanent. If #P is not part of FP, then such algorithm does not exist.

Below we introduce the work of Valiant on how to reduce #3SAT to computing the permanent of a matrix containing only values -1, 0, 1, 2 and 3. Any $n \times n$ square matrix A can be regarded as the adjacency matrix of an edge-weighted, directed graph, $\vec{G}$ with n vertices, where $a_{i,j}$ is the weight of the edge going from v_i to v_j. $\vec{G}$ might contain loops and antiparallel edges with different weights. It is easy to see that any permutation is a cycle cover of the vertices. A loop is considered as a cycle of length one covering its vertex. We can define the weight of a cycle cover as the product of the edge weights in it. Having said this, it is clear that the permanent of a matrix A is the sum of the weights of the cycle covers in $\vec{G}$.

For any 3CNF Φ, let $t(\Phi)$ be twice the number of occurrences of the literals in Φ minus the number of clauses in Φ. We are going to construct a directed graph $\vec{G}$ such that $4^{t(\Phi)}$ times the number of satisfying assignments of Φ is the sum of the weights of the cycle covers in $\vec{G}$, that is, the permanent of the corresponding weighted adjacency matrix. Furthermore, $\vec{G}$ can be constructed from Φ in polynomial time. This construction proves that calculating the permanent is #P-hard, since calculating $t(\Phi)$ as well as dividing an integer with $4^{t(\phi)}$ are both easy.

The construction is such that each cycle cover corresponding to a satisfying assignment has a weight $4^{t(\Phi)}$ and all other "spurious" cycle covers cancel each other out. Let $\Phi = C_1 \wedge C_2 \wedge \ldots C_m$, where each $C_i = (y_{i,1} \vee y_{i,2} \vee y_{i,3})$ with $y_{i,j} \in \{x_1, \bar{x}_1, x_2, \bar{x}_2, \ldots, x_n, \bar{x}_n\}$. The graph is built up using the following gadgets:

(a) A *track* T_k for each variable x_k;

(b) an *interchange* R_i for each clause C_i;

(c) for each literal $y_{i,j}$ such that $y_{i,j}$ is either x_k or $\bar{x}_k$, a *junction* $J_{i,k}$ at which R_i and T_k meet. Interchanges also have *internal junctions* with the same structure.

Each junction is a 4-vertex, weighted, directed graph with the following weighted adjacency matrix X:

$$X := \begin{pmatrix} 0 & 1 & -1 & -1 \\ 1 & -1 & 1 & 1 \\ 0 & 1 & 1 & 2 \\ 0 & 1 & 3 & 0 \end{pmatrix}. \tag{4.8}$$

Each junction has external connections via the 1st and 4th vertices and not via the other two vertices. Let $X[\gamma; \delta]$ denote the submatrix obtained from X deleting the rows γ and columns δ. The following properties are easy to verify:

(a) $per(X) = 0$,

(b) $per(X[1; 1]) = 0$,

(c) $per(X[4; 4]) = 0$,

(d) $per(X[1, 4; 1, 4]) = 0$,

(e) $per(X[1; 4]) = per(X[4; 1]) =$ non-zero constant $(= 4)$.

Only the junctions have edges with weight other than 1. All other edges in the construction have weight 1.

A track T_k consists of 2 vertices, $v_{k,1}$ and $v_{k,2}$, and r_k junctions where r_k is the number of clauses in which literals x_k and $\bar{x}_k$ participate. There is an edge going from $v_{k,1}$ to $v_{k,2}$, and there are 1-1 paths from $v_{k,2}$ to $v_{k,1}$ picking up the first and last edges of the junctions. One of the paths picks up the junctions of clauses in which x_k is a literal and the other path picks up the junctions of clauses in which $\bar{x}_k$ is a literal. An example is shown on Figure 4.2.

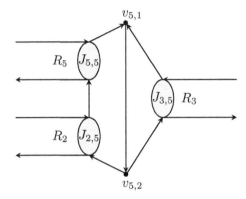

FIGURE 4.2: The track T_5 for the variable x_5 when x_5 is a literal in C_2 and C_5 and $\bar{x}_5$ is a literal in C_3.

Interchange R_i contains two vertices $w_{i,1}$ and $w_{i,2}$, and 3 junctions for the 3 literals and 2 internal junctions. They are wired as shown on Figure 4.3. The interchanges do not distinguish literals x_i and $\bar{x}_i$, the edges connecting the junctions are always the same.

We define a *route* in $\vec{G}$ as a set of cycle covers that contains the same edges outside the junctions. A route is *good* if every junction and internal junction is entered exactly once and left exactly once at the opposite end. A route might not be good for several reasons:

1. some junction and/or internal junction is not entered and left, or

2. it is entered and left on the same end, or

3. it is entered and left twice.

Due to the properties of the permanent of the submatrices of X, any route which is not good, contributes 0 to the permanent. Indeed, if a junction is not entered and left in a route, then the sum of the weights of the cycle covers in that route will be 0 due to property (a). Similarly, if a junction is entered and left on the same end or it is entered and left twice, the sum of the weights of cycle covers will be 0 due to conditions (b)–(d). On the other hand, condition (e) ensures that any good route contributes $4^{t(\Phi)}$ to the permanent: the number of junctions and internal junctions is indeed $t(\Phi)$.

We have to show that the good routes are in a bijection with the satisfying assignments of Φ. Observe the following:

1. Any good route in any track T_k either "picks up" the junctions corresponding to the x_k literal or picks up the junctions corresponding to the $\bar{x}_k$ literal. Assign the logical TRUE value to x_k in the first case and the logical FALSE value in the second case.

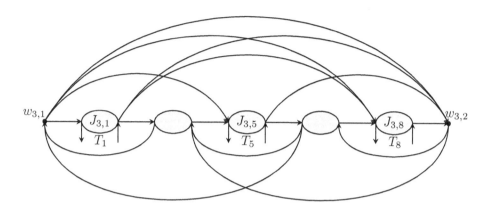

FIGURE 4.3: The interchange R_3 for the clause $C_3 = (x_1 \vee \overline{x}_5 \vee x_8)$. Note that interchanges do not distinguish literals x_i and $\overline{x}_i$. Each edge in and above the line of the junctions goes from left to right, and each edge below the line of the junctions goes from right to left. Junctions without labels are the internal junctions.

2. The interchanges are designed so that a good route can pick up the two internal junctions and any subset of the junctions except all of them. Furthermore, it can pick up these subsets in exactly one way. Therefore, in any good route, each interchange picks up the junctions corresponding to the literals that do not satisfy the clause corresponding to the interchange. Since it cannot pick up all the literals, some of them will satisfy the clause. This is true for each interchange, namely, the assignment we defined based on the good route is a satisfying assignment.

3. It is clear that different good routes define different satisfying assignments, thus, the mapping we defined is an injection. On the other hand, each satisfying assignment defines a good route, and it is just the inverse of the mapping we defined.

Since there is a bijection between good routes and satisfying assignments, each good route contributes with $4^{t(\Phi)}$ to the permanent and any route that is not good contributes 0 to the permanent, we can conclude that the permanent of the adjacency matrix of $\vec{G}$ is indeed $4^{t(\Phi)}$ times the number of satisfying assignments. This proves that calculating the permanent of an arbitrary matrix is #P-hard. Actually, finding the permanent of a matrix remains #P-hard if we restrict the cases to matrices containing only values -1, 0, 1, 2 and 3. Restricting the values to non-negative numbers is particularly tricky. The proof that finding the permanent remains #P-hard for non-negative entries is based on a polynomial reduction that contains some computational steps that do not preserve the relative error. Later we will see that a polynomial reduction containing only computational steps that preserve the relative error does not exist unless RP = NP.

4.1.3   Counting the most parsimonious substitution histories on an evolutionary tree

In this subsection, we are going to introduce a #P-completeness proof based on the work of Miklós, Tannier and Kiss [135] and Miklós and Smith [133]. The small parsimony problem is defined in the following way.

Problem 7.
Name: SP-Tree.
Input: a finite alphabet Γ, a rooted binary tree $T = (V, E)$, where $L \subset V$ denotes the leaves of the tree and $f : L \to \Gamma^k$, a function assigning a sequence of length k to each leaf of the tree.
Output: a function $g : V \to \Gamma^k$ such that $g(v) = f(v)$ for all $v \in L$, and the score

$$\sum_{(u,v) \in E} H(g(u), g(v)) \qquad (4.9)$$

is minimized, where H denotes the Hamming distance of the sequences, that is, the number of positions in which the two sequences differ.

We learned in Chapter 2 that the SP-Tree problem is an easy optimization problem, and the number of functions g minimizing the score in Equation (4.9) can be found in polynomial time. However, the counting problem becomes hard if we would like to obtain the number of most parsimonious scenarios instead of labelings of the internal nodes. There are $H(g(u), g(v))!$ number of ways to transform the sequence labeling vertex u to the sequence labeling vertex v using $H(g(u), g(v))$ substitutions (a substitution changes one character in a sequence). Therefore, if $\mathcal{G}$ denotes the set of functions minimizing the score in Equation (4.9), then we would like to compute

$$\sum_{g \in \mathcal{G}} \prod_{(u,v) \in E} H(g(u), g(v))! \qquad (4.10)$$

We will denote this counting problem #SPS-Tree (small parsimony scenario on trees). We are going to show that computing this number is #P-complete even if the problem is restricted to an alphabet of size 2. First we need the following lemma.

Lemma 7. *For any 3CNF Φ, there exists a 3CNF Φ' such that the following hold:*

1. *Φ' can be constructed in polynomial time, particularly, the size of Φ' is a polynomial function of the size Φ.*

2. Φ and Φ' have the same number of satisfying assignments.

3. Φ' contains an even number of variables, and in any satisfying assignments of Φ', exactly half of the variables have a logical value TRUE.

Proof. Let $x_1, x_2, \ldots x_n$ be the variables in Φ. Φ' contains the variables $x_1, x_2, \ldots, x_n, y_1, y_2, \ldots y_n$, and is defined as

$$\Phi' := \Phi \land \bigwedge_{i=1}^{n} \quad ((x_i \lor y_i \lor x_{i+1}) \land (\overline{x}_i \lor \overline{y}_i \lor x_{i+1}) \land$$

$$(x_i \lor y_i \lor \overline{x}_{i+1}) \land (\overline{x}_i \lor \overline{y}_i \lor \overline{x}_{i+1})) \tag{4.11}$$

where the index of x_{i+1} is taken modulo n, that is, x_{n+1} is defined as x_1.

The assignments of the x_i variables in any satisfying assignment of Φ' also satisfies Φ, therefore, it is sufficient to show that any satisfying assignment of Φ can be extended to exactly one satisfying assignment of Φ'. But this is obvious: the conjunctive form in Equation (4.11) forces that y_i must take the value of $\overline{x}_i$. This provides also that in any satisfying assignment of Φ', exactly half of the values will take the TRUE value. $\square$

To prove that computing the quantity in Equation (4.10) is #P-complete, we give a polynomial running time algorithm which for any 3CNF formula Φ, constructs a problem instance $p \in \#SPS - TREE$ with the following property: The number of solutions of p can be written as $a + by$, where y is the number of satisfying assignments of Φ, b is an easy-to-calculate positive integer, and $0 < a \ll b$. Thus, if s is the number of solutions of p, then $\lfloor \frac{s}{b} \rfloor$ is the number of satisfying assignments of Φ.

Let Φ be a 3CNF, and let Φ' be the 3CNF that has as many satisfying assignments as Φ and in all satisfying assignments, the number of TRUE and FALSE values are the same. Let n denote the number of logical variables in Φ' and let k denote the number of clauses in Φ'. We are going to construct a tree denoted by T'_Φ, and to label its leaves with sequences over the alphabet $\{0, 1\}$. The first n characters of each sequence correspond to the logical variables x_i, and there are further, auxiliary characters. The number of auxiliary characters are $148k \left(\left\lceil (k \log(n!) + n \log(2)) / \log(\frac{2^{20}}{3^{12}}) \right\rceil + 1 \right)$. The construction is such that there will be 2^n most parsimonious labelings of the internal nodes, one for each possible logical assignment. Each labeling is such that the labeling of the root completely determines the labelings at the other internal nodes. The corresponding assignment is such that the value of the logical variable x_i is TRUE if there is a character 1 in the sequence at the root of the tree in position i. The characters in the auxiliary positions are 0 in all the most parsimonious labelings.

If an assignment is a satisfying assignment, then the corresponding labeling has many more scenarios than the labelings corresponding to non-satisfying assignments. Furthermore, for each satisfying assignment, the corresponding labelings have the same, easy-to-compute number of scenarios.

For each clause c_j, we construct a subtree T_{c_j}. The construction is done in three phases, illustrated on Figure 4.4. First, we create a constant-size subtree, called the unit subtree, using building blocks we call elementary subtrees. Then in the blowing-up phase, this unit subtree is repeated several times, and in the third phase it is amended with another constant-size subtree. The reason for this construction is the following: the unit subtree is constructed in such a way that if a clause is satisfied, the number of scenarios on this subtree is large, and is always the same number not depending on how many literals provide satisfaction of the clause. When the clause is not satisfied, the number of scenarios is a smaller number. The blowing up is necessary for sufficiently separating the number of solutions for satisfying and non-satisfying assignments. Finally, the amending is necessary for achieving 2^n most parsimonious labelings on each T_{c_j} and to guarantee that the number of most parsimonious scenarios is the same for each satisfying assignment. The amending is slightly different for those clauses that come from Φ and those that are in $\Phi' \setminus \Phi$.

We detail the construction of the subtree for the clause $c_j = x_1 \vee x_2 \vee x_3$, denoted by T_{c_j}. Subtrees for the other kinds of clauses are constructed similarly. The unit subtree is built from 76 smaller subtrees that we will call *elementary subtrees*. On each elementary subtree, the sequences labeling the leaves contains 0 at almost all the positions, except the positions corresponding to the literals of the clause and the position of the possible auxiliary character. Only 14 different types of elementary subtrees are in a unit subtree, but several of them have given multiplicity, and the total count of them is 76, see also Table 4.1. Some of the elementary subtrees are cherry motives (two leaves connected via an internal node, see also Fig. 4.5.a)) for which we arbitrarily identify a left and a right leaf. For some of these cherries, we introduce one or more auxiliary characters, which are 1 only on the indicated leaf of the cherry and 0 everywhere else in the tree. So the edges connecting these leaves to the rest of the entire tree $T_{\Phi'}$ will contain one or more additional substitutions in all the most parsimonious solutions.

The constructed unit subtree will be such that if the clause is not satisfied, the number of possible most parsimonious scenarios for the corresponding labeling on this unit subtree is $2^{136} \times 3^{76}$, and if the clause is satisfied, then the number of possible most parsimonious scenarios for each corresponding labeling is $2^{156} \times 3^{64}$. The ratio of the two numbers is $2^{20}/3^{12} > 1$. We will denote this number by γ.

Below we detail the construction of the elementary subtrees and also give the number of most parsimonious scenarios on them since the number of scenarios on the unit subtree is simply the product of these numbers. This part is quite technical, however, the careful reader might observe the following. The number of scenarios for a fixed labeling on a unit tree is the product of the number of scenarios on the elementary trees. These numbers are always in $2^x 3^y$ form, and we need a (linear) combination of unit trees such that the sum of the exponents both on 2 and 3 is the same for all satisfying assignments and different for the non-satisfying assignment; furthermore, the number of

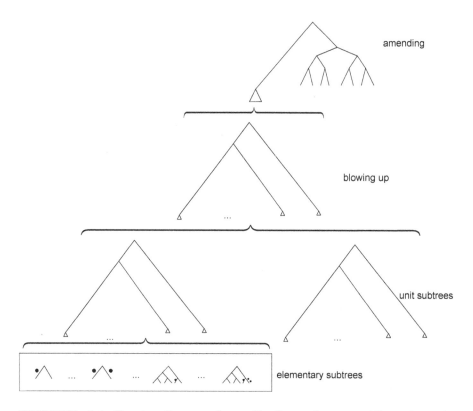

FIGURE 4.4: Constructing a subtree T_{c_j} for a clause c_j. The subtree is built in three phases. First, elementary subtrees are connected with a comb to get a unit subtree. In the second phase the same unit subtree is repeated several times, "blowing up" the tree. In the third phase, the blown-up tree is amended with a constant size, depth 3 fully balanced tree. The smaller subtrees constructed in the previous phase are denoted with a triangle in the next phase. See also text for details.

solutions for the non-satisfying assignment is smaller than for any of the satisfying assignments. Such combinations can be found by some linear algebraic considerations not presented here; below we just show one possible solution.

For each elementary tree, we give the characters at positions of the three literals. The elementary trees which are cherries are the following:

- There are four cherries on which the left leaf contains 1 in an extra position, and the characters in the positions of the three literals on the left and right leaf are given by

  011, 100
  101, 010
  110, 001
  000, 111.

  The first column shows the characters in the positions corresponding to the literals on the left leaf, while the second column shows those characters on the right leaf. Observe that there are 16 most parsimonious labelings of the root of the cherry motif and each needs 4 substitutions. However, 8 of them contain a character 1 in the auxiliary position, which will not be a most parsimonious labeling on $T_{\Phi'}$, as we discussed. So we have to consider only the other 8 labelings, where the characters are all 0, except the positions corresponding to the literals.

  The number of scenarios on one cherry is 24 if the sequence at the root of the cherry is the same as on the right leaf. Indeed, in that case, 4 substitutions are necessary on the left edge, and they can be performed in any order. If the number of substitutions are 3 and 1, respectively, on the left and right edges, or *vice versa*, the number of solutions is 6. Finally, if both edges have 2 SCJ operations, then the number of solutions is 4.

- There is one cherry motif without any extra adjacency, and the characters in the positions corresponding to the literals are

  000, 111.

  There are 8 most parsimonious labelings at the root, and each needs 3 substitutions. If the labeling at the root corresponds to a non-satisfying assignment, the number of scenarios on this cherry is 6; if all logical values are true, the number of scenarios is still 6; in any other case, the number of scenarios is 2.

  This elementary subtree is repeated 3 times.

- Finally, there are 3 types of cherry motifs with a character 1 at one-one auxiliary positions on both leaves. These are two different adjacencies, so both of them need one extra substitution on their incoming edge. The characters in the positions corresponding to the literals are

011, 100
101, 010
110, 001.

There are 8 possible labelings of the root which are most parsimonious in $T_{\Phi'}$, and each needs 5 substitutions. If all substitutions at the positions corresponding to the 3 literals falls onto one edge, then the number of scenarios is 24, otherwise the number of solutions is 12.

Each of these elementary subtrees is repeated 15 times.

The remaining elementary subtrees contain 3 cherry motifs connected with a comb, that is, a completely unbalanced tree, see also Figure 4.5. For the cherry at the right end of this elementary subtree, there is one or more auxiliary positions that have character 1 at one of the leaves and 0 everywhere else in T_Φ.

There are 3 elementary subtrees of this type which have only one auxiliary position. On these trees, the sequence at the right leaf of the rightmost cherry is all 0, and the sequence at the left leaf of the rightmost cherry motif is all 0 except at the auxiliary position and exactly 2 positions amongst the 3 positions corresponding to the literals.

The remaining leaves of these elementary subtrees are constructed in such a way that there are 8 most parsimonious labelings, each needing 7 substitutions, see the example in Figure 4.5. The number of substitutions is 0 or 1 at each edge except the two edges of the rightmost cherry motif. Here the number of substitutions might be 3 and 0, 2 and 1, or 1 and 2, yielding 6 or 2 scenarios, see also Table 4.1.

Each of these elementary subtrees are repeated 3 times.

Finally, there are 3 elementary subtrees of this type which have one auxiliary position for the left leaf of the rightmost cherry motif, and there are 2 auxiliary positions for the right leaf of the rightmost cherry motif. The sequence at the right leaf of the rightmost cherry is all 0 except at the 2 auxiliary positions, and the sequence at the left leaf of the rightmost cherry motif is all 0 except at the auxiliary positions and exactly 2 positions amongst the 3 positions corresponding to the literals.

The remaining leaves of these elementary subtrees are constructed in such a way that there are 8 most parsimonious labelings, each needing 9 substitutions, see the example in Figure 4.5. The number of substitutions is 0 or 1 on each edge except the two edges of the rightmost cherry motif. Here the number of substitutions might be 1 and 4, 2 and 3, or 3 and 2, yielding 24 or 12 scenarios, see also Table 4.1.

Each of these elementary subtrees are repeated 5 times.

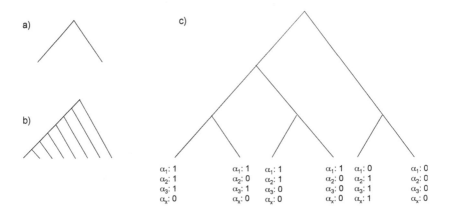

FIGURE 4.5: a) A cherry motif, i.e., two leaves connected with an internal node. **b)** A comb, i.e., a fully unbalanced tree. **c)** A tree with 3 cherry motifs connected with a comb. The assignments for 4 adjacencies, α_1, α_2, α_3 and α_x are shown at the bottom for each leaf. α_i, $i = 1, 2, 3$ are the adjacencies related to the logical variables b_i, and α_x is an extra adjacency. Note that Fitch's algorithm gives ambiguity for all adjacencies α_i at the root of this subtree.

	011	101	110	000	011	101	110	000	011	101	110	011	101	110
#	1	1	1	1	3	3	3	3	5	5	5	15	15	15
000	6	6	6	6	6^3	6^3	6^3	6^3	12^5	12^5	12^5	12^{15}	12^{15}	12^{15}
100	24	4	4	4	6^3	2^3	2^3	2^3	12^5	12^5	12^5	24^{15}	12^{15}	12^{15}
010	4	24	4	4	2^3	6^3	2^3	2^3	12^5	12^5	12^5	12^{15}	24^{15}	12^{15}
110	6	6	6	6	2^3	2^3	2^3	2^3	12^5	12^5	24^5	12^{15}	12^{15}	24^{15}
001	4	4	24	4	2^3	2^3	6^3	2^3	12^5	12^5	12^5	12^{15}	12^{15}	24^{15}
101	6	6	6	6	2^3	2^3	2^3	2^3	12^5	24^5	12^5	24^{15}	24^{15}	12^{15}
011	6	6	6	6	2^3	2^3	2^3	2^3	24^5	12^5	12^5	24^{15}	12^{15}	12^{15}
111	4	4	4	24	2^3	2^3	2^3	6^3	24^5	24^5	24^5	12^{15}	12^{15}	12^{15}

TABLE 4.1: The number of scenarios on different elementary subtrees of the unit subtree of the subtree T_{c_j} for clause $c_j = x_1 \lor x_2 \lor x_3$. Columns represent the 14 different elementary subtrees, the topology of the elementary subtree is indicated on the top. The black dots mean extra substitutions on the indicated edge due to the characters in the auxiliary positions; the numbers represent the presence/absence of adjacencies on the left leaf of a particular cherry motif, see text for details. The row starting with # indicates the number of repeats of the elementary subtrees. Further rows represent the logical true/false values of the literals, for example, 001 means $x_1 = $ FALSE, $x_2 = $ FALSE, $x_3 = $ TRUE. The values in the table indicate the number of scenarios, raised to the appropriate power due to multiplicity of the elementary subtrees. It is easy to check that the product of the numbers in the first line is $2^{136} \times 3^{76}$ and in any other lines is $2^{156} \times 3^{64}$.

In this way, the roots of all 76 elementary subtrees have 8 most parsimonious labelings corresponding to the 8 possible assignments of the literals in the clause. We connect the 76 elementary subtrees with a comb, and thus, there are still 8 most parsimonious labelings at the root of the entire subtree, which is the unit subtree. If the labeling at the root corresponds to a satisfying assignment of the clause, the number of scenarios is $2^{156} \times 3^{64}$, if the clause is not satisfied, the number of scenarios is $2^{136} \times 3^{76}$, as can be checked on Table 4.1. The ratio of them is indeed $2^{20}/3^{12} = \gamma$. The number of leaves on this unit subtree is 248, and 148 auxiliary positions are introduced.

This was the construction of the constant size unit subtree. In the next step, we "blow up" the system. Similar blowing up can be found in the seminal paper by Jerrum, Valiant and Vazirani [103], in the proof of Theorem 5.1. We repeat the above described unit subtree $\lceil (k \log(n!) + n \log(2))/ \log(\gamma) \rceil + 1$ times, and connect all of them with a comb (completely unbalanced tree). It is easy to see that there are still 8 most parsimonious labelings. For a solution satisfying the clause, the number of scenarios on this blown-up subtree is

$$X = \left(2^{156} \times 3^{64}\right)^{\left\lceil \frac{k \log(n!) + n \log(2)}{\log(\gamma)} \right\rceil + 1} \tag{4.12}$$

and the number of scenarios if the clause is not satisfied is

$$Y = \left(2^{136} \times 3^{76}\right)^{\left\lceil \frac{k \log(n!) + n \log(2)}{\log(\gamma)} \right\rceil + 1}. \tag{4.13}$$

It is easy to see that

$$\frac{X}{Y} = \gamma^{\left\lceil \frac{k \log(n!) + n \log(2)}{\log(\gamma)} \right\rceil + 1} \geq \gamma^{\frac{\log(n!)^k) + \log(2^n)}{\log(\gamma)}} = n!^k 2^n, \tag{4.14}$$

and

$$\frac{X}{Y} = \gamma^{\left\lceil \frac{k \log(n!) + n \log(2)}{\log(\gamma)} \right\rceil + 1} \leq \gamma^{\frac{\log(n!)^k) + \log(2^n)}{\log(\gamma)} + 2} = n!^k 2^n \gamma^2, \tag{4.15}$$

since for any positive number x,

$$x^{\frac{1}{\log(x)}} = e. \tag{4.16}$$

Let all adjacencies not participating in the clause be 0 on this blown-up subtree.

We are close to the final subtree T_{c_j} for one clause, c_j. In the third phase, we amend the so-far obtained tree with a constant-size subtree. The amending is slightly different for clauses coming from Φ and for those that are in $\Phi' \setminus \Phi$. We detail the amending for both cases.

If the clause contains only x logical variables, say, the clause is $x_1 \vee x_2 \vee x_3$, then construct two copies of a fully balanced depth 6 binary tree, on which the root has 64 most parsimonious labelings corresponding to the 64 possible assignments of the literals participating in the clause and their corresponding logical variables of the y type (namely, y_1, y_2 and y_3). This can be done with a construction similar to the left part of the tree on Figure 4.5.c).

In one of the copies, all other characters corresponding to logical variables not participating in the clause are 1 on all leaves, and thus, in each most parsimonious labelings of the root. In the other copy, those characters should be all 0.

In the copy, where all other characters are 0, the construction should be done in such a way that going from the root of the tree, first the y logical variables must be separated, then the x ones. Namely, characters at the position corresponding to y_1 should be the same (say, 0) on each leaf of the left subtree of the root and should be the other value on each leaf of the right subtree of the root. Similalry, for each of the four grandchildren of the root, the leaves must take the same value at the position corresponding to y_2, and these values must be different for the siblings. The same rule must be applied for the grand-grandchilden of the root. There is an internal node of this subtree such that on all of its leaves, each character at each position corresponding to y variables is 0. Replace the subtree at this position with the blown-up subtree. Connect the two copies with a common root. The so obtained tree is T_{c_j}.

Observe that there are 2^n possible most parsimonious labelings of T_{c_j}. We have the following lemma on them.

Lemma 8. *For any most parsimonious labelings, if Φ' and thus, particularly, the clause c_j is satisfied, then the number of scenarios on T_{c_j} is*

$$X \times \left(\left(\frac{n-6}{2} \right)! \right)^2 \geq Y \times (n!)^k \times 2^n \times \left(\left(\frac{n-6}{2} \right)! \right)^2. \qquad (4.17)$$

If the clause c_j is not satisfied, then the number of scenarios is at most $Y \times (n-6)!$. If the clause c_j is satisfied, however, Φ' is not satisfied, then the number of scenarios is at most $X \times (n-6)!$.

Proof. There are 3 logical x variables in the clause and there are 3 corresponding y variables. For the remaining $n-6$ variables, there are $n-6$ substitutions on the two edges of the root. If Φ' is satisfied, then for each i, exactly one in the couple (x_i, y_i) has the TRUE value and the other has the FALSE value. Therefore, there are $\frac{n-6}{2}$ substitutions on both edges of the root. On all remaining edges of the amending, there is either 0 or 1 substitution. Finally, the number of scenarios on the blown-up tree is X. Therefore, the number of scenarios is indeed $X \times \left(\left(\frac{n-6}{2} \right)! \right)^2$ if Φ' is satisfied. The inequality in Equation (4.18) comes from Equation (4.14).

If the clause is not satisfied, then the number of scenarios on the blown-up tree is Y. The substitutions on the two edges of the root might be arbitrarily distributed, however, in any cases, the number of scenarios is at most $(n-6)!$. This extremity is taken when all the substitutions fall onto the same edge.

If c_j is satisfied, however, Φ' is not, then the number of scenarios on the blown-up tree is X, and the number of scenarios on the two edges of the root is at most $(n-6)!$. $\qquad \square$

If the clause is in the form $x_i \lor y_i \lor x_{i+1}$ (some of the literals might be negated), then the amending is the following. Construct two copies of a fully balanced depth 4 binary tree, on which the root has 16 most parsimonious labelings corresponding to the 16 possible assignments of logical variables x_i, y_i, x_{i+1} and y_{i+1}. On one of the copies, all other characters are 0, while on the other copy, all other characters must be 1. On the copy, where all other characters are 0, the construction should be such that there must be an internal node such that at all of its leaves, the characters at the position corresponding to y_{i+1} are 0 and the subtree has depth 3. Replace this subtree with the blown-up tree. Connect the two copies with a common root. This is the final T_{c_j} tree.

For this tree, a lemma similar to Lemma 8 can be proved.

Lemma 9. *For any most parsimonious labelings, if Φ' and thus, particularly, the clause c_j is satisfied, then the number of scenarios on T_{c_j} is*

$$X \times \left(\left(\frac{n-4}{2} \right)! \right)^2 \geq Y \times (n!)^k \times 2^n \times \left(\left(\frac{n-4}{2} \right)! \right)^2. \qquad (4.18)$$

If the clause c_j is not satisfied, then the number of scenarios is at most $Y \times (n-4)!$. If the clause c_j is satisfied, however, Φ' is not satisfied, then the number of scenarios is at most $X \times (n-4)!$.

Proof. The proof is similar to the proof of Lemma 8, just now there are $n-4$ substitutions that must be distributed on the two edges of the root. $\qquad \square$

For all k clauses, construct such a subtree and connect all of them with a comb. This is the final tree $T_{\Phi'}$ for the 3CNF Φ'. It is easy to see that $T_{\Phi'}$ has 2^n most parsimonious labelings corresponding to the 2^n possible assignments of the logical variables. For these labelings, we have the following theorem.

Theorem 45. *If a labeling corresponds to a satisfying assignment, then the number of scenarios is*

$$X^k \times \left(\left(\frac{n-4}{2} \right)! \right)^{2n} \times \left(\left(\frac{n-6}{2} \right)! \right)^{k-2n} \geq$$
$$Y^k \times \left(n!^k \times 2^n \right)^k \times \left(\left(\frac{n-4}{2} \right)! \right)^{2n} \times \left(\left(\frac{n-6}{2} \right)! \right)^{k-2n}. \qquad (4.19)$$

If a labeling corresponds to a non-satisfying assignment, then the number of scenarios is at most

$$X^{k-1} \times Y \times (n-4)!^{2n} \times (n-6)!^{k-2n} \leq$$
$$Y^k \times \left(n!^k \times 2^n \times \gamma^2 \right)^{k-1} \times (n-4)!^{2n} \times (n-6)!^{k-2n}. \qquad (4.20)$$

Particularly, the number of scenarios corresponding to non-satisfying assignments is at most

$$Y^k \times \left(n!^k \times \gamma^2\right)^{k-1} \times (2^n)^k \times (n-4)!^{2n} \times (n-6)!^{k-2n} \le$$
$$Y^k \times \left(n!^k \times 2^n\right)^k \ll$$
$$Y^k \times \left(n!^k \times 2^n\right)^k \times \left(\left(\frac{n-4}{2}\right)!\right)^{2n} \times \left(\left(\frac{n-6}{2}\right)!\right)^{k-2n}. \quad (4.21)$$

Proof. If Φ' contains n logical variables, then there are $2n$ clauses in $\Phi' \setminus \Phi$ and $k - 2n$ clauses in Φ. Based on this, the number of scenarios for any labelings corresponding to a satisfying assignment can be easily calculated from Lemmas 8 and 9.

If Φ' is not satisfied, then at least one of the clauses is not satisfied causing a smaller number of scenarios on the corresponding subtree. However, the number of scenarios on other subtrees corresponding to other clauses might be higher due to the uneven distribution of the substitutions falling onto the two edges of the root of the subtrees. The upper bounds are based on Equation (4.15) considering that $\gamma = \frac{2^{20}}{3^{12}} < 2$ and $n \ge 6$. $\qquad \square$

What follows is that

$$\left\lfloor \frac{s}{X^k \times \left(\left(\frac{n-4}{2}\right)!\right)^{2n} \times \left(\left(\frac{n-6}{2}\right)!\right)^{k-2n}} \right\rfloor \qquad (4.22)$$

is the number of satisfying assignments of Φ where s is the number of most parsimonious scenarios on $T_{\Phi'}$. Since both the size of the tree $T_{\Phi'}$ and the length of the sequences labeling the leaves of $T_{\Phi'}$ is a polynomial function of size Φ, furthermore, $T_{\Phi'}$ together with the sequences labeling its leaves can be constructed in polynomial time, we get the following theorem.

Theorem 46. *The counting problem #SPS-TREE is in #P-complete.*

4.1.4   #IS and #Mon-2SAT

Martin Dyer, Leslie Ann Goldberg, Catherine Greenhill and Mark Jerrum proved that counting the number of independent sets is in #P-complete and it does not have an FPRAS unless RP = NP [57]. More surprisingly, the same holds for counting the satisfying assignments of monotone 2CNFs, although deciding if a monotone 2CNF has a satisfying assignment is trivial. A CNF is monotone if it does not contain any negated literal. Indeed, any monotone 2CNF is satisfiable.

First, we define the independent set problem.

Problem 8.
Name: #IS.

Input: a simple graph $G = (V, E)$.
Output: the number of independent sets of G. That is the number of $V' \in V$ such that for all $v_1, v_2 \in V'$, $(v_1, v_2) \notin E$.

The inapproximability is proved by reducing the problem of large independent sets to it. We learned that finding a large independent set is NP-complete, see Theorem 4. The reduction is the following. Let m be a positive integer, and let $G = (V, E)$ be a graph in which any independent set has a size at most m. We construct the following graph $G' = (V', E')$. The vertices V' are $V \times \{1, 2, \ldots, n+2\}$ and the edges are $E' = \{((v_1, r_1), (v_2, r_2)) | v_1, v_2 \in V$ and $r_1, r_2 \in \{1, 2, \ldots, r\}\}$, where $n = |V|$. Informally, each vertex $v \in V$ is replaced with a set of $n + 2$ vertices and any edge $(v_1, v_2) \in E$ is replaced with the complete bipartite graph $K_{n+2,n+2}$. There is a natural mapping of independent sets $I' \subseteq V'$ to the independent sets $I \subseteq V$. Indeed, if I' is an independent set, then

$$I - \varphi(I') := \{v | \exists r \ (v, r) \in I'\} \qquad (4.23)$$

is also an independent set. It is clear that φ is a surjection and the inverse image of any independent set $I \subseteq V$ has size $(2^{n+2} - 1)^k$, where $k = |I|$. We get that the number of independent sets in G' is

$$(2^{n+2} - 1)^m |\mathcal{I}_m(G)| + b, \qquad (4.24)$$

where $\mathcal{I}_m(G)$ is the set of independent sets in G of size m, and b is the number of independent sets in G' whose inverse image has size at most $m - 1$. Since there are at most 2^n independent sets in G, we get that

$$b \leq (2^{n+2} - 1)^{m-1} 2^n. \qquad (4.25)$$

Therefore

$$\left\lfloor \frac{|\mathcal{I}(G')|}{(2^{n+2} - 1)^m} \right\rfloor = \left\lfloor \frac{(2^{n+2} - 1)^m |\mathcal{I}_m(G)| + b}{(2^{n+2} - 1)^m} \right\rfloor = |\mathcal{I}_m(G)|, \qquad (4.26)$$

where $\mathcal{I}(G')$ is the set of independent sets of G'. It follows that if we could compute the number of independent sets in a graph, we could also compute the number of large independent sets in a graph. Particularly, we could decide if there is an independent set of size m, which is an NP-complete decision problem. Even if we could approximate the number of independent sets with an FPRAS having *epsilon* $= \frac{1}{2}$ and $\delta = \frac{2}{3}$, we would have a BPP algorithm for deciding if there is an independent set of size m. This would imply that RP = NP.

Dyer and his colleagues proved that #LARGEIS is in #P-complete by finding a parsimonious reduction from #SAT to #LARGEIS [57]. Thus #IS is not only in NP-hard, but also in #P-complete.

To prove that there is no FPRAS for counting the satisfying assignments of a monotone 2CNF, we find a parsimonious reduction from #IS to #MON-2SAT. Let $G = (V, E)$ be a graph. Define

$$\Phi := \bigwedge_{(v_i, v_j) \in E} (\overline{x}_i \vee \overline{x}_j). \qquad (4.27)$$

Clearly, Φ is monotone 2CNF in the variables $y_i := \overline{x}_i$, and there is a bijection between its satisfying assignments and the independent sets of G. The bijection is the following. If $X' \subseteq \{x_1, x_2, \ldots, x_n\}$ is a satisfying assignment, then $I := \{v_i | x_i = TRUE\}$ is an independent set.

4.2   #P-complete proofs not preserving the relative error

4.2.1   #DNF, #3DNF

First we define the disjunctive normal forms.

Definition 49. *A conjunctive clause is a logical expression of literals and AND operators ($\wedge$). A disjunctive normal form or DNF is a disjunction of conjunctive clauses, that is, conjunctive clauses connected with the logical OR ($\vee$) operator. The decision problem if there is a satisfying assignment of a disjunctive normal form is denoted by DNF and the corresponding counting problem is denoted by #DNF. Similar to 3CNF and 3SAT, we can define the 3DNF decision problem and its counting version #3DNF.*

Recall that a conjunctive normal form or CNF is the conjunction of disjunctive clauses. The relationship between CNFs and DNFs is provided by De Morgan's laws:

$$\overline{(x_{1,1} \vee \ldots \vee x_{1,m_1}) \wedge (x_{2,1} \vee \ldots \vee x_{2,m_2}) \wedge \ldots \wedge (x_{k,1} \vee \ldots \vee x_{k,m_k})} =$$
$$\overline{(x_{1,1} \vee \ldots \vee x_{1,m_1})} \vee \overline{(x_{2,1} \vee \ldots \vee x_{2,m_2})} \vee \ldots \vee \overline{(x_{k,1} \vee \ldots \vee x_{k,m_k})} =$$
$$(\overline{x}_{1,1} \wedge \ldots \wedge \overline{x}_{1,m_1}) \vee (\overline{x}_{2,1} \wedge \ldots \wedge \overline{x}_{2,m_2}) \vee \ldots \vee (\overline{x}_{k,1} \wedge \ldots \wedge \overline{x}_{k,m_k})$$

Namely, for any CNF Φ, we can generate in polynomial time a DNF Φ' such that any non-satisfying assignment for Φ is a satisfying assignment of Φ'. Therefore, if we can tell in polynomial time how many satisfying assignments a DNF has, we can also tell how many satisfying assignments a CNF has. Since this latter is a #P-complete counting problem, we just proved

Theorem 47. *The counting problem #DNF is in #P-complete.*

Observe that the negation of a 3CNF using the De Morgan's lows is a 3DNF. Since #3SAT is in #P-complete, the following theorem also holds.

Theorem 48. *The counting problem #3DNF is in #P-complete.*

4.2.2 Counting the sequences of a given length that a regular grammar can generate

In Chapter 2, we learned that counting the sequences of a given length that an unambiguous regular grammar can generate is in FP. Therefore, it might be surprising that the same problem for ambiguous grammars is hard [106].

The proof is based on reducing #DNF to this problem. Let Φ be a disjunctive normal form containing k clauses and n logical variables. We construct a regular grammar $G = (T, N, S, R)$ in the following way. $T = \{0, 1\}$. N contains $k(n-1)+1$ non-terminals, the start non-terminal S, and further $k(n-1)$ non-terminals $W_{i,j}$, where $i = 1, 2, \ldots, k$ and $j = 1, 2, \ldots, n-1$. For the start non-terminal we add the following rewriting rules. If neither x_1 nor $\overline{x}_1$ is a literal in the ith clause, then add the rewriting rules

$$S \rightarrow 0W_{i,1} \mid 1W_{i,1}. \tag{4.28}$$

If x_1 is a literal in the ith clause, then add the rewriting rule

$$S \rightarrow 1W_{i,1}. \tag{4.29}$$

If $\overline{x}_1$ is a literal in the ith clause, then add the rewriting rule

$$S \rightarrow 0W_{i,1}. \tag{4.30}$$

For $j = 1$ to $j = n - 2$, and for each i, if neither x_{j+1} nor $\overline{x}_{j+1}$ is a literal in the ith clause, then add the rewriting rules

$$W_{i,j} \rightarrow 0W_{i,j+1} \mid 1W_{i,j+1}. \tag{4.31}$$

If x_{j+1} is a literal in the ith clause, then add the rewriting rule

$$W_{i,j} \rightarrow 1W_{i,j+1}. \tag{4.32}$$

If $\overline{x}_{j+1}$ is a literal in the ith clause, then add the rewriting rule

$$W_{i,j} \rightarrow 0W_{i,j+1}. \tag{4.33}$$

Finally, for each i, if neither x_n nor $\overline{x}_n$ is a literal in the ith clause, then add the rewriting rules

$$W_{i,n-1} \rightarrow 0 \mid 1. \tag{4.34}$$

If x_n is a literal in the ith clause, then add the rewriting rule

$$W_{i,n-1}1. \tag{4.35}$$

If $\overline{x}_n$ is a literal in the ith clause, then add the rewriting rule

$$W_{i,n-1} \rightarrow 0. \tag{4.36}$$

We claim that there is a bijection between the language that the grammar G generates and the satisfying assignments of Φ. Specially, G generates n long 0-1 sequences $A = a_1 a_2 \ldots a_n$, and the image of that sequence is the assignment in which the logical variable x_i is TRUE if $a_i = 1$ and FALSE otherwise.

If A is part of the language, then there is a generation of it. Take any generation of A, and consider the first rewriting. It is either

$$S \to 0 W_{i,1}$$

or

$$S \to 1 W_{i,1}$$

for some i. For the literals in the ith clause of the DNF, the rewriting rules in the selected generation of A will be such that the corresponding assignments of the logical variables in the image of A satisfies the DNF. Therefore, any image is a satisfying assignment. Clearly different sequences have different images, thus, the mapping is an injection of the language to the satisfying assignments.

We can also inject the satisfying assignments to the language. Let X be a satisfying assignment, and assume that the ith clause satisfies the DNF. Let a_j be 1 if x_j is TRUE in the assignment S and let a_j be 0 otherwise. It is easy to see that

$$S \to a_1 W_{i,1} \to a_1 a_2 W_{i,2} \to \ldots \to a_1 a_2 \ldots a_n$$

is a possible generation in G, and it is easy to see that the image of the so-generated sequence is indeed X.

4.2.3 Computing the permanent of a non-negative matrix and counting perfect matchings in bipartite graphs

To prove that computing the permanent of a non-negative matrix is #P-hard, we need the following observation and theorem in number theory.

Observation 1. *Let μ be the largest absolute value in an $n \times n$ matrix A. Then*

$$|per(A)| \leq n! \mu^n. \tag{4.37}$$

The consequence of this observation is if we can calculate the permanent of A modulo p_i for each p_i in some set $\{p_1, p_2, \ldots, p_m\}$ whose product is at least $2n!\mu^n$, then we can compute the permanent A in polynomial time using the Chinese remainder theorem. However, we do know that the prime numbers are dense enough that their product is large enough.

Lemma 10. *There exists a constant d such that for any n and μ,*

$$\prod_{p \leq dn \log_2(\mu n)} p \geq 2n!\mu^n \tag{4.38}$$

where p is prime number.

This lemma comes from the well-known number theoretical result that the first Chebyshev function, defined as

$$\vartheta(n) := \sum_{p \le n} \log(p) \tag{4.39}$$

is asymptotically n.

We can find the list of prime numbers up to $dn \log_2(3n)$ in polynomial time using elementary methods (for example, the sieve of Eratosthenes). Note that the running time has to be polynomial in the value of n and not the number of digits necessary to write n. Let A be a matrix containing values $-1, 0, 1, 2$ and 3. Let p be a prime, and let A' be the matrix obtained from A by replacing each -1 with $p - 1$. Observe that

$$per(A) \equiv per(A') \pmod{p}. \tag{4.40}$$

Therefore, if we could compute the permanent of a non-negative matrix in polynomial time, then we could also compute the permanent of a matrix containing values $-1, 0, 1, 2$ and 3. Since this latter is a #P-hard computing problem, we get that

Theorem 49. *Computing the permanent of a non-negative matrix is in #P-hard.*

Note that the problem remains #P-hard if the entries are integers, and bounded by $O(n \log(n))$. This gives us the possibility to prove that computing the permanent of a 0-1 matrix is still #P-hard. Since any 0-1 matrix is an adjacency matrix of a bipartite graph, the permanent of a 0-1 matrix is the number of perfect matchings in a bipartite graph. Hence we can prove that counting the perfect matchings in a bipartite graph is #P-complete.

To prove that computing the permanent of a 0-1 matrix is #P-hard, we reduce computing the permanent of a non-negative integer matrix with $O(n \log(n))$ bounds on its entries to computing the permanent of a 0-1 matrix. Let A be a non-negative integer matrix. Consider the edge-weighted directed graph $\vec{G}$ whose adjacency matrix is A. Replace each edge $e = (v, w)$ in A with a weight $k > 1$ with a subgraph. The subgraph is illustrated in Figure 4.6 for the case $k = 3$. We get a new graph $\vec{G}'$. We claim that the permanent of the corresponding matrix A' is the permanent of A.

If (u, v) is not covered by a cycle in A, then there is only one way to cover the new vertices: a clockwise cycle. On the other hand, if (u, v) is covered by a cycle, then so must be the chain of edges from u to v in $\vec{G}'$. There are k ways to cover the remaining cycles, each containing one of the loops. Since any cycle covering of one of the subgraphs can be combined with any cycle covering of another subgraph, the number of cycle coverings in $\vec{G}'$ is the sum of the weights of cycle coverings in $\vec{G}$, that is, the permanent of A' is the permanent of A. Therefore, we get the following theorem:

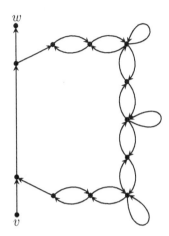

FIGURE 4.6: The unweighted subgraph replacing the edge (v, w) with a weight of 3. See text for details.

Theorem 50. *Computing the permanent of a 0-1 matrix is in #P-hard. The computational problem* #PERFECTMATCHING, *that is, counting the perfect matchings in bipartite graph, is in #P-complete. Counting the cycle covers in a directed graph is in #P-complete.*

4.2.4   Counting the (not necessarily perfect) matchings of a bipartite graph

To prove that #MATCHING, that is, finding the number of not necessarily perfect matchings of a bipartite graph is in #P-complete, we need the following lemma.

Lemma 11. *Let $f(x)$ be a polynomial of order of n in form $f(n) = \sum_{i=0}^{n} c_i x^i$. If the value of $f(x)$ is known in $n + 1$ rational points, $x_1, x_2, \ldots x_{n+1}$, then the coefficients of $f(x)$ can be computed in polynomial time in n, along with the size (number of digits) of the largest value and the size of the rational numbers. The size of a rational number $\frac{p}{q}$ is the number of digits in p and q, assuming that p and q are coprimes.*

Proof. Let A be the $(n + 1) \times (n + 1)$ matrix with $a_{i,j} = x_i^{j-1}$. Then A is a Vandermonde matrix, and thus, it can be inverted. The coefficients of $f(x)$ satisfy the matrix-vector equation

$$Ac = v \tag{4.41}$$

where c is the vector of coefficients and v is the vector of the values. However, A can be inverted since it is a Vandermonde matrix, and thus

$$c = A^{-1}v. \tag{4.42}$$

$\square$

We prove the #P-completeness of the number of matchings in a bipartite graph by reducing the number of perfect matchings to it. Let $G = (U, V, E)$ be a bipartite graph, such that $|U| = |V| = n$. Given a number k, generate a bipartite graph G_k in the following way. Add nk-nk vertices to both U and V, indexed by $u'_{i,l}$ and $v'_{j,l}$, $1 \le i, j \le n$, $1 \le l \le k$. Connect each $u'_{i,l}$ with v_i and connect each $v'_{j,l}$ with u_j.

Let m_r denote the number of matchings of G containing exactly $n-r$ edges. Each such matching will be contained in exactly $m_r(k+1)^{2r} = m_r(k^2+2k+1)^r$ matchings in G_k. Indeed, each vertex in G not covered by the matching might not be covered in G_k or might be covered with one of the new edges. Thus the number of matchings in G_k is

$$\sum_{r=0}^{n} m_r(k^2 + 2k + 1)^r = f(k^2 + 2k + 1). \tag{4.43}$$

Observe that for different ks, the number of matchings in G_k is the value of f at $k^2 + 2k + 1$. Hence, if we could calculate the matchings in G_k for $k = 0, 1, \ldots n$, then we could calculate the coefficients of f. Particularly, we could calculate m_0, that is, the number of perfect matchings in G. Therefore, we get

Theorem 51. *Computing the number of matchings in a bipartite graph is #P-complete.*

4.2.5 Counting the linear extensions of a poset

Brightwell and Winkler showed that counting the linear extensions of a partially ordered set is in #P-complete [26]. Recall that a linear extension of a poset $(P, \le_P)$ is a total ordering on P such that $a \le_P b$ implies that $a \le b$. We denote this counting problem by #LE.

The proof is based on reducing the #3SAT problem to #LE using modulo prime number calculations. We need the following lemma.

Lemma 12. *For any $n \ge 4$, the product of prime numbers strictly between n and n^2 is at least $n!2^n$.*

The proof of this lemma can be found in [26] and is based on the properties of the first and second Chebyshev functions.

The outline of proving the #P-completeness of #LE is the following. For

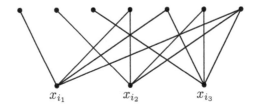

$$x_{i_1} \qquad\qquad x_{i_2} \qquad\qquad x_{i_3}$$

FIGURE 4.7: The Hasse diagram of a clause poset. See text for details.

any 3CNF Φ of k clauses and n variables, we are first going to construct a poset P_Φ of size $7k+n$. Let L_Φ denote the number of linear extensions of this poset. Next, we are going to find a set of primes between $7k+n$ and $(7k+n)^2$ such that none of them divides L_Φ, and their product is at least 2^n+1. Since $2^n+1 < 2^{7k+n}$ and $L_\Phi \leq (7k+n)!$, such set of primes exists according to Lemma 12. Then for each prime p in the so established set, we construct a poset $P_{\Phi,p}$ of size about $p(n+k)$ with the property that the number of linear extensions of $P_{\Phi,p}$ is $\alpha p + s(\Phi)\beta\gamma L_\Phi$, where α is a positive integer, β and γ are easily computable positive integers, none of them can be divided by p and $s(\Phi)$ is the number of satisfying assignments of Φ. Then we can compute $s(\Phi)$ modulo p. Using the Chinese remainder theorem, we can calculate the number of solutions modulo the product of the prime numbers, which is at least 2^n+1. Since the number of solutions may vary between 0 and 2^n, this means that we can actually exactly compute the number of solutions in polynomial time.

Let Φ be a 3CNF. The poset P_Φ is constructed in the following way. There are two types of vertices, n vertices for the n variables and 7 vertices for each clause. If x_{i_1}, x_{i_2} and x_{i_3} are the three logical variables that are literals in clause c_j, then each of the 7 vertices corresponding to c_j is placed above a different subset of $\{x_{i_1}, x_{i_2}, x_{i_3}\}$, see Figure 4.7. There are no other comparabilities in P_Φ.

Let p be a prime between $7k+n$ and $(7k+n)^2$ such that p does not divide L_Φ. We construct the poset $P_{\Phi,p}$ in the following way. There are two special vertices in the poset, a and b. Below a, there is an antichain of size $(p-1)(n+1)$, divided into $n+1$ parts. There is a subset of vertices U_i for each variable x_i, and there is an additional set U_0. Between a and b, there is an antichain of size $(k+1)(p-1)$. These are again divided into $k+1$ subsets, each of size $p-1$. For each clause c_j, there is a subset V_j and there is an additional subset V_0.

Finally, there are literal and clause vertices. For each variable x_i, there are two vertices x_i and $\bar{x}_i$. Namely, to simplify notations, we abuse notation by using the same symbol x_i and $\bar{x}_i$ for the literals and the vertices corresponding to them. There are 8 vertices for each clause, and the vertices for clause c_j are denoted by $c_{j,l}$, $l \in \{0, 1, \ldots, 7\}$.

Both x_i and $\bar{x}_i$ are above V_i. If clause x_j contains variables x_{i_1}, x_{i_2} and

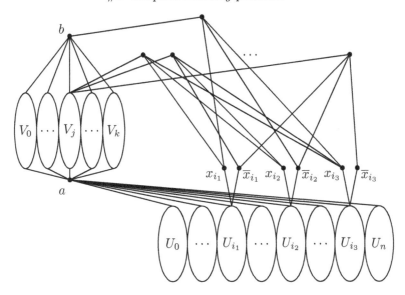

FIGURE 4.8: The poset $P_{\Phi,p}$. Ovals represent an antichain of size $p-1$. For sake of clarity, only the literal and some of the clause vertices for the clause $c_j = (x_{i_1} \vee \overline{x}_{i_2} \vee x_{i_3})$ are presented here. See also the text for details.

x_{i_3}, then each clause vertex $c_{j,l}$ is above a different triple of literal vertices $\{x_{i_1}, \overline{x}_{i_1}\}$, $\{x_{i_2}, \overline{x}_{i_2}\}$ and $\{x_{i_3}, \overline{x}_{i_3}\}$. The clause vertex which is above the triplets that actually constitutes c_j is above b and all other clause vertices are above V_j. There are no more comparabilities in $P_{\Phi,p}$. The poset contains

$$2 + (p-1)(k+n+2) + 2n + 8k = (p+7)k + (p+1)n + p \qquad (4.44)$$

vertices.

To count the linear extensions of $L_{\Phi,p}$, we partition them based on configurations that we define below. A configuration λ is a partition of the literal and clause vertices into 3 sets, B^λ, M^λ and T^λ, called the base, middle and top sets, respectively. We say that a linear extension respects a configuration $\lambda = \{B^\lambda, M^\lambda, T^\lambda\}$ if $B^\lambda \leq a \leq M^\lambda \leq b \leq T^\lambda$ in the linear extension. The set of linear extensions respecting a configuration λ is denoted by L^λ. A configuration λ is consistent if L^λ is not empty. It is easy to see that for such cases $|L^\lambda|$ is the product of the number of linear extensions of three posets, restricting $P_{\Phi,p}$ to the base, middle and top sets. We denote these posets by $P_{\Phi,p|B^\lambda}$, $P_{\Phi,p|M^\lambda}$, and $P_{\Phi,p|T^\lambda}$. The number of linear extensions respecting a configuration can be divided by p if and only if the number of linear extensions of any of these three posets can be divided by p. Therefore, in the following, we infer the cases when p does not divide any of these numbers of linear extensions.

The poset $P_{\Phi,p|B^\lambda}$ consists of an antichain of size $(p-1)(n+1)$ and some of the literal vertices. The set U_0 consists of isolated vertices and also those vertices in U_i for which neither x_i nor $\overline{x}_i$ is in $P_{\Phi,p|B^\lambda}$. The isolated vertices can be put into the linear extension to any position in any order, and for any such selection of positions and order, the remaining vertices can have an arbitrary linear extension. What follows is that the number of linear extensions of $P_{\Phi,p|B^\lambda}$ can be divided by $\binom{m}{r}r! = m(m-1)\ldots(m-r+1)$ where m is the size of the poset and r is the number of isolated vertices. Since $r \geq p-1$, this cannot be divided by p only if $r = p-1$ and $m \equiv -1 \mod p$. Since m is between $(p-1)(n+1)$ and $(p-1)(n+1)+2n$ and $p > n$, this implies that there are exactly n literal vertices in $P_{\Phi,p|B^\lambda}$, one for each logical variable (since the number of isolated vertices is $p-1$).

We claim that in this case, the number of linear extensions of $P_{\Phi,p|B^\lambda}$ is

$$\frac{(p(n+1)-1)!}{p^n},\tag{4.45}$$

which cannot be divided by p. Indeed, the size of the poset is $np+(p-1) = p(n+1)-1$. Any linear extension can be obtained by first selecting a place for each set of vertices U_i and its literal vertex x_i or $\overline{x}_i$ and also for the set of vertices U_0. This can be done in

$$\binom{p(n+1)-1}{p,p,\ldots p-1}\tag{4.46}$$

way. Then for each i, the vertices in U_i can be put into arbitrary order. This can be done in $(p-1)!$ way for each i, thus the number of linear extensions is indeed

$$\binom{p(n+1)-1}{p,p,\ldots p-1}(p-1)!^{n+1} = \frac{(p(n+1)-1)!}{p^n}.\tag{4.47}$$

It is trivial to see that this number cannot be divided by p since p is greater than $n+1$.

Similar analysis can be done for $P_{\Phi,p|M^\lambda}$. It contains an antichain of size $(p-1)(k+1)$ and some literal and clause vertices. It is easy to see that its number of linear extensions cannot be divided by p if and only if it contains none of the literal vertices and exactly one of the clause vertices for each V_j. In such a case, the number of linear extensions of $P_{\Phi,p|M^\lambda}$ is

$$\frac{(p(k+1)-1)!}{p^k}\tag{4.48}$$

which cannot be divided by p since $p > k$.

If the number of linear extensions cannot be divided by p for $P_{\Phi,p|B^\lambda}$ or for $P_{\Phi,p|M^\lambda}$, then $P_{\Phi,p|T^\lambda}$ contains one literal vertex for each logical variable and seven clause vertices for each literal. We are going to show that these configurations correspond to satisfying assignments. Indeed, in any such configuration, one of the literal vertices is in $P_{\Phi,p|B^\lambda}$ and the other is in $P_{\Phi,p|T^\lambda}$. Let the assignment be such that the literals in $P_{\Phi,p|T^\lambda}$ are TRUE. Then for each clause, the clause vertex which is in $P_{\Phi,p|M^\lambda}$ corresponds to a combination of literals that would build up a clause that the current assignment does not satisfy. However, this cannot be the clause vertex that corresponds to the combination of the literals which actually constitutes the clause, since that clause vertex is above b. Therefore the assignment satisfies each clause.

Also, if an assignment is a satisfying assignment, it determines a configuration for which the number of linear extensions cannot be divided by p for $P_{\Phi,p|B^\lambda}$ or for $P_{\Phi,p|M^\lambda}$.

The last piece of observation is that for any such assignment, $P_{\Phi,p|T^\lambda}$ is isomorphic to P_Φ. Indeed, $P_{\Phi,p|T^\lambda}$ contains, for each clause, 3 literal vertices and 7 clause vertices. The missed clause vertex is not comparable with any of the 3 literal vertices, therefore the presented 7 clause vertices correspond to the possible non-empty subsets of literal vertices. Then the number of linear extensions of $P_{\Phi,p|T^\lambda}$ is indeed L_Φ.

What we get is that the number of linear extensions of $P_{\Phi,p}$ is

$$\alpha p + s(\Phi)\frac{(p(n+1)-1)!}{p^n}\frac{(p(k+1)-1)!}{p^k}L_\Phi. \tag{4.49}$$

From this, $s(\Phi)$ can be calculated modulo p. If the number of linear extensions of the given posets are known, then building up the posets, finding the appropriate set of prime numbers, and computing $s(\Phi)$ modulo these prime numbers can all clearly be done in polynomial time. Therefore, we get the following theorem.

Theorem 52. *The counting problem #LE is in #P-complete.*

4.2.6 Counting the most parsimonious substitution histories on a star tree

We learned in Subsection 4.1.3 that counting the most parsimonious scenarios on an evolutionary tree is in #P-complete. Here we show that the problem remains #P-complete if the binary tree is replaced to a star tree. Below we define this problem.

Problem 9.
Name: #SPS-STAR.
Input: a multiset of sequences of the same length over the same alphabet $\mathcal{S} = \{A_1, A_2, \ldots A_n\}$.
Output: the value defined as

$$\sum_{M\in\mathcal{M}}\prod_{i=1}^{n}H(A_i, M)! \tag{4.50}$$

where $\mathcal{M}$ is the set of sequences that minimizes the sum of Hamming distances from the sequences in $\mathcal{S}$. Namely, for any $M \in \mathcal{M}$,

$$\sum_{i=1}^{n} H(A_i, M) \tag{4.51}$$

is minimal.   Surprisingly, this problem is #P-complete, even if the size of the alphabet is 2, although finding the size of $\mathcal{M}$ is trivial. It is easy to see that $\mathcal{M}$ consists of the sequences that contain the majority character for each position. The majority character might not be unique; the size of $\mathcal{M}$ is the product of the number of majority characters in each position. If the size of the alphabet is 2, say, it is $\{0,1\}$, then the size of $\mathcal{M}$ is 2^m, where m is the number of positions where half of the sequences in $\mathcal{S}$ contain 0 and the other half of them contain 1. Sequences in M are called *median sequences*. For each median sequence M, $\prod_{i=1}^{n} H(A_i, M)!$ is the number of corresponding scenarios.

Below we present a proof that #SPS-STAR is in #P-complete based on the work of Miklós and Smith [133]. The proof is based on reducing #3SAT to #SPS-STAR using modulo prime number calculations.

Let Φ be a 3CNF with n variables and k clauses, and let p be a prime number between min 300, $n + 5$ and 5 min 300, $n + 5$. We are going to construct a multiset $\mathcal{S}$ containing $2 + 2n + 50k$ sequences, each of them of length $2n + 2(q + 4) + 2n(q + 3) + k(75 + 50)$, where $q = p - n + 5$. Each sequence is in the form

$$a_1b_1a_2b_2 \ldots a_nb_ne_1e_2 \ldots e_{t(p)}, \tag{4.52}$$

where $t(p) = 2(q + 4) + 2n(q + 3) + k(75 + 50q)$. The a_i and b_i characters correspond to the logical variable x_i in Φ, and the e_j characters are additional characters. In these additional positions, all sequences contain character 0 except one of them. We will say that a sequence *contains x additional ones*, which means that there are x additional positions where the sequence contains 1s. The sequences come in pairs such that they are the complement of each other in the first $2n$ positions. What follows is that there are 2^{2n} median sequences. The sequences are the following.

1. There is a sequence that contains all 0 characters in the first $2n$ positions and has $q + 4$ additional ones. Furthermore, there is a sequence that contains all 1s in the first $2n$ positions and contains $q + 4$ additional ones. We denote these sequences as A and $\overline{A}$.

2. For each index $i = 1, \ldots n$, there are a couple of sequences. For one of them $a_i = b_i = 1$, and for all other $j \neq i$ $a_i = b_i = 0$, and the sequence contains $q + 3$ additional ones. The other sequence is the complement of the first one in the first $2n$ positions and also contains $q + 3$ additional ones. We denote these sequences by A_i and $\overline{A}_i$.

3. For each clause, there are 50 sequences, see Table 4.2. Each sequence

differs in the characters corresponding the logical variables participating in the clause, in the characters corresponding to other logical variables and in the number of additional ones. In Table 4.2, column **A** gives the characters $a_{i_1}, b_{i_1}, a_{i_2}, b_{i_2}, a_{i_3}$ and b_{i_3} for each sequence if the clause is $(x_{i_1} \vee x_{i_2} \vee x_{i_3})$. If some of the literals are negated, the corresponding a and b values must be swapped. In each sequence, for all $j \neq i_1, i_2, i_3$, all characters a_j and b_j are the same. Column **B** tells if it is 1 or 0. Each sequence has q additional ones plus the number that can be found in column **C**. These 3 columns completely describe the sequences; the remaining columns in the table are explained later.

It is easy to see that there are 2^{2n} median sequences; the first $2n$ characters might be arbitrary, and the characters in the additional positions must be 0. We set up three properties on the medians.

Property 1. Exactly n of the characters are 1 in the median.

Property 2. For each i, $a_i + b_1 = 1$.

Property 3. For each i, $a_i + b_i = 1$, and the assignment

$$x_i = \begin{cases} TRUE & \text{if } a_i = 1 \\ FALSE & \text{if } a_i = 0 \end{cases} \tag{4.53}$$

satisfies Φ.

It is easy to see that these properties are nested, namely, if a median sequence has Property i, then it also has Property j for each $j < i$. We prove the following on the median sequences.

If a median sequence M does not have Property 1, then the number of corresponding scenarios can be divided by p. Indeed, in such a case, either $H(M, A) \geq p$ or $H(M, \overline{A}) \geq p$ and thus either $H(M, A)!$ or $H(M, \overline{A})!$ can be divided by p.

If a median sequence M has Property 1, but does not have Property 2, then the number of corresponding scenarios can be divided by p. Indeed, let i be such that $a_i + b_i = 0$ or $a_i + b_i = 2$. Then either $H(M, A_i) = p$ or $H(M, \overline{A}_i) = p$, making the corresponding factorial dividable by p.

If a median sequence M has Properties 1 and 2, but does not have Property 3, then the number of corresponding scenarios can be divided by p. Assume that $c_j = (x_{i_1} \vee x_{i_2} \vee x_{i_3})$ is the clause that is not satisfied by the assignment defined in Equation (4.53). Then $a_{i_1} = a_{i_2} = a_{i_3} = 0$ and $b_{i_1} = b_{i_2} = b_{i_3} = 1$. In that case, the Hamming distance between M and the sequence that is defined for clause c_j in the last row of Table 4.2 is p. It follows that the number of corresponding scenarios can be divided by p. If some of the literals are negated in a clause not satisfied by the assignment defined in Equation (4.53), the same arguing holds, since both in the constructed sequences and in M, some of the a and b values are swapped.

A	B	C	M1 111	M2 110	M3 101	M4 011	M5 100	M6 010	M7 001	M8 000
01 00 00	0	+3	$p-1$	$p-1$	$p-1$	$p-3$	$p-1$	$p-3$	$p-3$	$p-3$
00 01 00	0	+3	$p-1$	$p-1$	$p-3$	$p-1$	$p-3$	$p-1$	$p-3$	$p-3$
00 00 01	0	+3	$p-1$	$p-3$	$p-1$	$p-1$	$p-3$	$p-3$	$p-1$	$p-3$
10 11 11	1	+0	$p-6$	$p-6$	$p-6$	$p-4$	$p-6$	$p-4$	$p-4$	$p-4$
11 10 11	1	+0	$p-6$	$p-6$	$p-4$	$p-6$	$p-4$	$p-6$	$p-4$	$p-4$
11 11 10	1	+0	$p-6$	$p-4$	$p-6$	$p-6$	$p-4$	$p-4$	$p-6$	$p-4$
10 10 00	0	+2	$p-5$	$p-5$	$p-3$	$p-3$	$p-3$	$p-3$	$p-1$	$p-1$
10 00 10	0	+2	$p-5$	$p-3$	$p-5$	$p-3$	$p-3$	$p-1$	$p-3$	$p-1$
00 10 10	0	+2	$p-5$	$p-3$	$p-3$	$p-5$	$p-1$	$p-3$	$p-3$	$p-1$
10 10 00	0	+2	$p-5$	$p-5$	$p-3$	$p-3$	$p-3$	$p-3$	$p-1$	$p-1$
10 00 01	0	+2	$p-3$	$p-5$	$p-3$	$p-1$	$p-5$	$p-3$	$p-1$	$p-3$
00 10 01	0	+2	$p-3$	$p-5$	$p-1$	$p-3$	$p-3$	$p-5$	$p-1$	$p-3$
10 01 00	0	+2	$p-3$	$p-3$	$p-5$	$p-1$	$p-5$	$p-1$	$p-3$	$p-3$
10 00 10	0	+2	$p-5$	$p-3$	$p-5$	$p-3$	$p-3$	$p-1$	$p-3$	$p-1$
00 01 10	0	+2	$p-5$	$p-1$	$p-5$	$p-3$	$p-3$	$p-1$	$p-5$	$p-3$
01 10 00	0	+2	$p-3$	$p-3$	$p-1$	$p-5$	$p-1$	$p-5$	$p-3$	$p-3$
01 00 10	0	+2	$p-3$	$p-1$	$p-3$	$p-5$	$p-1$	$p-3$	$p-5$	$p-3$
00 10 10	0	+2	$p-5$	$p-3$	$p-3$	$p-5$	$p-1$	$p-3$	$p-3$	$p-1$
10 01 00	0	+2	$p-3$	$p-3$	$p-5$	$p-1$	$p-5$	$p-1$	$p-3$	$p-3$
10 00 01	0	+2	$p-3$	$p-5$	$p-3$	$p-1$	$p-5$	$p-3$	$p-1$	$p-3$
00 01 01	0	+2	$p-1$	$p-1$	$p-3$	$p-1$	$p-5$	$p-3$	$p-3$	$p-5$
01 10 00	0	+2	$p-3$	$p-3$	$p-1$	$p-5$	$p-1$	$p-5$	$p-3$	$p-3$
01 00 01	0	+2	$p-1$	$p-3$	$p-1$	$p-3$	$p-3$	$p-5$	$p-3$	$p-5$
00 10 01	0	+2	$p-3$	$p-5$	$p-1$	$p-3$	$p-3$	$p-5$	$p-1$	$p-3$
01 01 00	0	+2	$p-1$	$p-1$	$p-3$	$p-3$	$p-3$	$p-3$	$p-5$	$p-5$
01 00 10	0	+2	$p-3$	$p-1$	$p-3$	$p-5$	$p-1$	$p-3$	$p-5$	$p-3$
00 01 10	0	+2	$p-3$	$p-1$	$p-5$	$p-3$	$p-3$	$p-1$	$p-5$	$p-3$
10 10 11	1	+1	$p-6$	$p-6$	$p-4$	$p-4$	$p-4$	$p-4$	$p-2$	$p-2$
10 11 01	1	+1	$p-4$	$p-6$	$p-4$	$p-2$	$p-6$	$p-4$	$p-2$	$p-4$
11 10 01	1	+1	$p-4$	$p-6$	$p-2$	$p-4$	$p-4$	$p-6$	$p-2$	$p-4$
10 01 11	1	+1	$p-4$	$p-4$	$p-6$	$p-2$	$p-6$	$p-2$	$p-4$	$p-4$
10 11 10	1	+1	$p-6$	$p-4$	$p-6$	$p-4$	$p-4$	$p-2$	$p-4$	$p-2$
11 01 10	1	+1	$p-4$	$p-2$	$p-6$	$p-4$	$p-4$	$p-2$	$p-6$	$p-4$
01 10 11	1	+1	$p-4$	$p-4$	$p-2$	$p-6$	$p-2$	$p-6$	$p-4$	$p-4$
01 11 10	1	+1	$p-4$	$p-2$	$p-4$	$p-6$	$p-2$	$p-4$	$p-6$	$p-4$
11 10 10	1	+1	$p-6$	$p-4$	$p-4$	$p-6$	$p-2$	$p-4$	$p-4$	$p-2$
10 01 11	1	+1	$p-4$	$p-4$	$p-6$	$p-2$	$p-6$	$p-2$	$p-4$	$p-4$
10 11 01	1	+1	$p-4$	$p-6$	$p-4$	$p-2$	$p-6$	$p-4$	$p-2$	$p-4$
11 01 01	1	+1	$p-2$	$p-4$	$p-4$	$p-2$	$p-6$	$p-4$	$p-4$	$p-6$
01 10 11	1	+1	$p-4$	$p-4$	$p-2$	$p-6$	$p-2$	$p-6$	$p-4$	$p-4$
01 11 01	1	+1	$p-2$	$p-4$	$p-2$	$p-4$	$p-4$	$p-6$	$p-4$	$p-6$
11 10 01	1	+1	$p-4$	$p-6$	$p-2$	$p-4$	$p-4$	$p-6$	$p-2$	$p-4$
01 01 11	1	+1	$p-2$	$p-2$	$p-4$	$p-4$	$p-4$	$p-4$	$p-6$	$p-6$
01 11 10	1	+1	$p-4$	$p-2$	$p-4$	$p-6$	$p-2$	$p-4$	$p-6$	$p-4$
11 01 10	1	+1	$p-4$	$p-2$	$p-6$	$p-4$	$p-4$	$p-2$	$p-6$	$p-4$
01 01 11	1	+1	$p-2$	$p-2$	$p-4$	$p-4$	$p-4$	$p-4$	$p-6$	$p-6$
01 11 01	1	+1	$p-2$	$p-4$	$p-2$	$p-4$	$p-4$	$p-6$	$p-4$	$p-6$
11 01 01	1	+1	$p-2$	$p-4$	$p-4$	$p-2$	$p-6$	$p-4$	$p-4$	$p-6$
01 01 01	0	+1	$p-2$	$p-3$	$p-3$	$p-1$	$p-5$	$p-5$	$p-5$	$p-7$
10 10 10	1	+2	$p-6$	$p-4$	$p-4$	$p-4$	$p-2$	$p-2$	$p-2$	p

TABLE 4.2: Constructing the 50 sequences for a clause. See text for explanation.

If a median sequence M satisfies Property 3, then the number of corresponding scenarios are

$$(p-6)!^{7k}(p-5)!^{6k}(p-4)!^{12k}(p-3)!^{12k}(p-2)!^{6k+2n}(p-1)!^{7k+2} \quad (4.54)$$

which cannot be divided by p. Indeed, if Property 3 holds, then $H(M, A) = H(M, \overline{A}) = p-1$ and for each i, $H(M, A_i) = H(M, \overline{A}_i) = p-2$. Since a clause $c_j = (x_{i_1} \vee x_{i_2} \vee x_{i_3})$ is satisfied, characters a_{i_1}, a_{i_2} and a_{i_3} in M are one of the combinations that can be found in the 7 columns in Table 4.2 labeled by **M1–M7** and the corresponding b characters are the complements. Then the Hamming distances between M and the 50 sequences defined for c_j are the values indicated in the appropriate column. It is easy to verify that in each column from **M1** to **M7**, there are $7 - 7$ $(p-6)$ and $(p-1)$, $6-6$ $(p-5)$ and $(p-2)$, and $12 - 12$ $(p-4)$ and $(p-3)$. If some of the literals are negated in a clause, then both in the constructed sequences and in the median sequences, the corresponding a and b values are swapped, and the same reasoning holds.

What follows is that only those median sequences contribute to the number of scenarios modulo p that have Property 3. Therefore,

$$\sum_{M \in \mathcal{M}} \prod_{A \in \mathcal{S}} H(M, A)! \equiv s(\Phi)(p-6)!^{7k}(p-5)!^{6k}(p-4)!^{12k} \times$$
$$\times (p-3)!^{12k}(p-2)!^{6k+2n}(p-1)!^{7k+2} \quad \bmod \ p \quad (4.55)$$

where $s(\Phi)$ is the number of satisfying assignments of Φ. Since all the calculations in the reduction can be done in polynomial time, we get the following theorem.

Theorem 53. *The counting problem #SPS-STAR is in #P-complete.*

4.2.7 Counting the (not necessarily perfect) matchings in a planar graph

While the number of perfect matchings in a planar graph can be calculated in polynomial time, the counting problem becomes hard if all matchings are to be calculated. This quite surprising result was proved by Mark Jerrum [97]. The reduction is via (vertex) weighted matchings that we define below.

Problem 10.
Name: #PL-W-MATCHING.
Input: a simple graph $G = (V, E)$, and a weight function $w : V \to \mathbb{Z}$.
Output: the value defined as

$$\sum_{M \in \mathcal{M}(G)} \prod_{v \notin M} w(v), \quad (4.56)$$

where $\mathcal{M}(G)$ is the set of the (not necessarily perfect) matchings of G. Later on, we are going to introduce weighted graphs where the weight function maps to the multivariate polynomial ring $\mathbb{Z}[x_1, x_2, \ldots, x_k]$. In such cases, the sum in Equation (4.56) will be called the *matching polynomial*.

Clearly, #W-MATCHING is a #P-complete problem, since the problem reduces to the #PERFECTMATCHING by choosing the weight function to be the constant 0 function. Indeed, then only the perfect matchings contribute to the sum in Equation (4.56), and all of them with 1 (recall that the empty product is defined as 1). It is also clear that the #W-MATCHING problem remains in #P-complete if only weights 1 and 0 are used.

First we show that #PL-W-MATCHING is also in #P-complete by reducing #W-MATCHING to it. Let $G = (V, E)$ be a simple graph with weights $w : V \to \mathbb{Z}$. We can draw G on a plane in polynomial time such that there are no points where more than 2 edges cross each other. The number of crosses is $O(n^4)$, where $n = |V|$. We are going to replace each crossing to a constant size planar gadget such that for the so obtained planar graph $G' = (V, E')$ with weights $w' : V' \to \mathbb{Z}$, the equality

$$\sum_{M' \in \mathcal{M}(G')} \prod_{v' \notin M'} w'(v') = 8^c \sum_{M \in \mathcal{M}(G)} \prod_{v \notin M} w(v) \qquad (4.57)$$

holds, where c is the number of crossings in the drawing of G. The gadget is built up using the two units in Figures 4.9 and 4.10. These gadgets contain indeterminants labeling some vertices. These indeterminants can be considered as monomials in the multivariate polynomial ring. It is easy to see that the matching polynomial is

$$1 + xy + yz + zx \qquad (4.58)$$

for the gadget Δ_1 in Figure 4.9, and

$$2(1 + xyz) \qquad (4.59)$$

for the gadget Δ_2 on Figure 4.10.

Call the degree 1 vertices "external" and all other vertices "internal" in Δ_1 and Δ_2. Similarly, let an edge be "internal" if both of its vertices are internal, otherwise external. The matching polynomials can be interpreted by considering the gadgets in a larger graph H, where only the external vertices have connections toward the remaining part of H_1. The matching polynomial of Δ_1 tells us that Δ_1 contributes to the matching polynomial of H_1 only when 1 or 3 of the external vertices are covered by edges outside of Δ_1, and thus, 3 or 2 external edges are used in a matching. The contribution in each case is 1. Namely, if MP denotes the matching polynomial, then

$$MP(H_1) = MP((H_1 \setminus \Delta_1) \cup \{x, y, z\}) + yzMP((H_1 \setminus \Delta_1) \cup \{x\}) +$$
$$xzMP((H_1 \setminus \Delta_1) \cup \{y\}) + xyMP((H_1 \setminus \Delta_1) \cup \{z\}). \qquad (4.60)$$

Here we slightly abused the notations that some of the vertices in Δ_1 are denoted by their assigned weights.

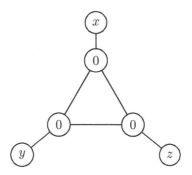

FIGURE 4.9: The gadget component Δ_1 for replacing a crossing in a non-planar graph. See text for details.

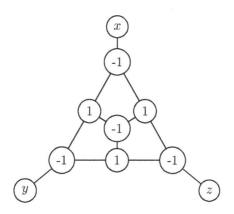

FIGURE 4.10: The gadget component Δ_2 for replacing a crossing in a non-planar graph. See text for details.

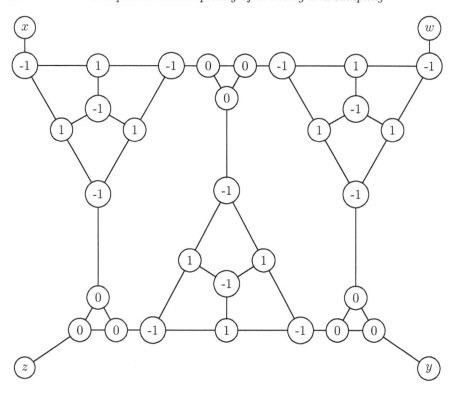

FIGURE 4.11: The gadget Γ replacing a crossing in a non-planar graph. See text for details.

It is easy to see that a similar argument holds for Δ_2, when it is a part in a larger graph H_2, then

$$MP(H_2) = 2MP((H_2 \setminus \Delta_2) \cup \{x, y, z\}) + 2xyzMP(H_2 \setminus \Delta_2). \qquad (4.61)$$

The complete gadget Γ is in Figure 4.11. It is easy to observe that Γ is constructed from three copies of Δ_1 and three copies of Δ_2. Only those matchings will contribute to the matching polynomial in which 3 or 2 external edges are used in each copy of Δ_1 and 3 or 0 external edges are used in each copy of Δ_2. From this, a straightforward calculation reveals that the matching polynomial of Γ is

$$8(1 + xy + wz + wxyz). \qquad (4.62)$$

Namely, Γ has exactly the properties required to substitute a crossover. The diametrically opposed external edges in Γ are forced to act in the same way in any matching: either both of them are in the matching or none of them. What follows is that if a crossover is replaced with Γ and all w, x, y, and z are replaced with 1, then the matching polynomial becomes 8 times the matching polynomial of the original graph. By replacing each matching in G with Γ and substituting each non-determinant by 1, the matching polynomial becomes 8^c times the matching polynomial of G, where c is the number of crossings in G. It is clear that the whole reduction is polynomial computable. Since #W-MATCHING is in #P-complete, we get that #PL-W-MATCHING is also in #P-complete. It is clear that #PL-W-MATCHING remains in #P-complete if only weights -1, 0 and 1 are used.

In the last step, we prove that #PL-MATCHING, that is, counting the matchings in a planar graph, is also in #P-complete by showing that #PL-W-MATCHING is polynomial reducible to it. This reduction uses polynomial interpolation twice. In the first step, we show that #PL-W-MATCHING with weights -1, 0 and 1 is polynomial reducible to the #PL-W-MATCHING problem using only weights -1 and 1. In the second step, we show that the #PL-W-MATCHING problem using only weights -1 and 1 is polynomial reducible to the #PL-MATCHING problem.

Let G be a planar graph with vertex weights from the set $\{-1, 1, 0\}$. Replace each 0 with x. Then the sum of the weighted matchings is the matching polynomial

$$MP(G) := \sum_{i=0}^{k} a_i x^i \tag{4.63}$$

evaluated at $x = 0$, where k is the number of vertices labeled by 0. Generate graphs $G_0, G_1, \ldots, G_k$ by attaching $0, 1, \ldots k$ number of auxiliary vertices to each vertex labeled by x. Assign weight 1 to each auxiliary vertex and replace each x with 1. Then the sum of the weighted matchings in G_j is

$$\sum_{i=0}^{k} a_i (j+1)^i. \tag{4.64}$$

To see this, observe that a_i is the sum of weighted matchings over those matchings that avoid exactly i vertices having weight x in G. Any such matching can be extended to a matching in G_j by j ways adding an edge to an avoided vertex and one way by not adding any edge. Thus the sum of the weighted matchings in G_j is the matching polynomial of G evaluated at $x = j + 1$. Evaluating $PM(G)$ at $k + 1$ different points, we can compute the coefficients of it in polynomial time. Particularly, we can compute a_0, that is, the sum of weighted matchings of G.

Similar reduction can be done in the second step. Let G be a planar graph with vertex weights from the set $\{-1, 1\}$. Replace each -1 with x. Then the sum of the weighted matchings is the matching polynomial

$$MP(G) := \sum_{i=0}^{k} a_i x^i \tag{4.65}$$

evaluated at $x = -1$, where k is the number of vertices labeled by -1. We can

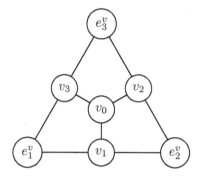

FIGURE 4.12: The gadget Δ replacing a vertex in a planar, 3-regular graph. See text for details.

again generate $k + 1$ graphs in which the number of matchings is equal to the matching polynomial evaluated at $k + 1$ different integer values. Therefore, the coefficients can be computed in polynomial time, and thus, the evaluation at $x = -1$ can also be done in polynomial time. This completes the proof of the following theorem.

Theorem 54. *The computational problem* #PL-MATCHING *is in* #P-*complete.*

4.2.8 Counting the subtrees of a graph

We learned that counting the spanning trees in a graph is easy. Surprisingly, counting all trees of a graph becomes #P-complete, even if the problem is restricted to planar graphs. The proof is based on reducing #PL-3REG-H-PATH, that is, counting the Hamiltonian paths in a planar, 3-regular graph to this problem. It is known that #PL-3REG-H-PATH is in #P-complete. We do not introduce the proof of the #P-completeness of the #PL-3REG-H-PATH problem; the proof can be found in [98] based on the work of Garey, Johnson and Tarjan [74], who proved that PL-3REG-H-PATH is in NP-complete.

The reduction is done via two intermediate problems. The #PL-S-K-SUBTREE is the problem of counting the subtrees of planar graphs containing exactly k edges and a subset S of vertices. The #PL-S-SUBTREE is the problem of counting the subtrees of a planar graph containing a subset S of vertices. The reductions, except the first one, use polynomial interpolations.

First, we show that #PL-S-K-SUBTREE is in #P-complete by reducing the #PL-3REG-H-PATH to it. Let $G = (V, E)$ be a planar, 3-regular graph; let n denote the number of vertices. We replace each vertex v of G incident to edges e_1, e_2 and e_3 with a gadget Δ_v shown in Figure 4.12. In the modified graph, G', the edges are incident to vertices e_1^v, e_2^v and e_3^v. In each gadget, the e_i^v vertices are called external vertices, and the v_j vertices are called internal vertices. The edges in $G' \subset G$ are called external edges. These edges connect the external vertices of the gadgets.

We ask for the number of trees in G' with exactly $k = 4n - 3$ edges containing the set $S = \{v_0 | v \in V\}$. We show that this is exactly four times the number of Hamiltonian paths in G. Let p be a Hamiltonian path in G, and consider the corresponding external edges in G'. This set can be extended to a tree containing k edges and the set S in exactly 4 different ways. For each adjacent edge in p, there is a unique way to connect e_i and e_{i+1} with 3 edges, (e_i^v, v_i), (v_i, v_0), and (v_i, e_{i+1}^v), involving vertex v_0. (The indexes are modulo 3.) At the end of the Hamiltonian path, which were vertices s and t in G, there are 2 possible ways to connect s_0 and t_0 to the tree in G' using exactly 2 edges. It is easy to see that the number of vertices is indeed $4n - 3$. There are $n - 1$ external edges, and there are 3 internal edges in each gadget except at the end of the Hamiltonian path, where the number of internal edges are 2. Thus, the number of edges is $n - 1 + 3(n - 2) + 4 = 4n - 3$.

We are going to show that these are the only subtrees with k edges and covering the subset S by proving that any minimal subtree covering S contains exactly k edges if and only if its external edges form a Hamiltonian path in G. Let T be a minimal subtree covering S. Then for each gadget, the number of external vertices in the subtree is either 1, 2 or 3. It is easy to show that the number of internal edges for these three cases is 2, 3 or 5, respectively. The external edges form a spanning tree in G, thus, the number of external edges is $n - 1$ and the sum of the external vertices in the gadgets are $2n - 2$. Then the number of edges is k only if the external edges form a Hamiltonian path in G. Indeed, if there are m gadgets with 3 external vertices, then there are $m + 2$ gadgets with 1 external vertex, and $n - 2m - 2$ gadgets with 2 external vertices. Then the total number of edges is

$$n - 1 + 5m + 2(m + 2) + 3(n - 2m - 2) = 4n - 3 + m. \tag{4.66}$$

It is $2n - 3$ only if $m = 0$. That is, the external edges form a Hamiltonian path.

To show that #PL-S-SUBTREE is in #P-complete, consider a planar, graph $G = (V, E)$, and a subset of vertices $S \subseteq V$. We introduce the following polynomial:

$$TP(G) := \sum_{i=1}^{n} a_i x^i, \tag{4.67}$$

where a_i is the number of subtrees of G that contain i vertices and cover the subset S. Introduce a series of graphs $G_0, G_1, \ldots, G_{n-1}$, by attaching $0, 1, \ldots, n - 1$ auxiliary vertices to each vertex. Then the number of subtrees that cover the subset S in G_j is

$$\sum_{i=1}^{n} a_i (j + 1)^i, \tag{4.68}$$

which is $TP(G)$ evaluated at $x = j + 1$. Indeed, any subtree of G containing i vertices can be extended to a subtree of G_j in $(j + 1)^i$ different ways by not attaching or attaching one of the j auxiliary vertices to each of the vertices in the subtree. Thus, if we can compute the number of subtrees covering S in each graph $G_0, G_1, \ldots, G_{n-1}$, then we can calculate the coefficients of $TP(G)$. In particular, we can calculate a_{k+1}, the number of subtrees covering S and containing k vertices. That is, containing exactly k edges.

Finally, we can prove that #PL-SUBTREE is in #P-complete by reducing #PL-S-SUBTREE to it. Polynomial interpolation is applied again in the reduction. Let $G = (V, E)$ be a planar graph, let $S \subseteq V$ be a subset of vertices, and let k denote the size of S. Generate graphs $G_0, G_1, \ldots G_k$ by attaching $0, 1, \ldots, k$ additional vertices to each vertex in S. Then the number of subtrees in G_j is

$$\sum_{i=0}^{k} b_i(j + 1)^i \qquad (4.69)$$

where b_i is the number of subtrees in G that cover i vertices from S. By computing the number of subtrees in each G_j, we can evaluate the polynomial

$$\sum_{i=0}^{k} k b_i x^i \qquad (4.70)$$

at $k + 1$ different points. This allows the estimation of the coefficients in polynomial time. Particularly, we can compute b_k, the number of subtrees in G covering each vertex in S. This completes the proof of the following theorem.

Theorem 55. *The counting problem #PL-SUBTREE is in #P-complete.*

4.2.9 Number of Eulerian orientations in a Eulerian graph

Problem 11.
Name: #EULERIANORIENTATION.
Input: a Eulerian graph $G = (V, E)$, with possible parallel edges.
Output: the number of Eulerian Orientations of G. That is, the number of orientations of the edges such that for every vertex $v \in V$, the number of incoming and outgoing edges of v is the same.

Milena Michail and Peter Winkler showed that counting the Eulerian orientation of a graph is #P-complete [129]. They reduce the number of perfect matchings to the number of Eulerian orientations. The reduction is the following. Let $G = (U, V, E)$ be a bipartite graph. Since we are interested in the number of perfect matchings in G, we can assume, without loss of generality, that each degree in G is greater than 1. Let n denote $|U| = |V|$, let m denote $\sum_{u \in U} d(u) = \sum_{v \in V} d(v)$, and let m' be $m - 2n$. We construct the following graphs. Let G' be the graph that amends G by adding two vertices s and t. Each $u \in U$ is connected to s with $d(u) - 2$ parallel edges, and each $v \in V$ is connected to t with $d(v) - 2$ parallel edges. Finally, let G_k denote the graph that amends G' by connecting s and t with k parallel edges.

It is clear that in all Eulerian orientations of $G_{m'}$ in which all edges between s and t are oriented from t to s, all edges connecting U with s must be oriented toward U, and all edges connecting V to t must be oriented toward t. What follows is that all edges between U and V are oriented from V to U except exactly one edge for each $u \in U$ and $v \in V$. These edges indicate a perfect matching between u and v. It is also easy to see that there is a bijection between those Eulerian orientations of $G_{m'}$ and the perfect matchings in G.

Let R_j denote the number of (not necessarily balanced) orientations of G' in which each vertex in U and V are balanced and there are exactly j edges from the m' ones connecting s with U that are oriented toward U and similarly, exactly j edges from the m' ones connecting t with V that are oriented toward t. It is clear that $|R_{m'}|$ is the number of perfect matchings in G. Observe that due to symmetry,

$$|R_j| = |R_{m'-j}| \tag{4.71}$$

since there is a bijection between R_j and $R_{m'-j}$ obtained by changing the orientation of each edge.

Now, let $P(G_k)$ denote the number of Eulerian orientations in G_k. Then we have that the equation

$$P(G_k) = \sum_{i=0}^{k} \binom{k}{i} \left| R_{\frac{m'+k}{2} - i} \right| \tag{4.72}$$

holds. Indeed, if there are i number of edges oriented from s to t, then there are $k - i$ edges oriented from s to t. To get a Eulerian orientation, there must be $\frac{m'+k}{2} - i$ edges oriented from s to U and also, there must be the same number of edges oriented from V to t. Note that G_k is a Eulerian graph if and only if the parity of k and m are the same. Now observe that Equation (4.72) for $k = 0, 2, \ldots, m'$ if m' is even and for $k = 1, 3, \ldots, m'$ if m' is odd together with Equation (4.71) for each $j = 0, 1, \ldots, \left\lfloor \frac{m'-1}{2} \right\rfloor$ define $m' + 1$ linear equations. These equations contain $m' + 1$ indeterminants, $|R_j|$, with $j = 0, 1, \ldots, m'$, assuming that we can compute $P(G_k)$ for each k. It is easy to see that this equation system has a unique solution (the corresponding determinant is non-zero). Therefore, each $|R_j|$, particularly, the number of perfect matchings in R, can be computed in polynomial time if each $P(G_k)$ is obtained. This completes the proof of the following theorem.

Theorem 56. *The counting problem* #EULERIANORIENTATION *is in #P-complete.*

4.3 Further reading and open problems

4.3.1 Further results

- Leslie Valiant was the first who showed that many computing problems are in #P-complete. In his paper [176], he gave a list of 14 problems which are #P-hard or #P-complete. This list includes many of the problems we already covered in this chapter: computing the permanent, the number of perfect matchings, the number of not necessarily perfect matchings, and the number of satisfying assignments of a monotone 2CNF. Further important counting problems in his lists are:

  - Number of prime indications of monotone 2CNFs. The input is a 2CNF Φ built up from the set of logical variables X, and the output is the number of subsets $X' \subseteq X$ such that the implication

$$\bigwedge_{x \in Y} x \Rightarrow \Phi \qquad (4.73)$$

  holds for $Y = X'$ but does not hold for any $Y \subset X'$. The proof is based on reducing the number of perfect matchings to this problem using polynomial interpolation.

  - Number of minimal vertex covers. The input is a simple graph $G = (V, E)$, and the output is the number of subsets $V' \subseteq V$ such that V' covers the edge set, and it is minimal. That is, the implication

$$(u, v) \in E \Rightarrow u \in A \lor v \in A \qquad (4.74)$$

  holds for $A = V'$ but does not hold for any $A \subset V'$. The proof is based on reducing the monotone prime implications to this problem using bijection between the number of solutions.

- Number of maximal cliques. The input is a simple graph $G = (V, E)$, and the output is the number of subsets $V' \subseteq V$ such that V' is a clique, and it is maximal. That is, the implication

$$u, v \in A \Rightarrow (u, v) \in E \tag{4.75}$$

holds for $A = V'$ but does not hold for any $A \supset V'$. The proof is based on the fact that the set of vertices in any maximal clique is the complement of a minimal vertex cover in the complement graph and vice versa.

- Directed trees in a directed graph. The input is a directed graph $\vec{G} = (V, E)$, and the output is the number of subsets $E' \subseteq E$ that form a rooted tree. Compare this with the fact that the number of directed spanning trees rooted into a given vertex is easy to compute! The proof is based on a series of reductions of intermediate problems using polynomial interpolations.

- Number of $s - t$ paths. The input is a directed graph $\vec{G} = (V, E)$, and two vertices $s, t \in V$ and the output is the number of paths from s to t. The proof is similar to proving the #P-completeness of counting cycles in a directed graph. That is, the number of $s - t$ paths cannot be approximated with an FPRAS unless RP = NP.

- J. Scott Provan and Michael O. Ball proved [144] that the following problems are in #P-complete:

  - Counting the independent sets in a bipartite graph.

  - Counting vertex covers in a bipartite graph.

  - Counting antichains in a partial order.

  - Counting the minimum cardinality $s-t$ cuts. The input is a directed graph $\vec{G} = (V, E)$ and two vertices $s, t \in V$, and the output is the number of subsets $V' \subseteq V \setminus \{s, t\}$ such that the size of V' is minimal, and there is no path from s to t in $\vec{G} \setminus V'$.

  - Computing the probability that a graph is connected.

- Nathan Linial also gave a list of hard counting problems in geometry and combinatorics [118]. His list is the following:

  - Number of vertices of a polytope. The input is a set of linear inequalities in the form of $Ax \leq b$. Each linear inequality is a half space, and the intersection of them defines a polytope $P \subset \mathbb{R}^n$. The output is the number of vertices of P. The proof is based on reducing the number of antichains to this problem using bijection between the number of solutions.

- Number of facets of a polytope. The input is a set of linear inequalities in the form of $Ax \leq b$ defining a polytope P and a fixed d. The output is the number of d-dimensional facets of P. The proof is based on reducing the number of vertices of a polytope to this problem using a linear equation system.

- Number of $d - 1$-dimensional facets.

- Number of acyclic orientations of a graph. The input is a simple graph G, and the output is the number of orientations of G not containing a (directed) cycle. The proof is based on reducing the computation of the chromatic polynomial to this problem using polynomial interpolation.

- Components of a slotted space. The input is a set of hyperplanes H_i of a Euclidian space $\mathbb{R}^n$ defined by linear equations, and the output is the number of components of $\{\mathbb{R}^n \setminus \cup_i H_i\}$. The proof is based on reducing the number of acyclic orientations of a graph to this problem finding a bijection between the solutions.

- Number of 3-colorings of a bipartite graph. The input is a bipartite graph G, and the output is the number of proper 3-colorings of the vertices of G. The proof is based on reducing the number of independent sets in a bipartite graph to this problem finding a many-to-one mapping between the solutions. See also Exercise 13.

- Number of satisfying assignments of an implicative Boolean formula. The input is a Boolean formula Φ in the form

$$\bigwedge (x_i \vee \overline{x}_j), \tag{4.76}$$

and the output is the number of satisfying assignments. The proof is based on reducing the number of antichains of a poset to this problem finding a bijection between the number of solutions.

• Leslie Ann Goldberg and Mark Jerrum showed that counting the non-isomorphic subtrees of a tree is in #P-complete. The proof is based on reducing the number of matchings in bipartite graphs to this problem using polynomial interpolation [78].

• We learned that counting the Eulerian circuits in a Eulerian directed graph is easy. The problem becomes hard for undirected graphs. The proof was given by Graham Brightwell and Peter Winkler [27]. The proof is based on reducing the number of Eulerian orientations to the number of Eulerian circuits using modulo prime number calculations. Patric John Creed proved #P-completeness for planar graphs [48] and Qi Ge and Daniel Štefankovič proved that counting Eulerian circuits remains #P-complete for 4-regular planar graphs [75].

- Salil Vadhan introduced an interpolation technique that preserves several properties of graphs like regularity or sparseness [173]. Using this technique, he was able to prove that many hard enumeration problems remain in #P-complete when restricted to regular and/or planar graphs, possibly further constrained to have small constant maximum degree. Catherine Greenhill also applied this technique to prove that counting the proper colorings of a graph remains #P-complete even if the maximum degree is 3 [82].

4.3.2   #BIS-complete problems

The #BIS-complete problems have been introduced by Dyer and his coworkers [57]. #BIS is the counting problem asking the number of independent sets in a bipartite graph. It is known that #BIS is in #P-complete, as was mentioned in the previous subsection [144]. We say that two counting problems #A and #B are AP-interreducible if there is an approximation preserving polynomial reduction from #A to #B and there is also such reduction from #B to #A. The #BIS-complete problems are those counting problems which are AP-interreducible with #BIS. A large number of relatively diverse counting problems are known to be #BIS-complete, and none of them are known to have an FPRAS. Therefore, it is conjectured that these problems do not have FPRAS, although we cannot prove it, even conditionally, that is, assuming that RP does not equal NP. Dyer and his coworkers gave a logical characterization of the #BIS class, based on the wok of Saluja, Subrahmanyam and Thakur [148], who gave a logical characterization of the #P class similar to Fagin's characterization of NP [67]. Dyer and his coworkers proved that several graph isomorphism problems, the #ANTICHAIN and the #1N1P-SAT problems, are all #BIS-complete. The counting problem #1N1P-SAT asks the number of satisfying assignments of a conjunctive normal form in which each clause contains at most one negated literal and at most one non-negated literal. The relative complexity of counting graph homomorphisms is a current hot research topic [71].

4.3.3   Open problems

The following problems are suspected to be in #P-complete. The corresponding decision/optimization problems are known to be solvable in polynomial time.

- Number of realizations of degree sequences. A graph G is a realization of a degree sequence $D = d_1, d_2, \ldots, d_n$, $d_i \in \mathbb{Z}^+$, if the degrees of the vertices of G are exactly D. Polynomial running time algorithms exist to decide if a degree sequence D has a realization [63, 92, 88].

- Number of most parsimonious DCJ scenarios. In the DCJ (double cut and join) model, genomes are represented as edge-labeled directed graphs consisting of (not necessarily directed) paths and cycles. A DCJ operation takes at most two vertices and combines the edge ends in it into at most two new vertices. Finding the minimum number of DCJ operations necessary to transform a genome into another can be done in polynomial time [185, 17]. The number of solutions can be computed in polynomial time in special cases [138, 127], but this counting problem is conjectured to be in #P-complete in the general case.

- Number of most parsimonious reversal scenarios. A signed permutation is a permutation of numbers $1, 2, \ldots, n$ together with a sign assigned to each number. A reversal flips a consecutive part of the permutation and also changes the signs of the numbers. A single number can be reverted, and in that case, only its sign is changed. Hannenhalli and Pevzner gave the first polynomial running time algorithm to find the minimum number of reversals necessary to transform a signed permutation into $+1, +2, \ldots + n$ [89]. In spite of the enormous work on faster algorithms to find a solution [107, 11, 166] and on exploring the solution space [4, 154, 23, 165], there is no polynomial running time algorithm that could count the most parsimonious reversal scenarios.

- Number of triangulations of a general polygon. We learned that computing the number of triangulations of a convex polygon is easy, and in fact, the number of triangulations of a convex n-gon is C_{n-2}, the $n - 2^{\text{nd}}$ Catalan number. However, there is no known polynomial time algorithm to compute the number of triangulations of a general simple polygon, although any simple polygon can be triangulated in linear time [40, 7].

- Number of evolutionary scenarios in perfect phylogenies. Let $\mathcal{S}$ be a set of sequences of the same length from the alphabet $\{0, 1\}$. Let S_i be the subset of sequences that contain 1 at position i. We say that $\mathcal{S}$ has a perfect phylogeny [90, 87] if for any S_i and S_j, the intersection of the two subsets is empty or one of the subsets contains the other. An evolutionary scenario is a series of events, and each event is one of the following two types:

    (a) Substitution. A substitution changes a character in one of the sequences in one of the positions.

    (b) Coalescent. Two identical sequences are merged into one.

It is easy to see that in case of perfect phylogeny, there is at most one substitution in each position. Finding an evolutionary scenario with a minimum number of events is easy. However, it seems to be hard to compute how many shortest evolutionary scenarios exist.

4.4 Exercises

1. Generate a 3CNF Φ for each of these logical expressions such that Φ has the same number of satisfying assignments as the logical expressions.

    (a) * $x_1 \rightarrow ((x_2 \wedge x_3) \vee (\overline{x}_2 \wedge x_4))$

    (b) $\circ$ $(x_1 \leftrightarrow x_2) \wedge (\overline{x}_1 \rightarrow x_3)$

    (c) $(x_1 \wedge \overline{x}_2) \rightarrow (x_3 \vee \overline{x}_4)$

2. * Prove that it can be decided in polynomial time if a bipartite graph has an even number of perfect matchings.

3. Construct the directed graph isomorphic to a junction defined by Valiant in the proof of #P-completeness of the permanent. Its weighted adjacency matrix is in Equation (4.8).

4. Find the cycles picking up 0, 1 and 2 junctions + the 2 internal junctions in an interchange. See Figure 4.3

5. Prove that there is no 3×3 matrix M that satisfies the following properties:

    (a) $per(M) = 0$

    (b) $per(M[1; 1]) = 0$,

    (c) $per(M[3; 3]) = 0$,

    (d) $per(M[1, 3; 1, 3]) = 0$,

    (e) $per(M[1; 3]) = per(M[3; 1]) = $ non-zero constant.

6. Prove that there is no 4×4 matrix M satisfying the following conditions:

    (a) $det(M) = 0$

    (b) $det(M[1; 1]) = 0$,

    (c) $det(M[4; 4]) = 0$,

    (d) $det(M[1, 4; 1, 4]) = 0$,

    (e) $det(M[1; 4]) = det(M[4; 1]) = $ non-zero constant.

7. $\circ$ Prove that #DNF remains a hard computational problem if it is restricted to monotone disjunctive normal forms. Namely, prove that #MON-DNF is in #P-complete.

8. * Prove the #P-completeness of #MON-2SAT by finding a parsimonious reduction from #MATCHING to it.

9. Compute the number of linear extensions of the poset in Figure 4.7.

10. ∘ The Steiner tree problem is to find the smallest subtree of a graph $G = (V, E)$ that covers a subset of vertices $S \subseteq V$. Show that the problem is NP-hard even if the problem is restricted to planar, 3-regular graphs.

11. ∘ Prove that the number of Eulerian orientations remains in #P-complete even if it is restricted to simple Eulerian graphs.

12. Prove that if V' is a minimal vertex cover in $G = (V, E)$, then $V \setminus V'$ forms a maximal clique in $\overline{G}$, the complement of G.

13. * Prove the #P-completeness of the number of 3-colorings of a planar graph by reducing the number of independent sets of a planar graph to it.

14. The Erdős-Gallai theorem is the following. Let $D = d_1 \geq d_2 \geq \ldots \geq d_n$ be a degree sequence. Then D is graphical (has at least one simple graph G whose degrees are exactly D) if and only if the sum of the degrees is even and for all $k = 1, 2, \ldots, n$

$$\sum_{i=1}^{k} d_i \leq k(k-1) + \sum_{j=k+1}^{n} \min\{k, d_j\}. \tag{4.77}$$

Prove the necessary direction of the Erdős-Gallai theorem. That is, if D is graphical, then the sum of the degrees is even and for all $k = 1, 2, \ldots, n$, the Erdős-Gallai inequalities (Equation (4.77)) hold.

4.5   Solutions

Exercise 1.

(a) The Boolean circuit of the function is

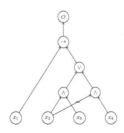

Introduce new variables for the internal nodes of the circuit. The 3CNF we are looking for is:

$$(\overline{x}_2 \vee \overline{x}_3 \vee y_1) \wedge (x_2 \vee \overline{x}_3 \vee y_1) \wedge (\overline{x}_2 \vee x_3 \vee y_1) \wedge (x_2 \vee x_3 \vee \overline{y}_1) \wedge$$
$$(x_2 \vee \overline{x}_4 \vee y_2) \wedge (\overline{x}_2 \vee \overline{x}_4 \vee y_2) \wedge (x_2 \vee x_4 \vee y_2) \wedge (\overline{x}_2 \vee x_4 \vee \overline{y}_2) \wedge$$
$$(\overline{y}_1 \vee \overline{y}_2 \vee y_3) \wedge (y_1 \vee \overline{y}_2 \vee \overline{y}_3) \wedge (\overline{y}_1 \vee y_2 \vee \overline{y}_3) \wedge (y_1 \vee y_2 \vee \overline{y}_3) \wedge$$
$$(x_1 \vee \overline{y}_3 \vee O) \wedge (\overline{x}_1 \vee \overline{y}_3 \vee O) \wedge (x_1 \vee y_3 \vee O) \wedge (\overline{x}_1 \vee y_3 \vee \overline{O}) \wedge$$
$$(O \vee O' \vee O") \wedge (O \vee O' \vee \overline{O"}) \wedge (O \vee \overline{O'} \vee O") \wedge (\overline{O} \vee O' \vee O") \wedge$$
$$(\overline{O} \vee \overline{O'} \vee O") \wedge (\overline{O} \vee O' \vee \overline{O"}) \wedge (O \vee \overline{O'} \vee \overline{O"})$$

(b) Observe that the logical form $x_1 \leftrightarrow x_2$ can be described in 3CNF as

$$(\overline{x}_1 \vee \overline{x}_2 \vee \overline{y}) \wedge (x_1 \vee \overline{x}_2 \vee y) \wedge (\overline{x}_1 \vee x_2 \vee y) \wedge (x_1 \vee x_2 \vee \overline{y}),$$

where y is the outcome of $x_1 \leftrightarrow x_2$.

Exercise 2. Observe that for any square matrix A,

$$per(A) \equiv det(A) \quad \mod (2).$$

Let A be the adjacency matrix of the bipartite graph G. Then, $per(A)$ is the number of perfect matchings, however it is $det(A)$ modulo 2. Since the determinant can be calculated in polynomial time, it can be decided in polynomial time if G contains an even or an odd number of perfect matchings.

Exercise 7. Observe that monotone satisfiability is in #P-complete, and in fact, restricted to #MON-2SAT, the counting problem is still in #P-complete. The complement of a monotone 2CNF can be expressed as a monotone 2DNF.

Exercise 8. Let $G = (U, V, E)$ be a bipartite graph with n vertices on both vertex classes. The 2CNF Φ that has as many satisfying assignments as the number of matchings in G contains a logical variable $x_{i,j}$ for each $(u_i, v_j) \in E$, and it is defined as

$$\Phi := \left(\bigwedge_{i=1}^{n} \bigwedge_{j_1 \neq j_2} (\overline{x}_{i,j_1} \vee \overline{x}_{i,j_2}) \right) \wedge \left(\bigwedge_{j=1}^{n} \bigwedge_{i_1 \neq i_2} (\overline{x}_{i_1,j} \vee \overline{x}_{i_2,j}) \right). \quad (4.78)$$

Indeed, the 2CNF forces there to be at most one edge on each vertex. It should be clear that Φ in Equation (4.78) is a monotone function of the logical variables $y_{i,j} = \overline{x}_{i,j}$. Therefore, we get that #MON-2SAT is in #P-complete.

Exercise 10. Let G be a planar, 3-regular graph. Replace each vertex with the gadget in Figure 4.12. Let this graph be denoted by G', and let S be the set of vertices containing the v_0 vertices of the gadgets. Show that from the solution of the Steiner tree problem of G' and S, it could be decided if G contains a Hamiltonian path.

Exercise 11. Replace each edge with a path of length 3 (containing two internal vertices).

Exercise 13. Reduce #BIS, the number of independent sets in bipartite graphs to this problem. Let $G = (U, V, E)$ be a bipartite graph. Extend it with two vertices a and b, and connect a to $V \cup \{b\}$ and connect b to $U \cup \{a\}$. The so-obtained G' is also a bipartite graph. Consider its 3-colorings. Then without loss of generality, a has color 1 and b has color 2. Then the vertices having color 3 form an independent set in G. Similarly, every independent set of G indicates a good coloring of G'. We obtain that the number of 3-colorings of G' is 6 times the number of independent sets of G. Indeed, there are 6 possible good colorings of vertices a and b.

Chapter 5

Holographic algorithms

The term *holographic reduction* was coined by Leslie Valiant. A holographic reduction is a many-to-many mapping between two finite sets, such that the sum of the fragments of elements in one set is mapped onto the sum of the fragments of elements in another set. However, the mapping is such that it proves the equality between the cardinality of the two sets. When the two sets are the solutions of problem instances in two different counting problems, the holographic reduction can prove that the two counting problems have the same computational complexity. If one of the problems can be solved in polynomial time, so can the other problem be solved. Similarly, the #P-completeness of one of the problems proves the #P-completeness of the other problem.

In classical computational complexity, the reductions are one-to-one or possibly many-to-one or one-to-many. An example of one-to-one mapping is the mapping of the satisfying assignments of a CNF Φ to the satisfying assignments of a 3CNF Φ', see Subsection 4.1.1. This mapping proves the #P-completeness of #3SAT (also, the 3CNF Φ' to the corresponding CNF Φ must be polynomial time computable). An example of a many-to-one reduction is the mapping of Eulerian circuits of a directed Eulerian graph $\vec{G}$ onto the arborescences of the same graph, see Section 3.3. This mapping proves that the number of Eulerian circuits of a directed Eulerian graph can be computed in polynomial time since the number of arborescences of a directed graph also can be computed in polynomial time and each arborescence in the mapping appears in the same number of times, $x_{\vec{G}}$, where $x_{\vec{G}}$ is also a polynomial time computable number depending only on $\vec{G}$ and not on the arborescence. The mapping of satisfying assignments of a 3CNF Φ onto the cycle covers of a weighted graph is a one-to-many mapping proving the #P-hardness of computing the permanent. Here a satisfying assignment is mapped onto a set of cycle covers whose summed weight is a constant x_Φ, which is polynomial time computable and depends on only Φ and not on the satisfying assignment.

In holographic reductions, the correspondences between individual solutions are no longer identifiable. This is a dramatically new computational primitive that has not appeared even implicitly in previous works in computer science. In his seminal paper [178], Valiant gave a list of computational problems solvable efficiently via holographic reductions. The introduced problems might look artificial and might be of little interest, however, all the introduced problems are in proximity of hard computational problems and cannot be solved efficiently without holographic reductions. The intriguing question is why the introduced problems are easy to compute and why the look-alike problems fall into the #P-complete class. Answering this question might be as hard as resolving the P vs. NP problem. As Valiant concluded, any P $\neq$ NP proof may need to explain, and not only imply, the absence of holographic reductions between polynomial time computable and #P-complete counting problems.

The theory of holographic reductions has been developed considerably in the recent years. Many counting problems have been proved to be in #P-complete via holographic reductions. Jin-Yi Cai and his coworkers developed dichotomy theories for counting problems. They published a comprehensive book on it [32], and interested readers are referred to that. In this book, we give only a brief introduction of the theory and a few simple examples.

5.1 Holographic reduction

Holographic reduction is a linear transformation between the solutions of a problem instance of a counting problem and the weighted perfect matchings of a planar graph. The planar graph consists of gadgets called *matchgates* and connecting edges. There are two types of matchgates, *recognizers* and *generators*. The generator matchgates have some distinguished nodes called *output* nodes, while the recognizers have some distinguished nodes called *input* nodes. One might consider a third type of matchgate, a *transducer* matchgate that has both input and output nodes, however, such mixed matchgates are not considered here. Any connecting edge connects an output vertex of a generator to an input vertex of a recognizer. Clearly, input and output vertices must be outer vertices in the planar embedding of matchgates. The number of input or output vertices is the *arity* of the matchgate. The entire graph is called a *matchgrid*.

Only the edges in the matchgates have non-trivial weights, and the weight of any edge in C is 1. The weights come from an arbitrary field, or even more generally, from an arbitrary commutative ring. Let G denote the weighted graph. We define

$$Z(G) := \sum_{M \in PM(G)} \prod_{e \in M} w(e) \tag{5.1}$$

where $PM(G)$ denotes the set of perfect matchings, and w is the weight function on the edges. Recall that computing this number, that is, the sum of weights of perfect matchings in a planar graph, needs only a polynomial number of ring operations for any commutative ring, see Theorems 41 and 31.

We define the *standard signature* of a matchgrid. Let X be a (generator or recognizer) matchgate. Let $e_{i_1}, e_{i_2}, \ldots, e_{i_l}$ be the edges that are incident to the input or output vertices of X. For any subset F of these edges, let $X \setminus F$ denote the graph obtained from X in which the vertices incident to any edge in F are omitted. We define the index of F as

$$ind(F) := 1 + \sum_{e_{i_j} \in F} 2^{j-1}.$$

The standard signature of X is a 2^l dimensional vector, for which the entry at index $ind(F)$ is $Z(X \setminus F)$.

We are going to rearrange the summation in Equation (5.1). Let X denote the set of generator matchgates, let Y denote the set of the recognizer matchgates, and let C denote the edges of G that connect the output nodes of the generator matchgates with the input nodes of the recognizer matchgates. Furthermore, for any $F \subseteq C$ and $X_i \in X$ (respectively, $Y_j \in Y$), let $X_i \setminus F$ (respectively, $Y_j \setminus F$) denote the planar graph in which vertices incident to any edge in F are omitted. We can partition the perfect matchings in G based on which edges coming from C are used in the perfect matching. Therefore we get that

$$Z(G) = \sum_{F \in 2^C} \prod_{X_i \in X} Z(X_i \setminus F) \prod_{Y_j \in Y} Z(Y_j \setminus F). \tag{5.2}$$

Observe that we can look at $Z(G)$ as a tensor of the standard signature vectors of the matchgates. Indeed, it is easy to see that $Z(G)$ is a linear function for each standard signature vector. The holographic reduction is a base change of this tensor.

For readers not familiar with tensor algebra, we derive a linear algebraic description which is more tedious, but does not need background knowledge of base changes in tensor algebra. Observe that the right-hand side of Equation (5.2) is the dot product of two vectors of dimension $2^{|C|}$. We can explicitly define these vectors in the following way. We are looking for the vectors in the forms

$$g := (g_1, g_2, \ldots, g_{2^{|C|}})$$

and

$$r^T := (r_1, r_2, \ldots, r_{2^{|C|}}).$$

That is, we are looking for a row vector g and a column vector r. We give an indexing of the edges in C in such a way that first we consider the edges of the last generator matchgate, then the edges of the next to last generator matchgate, etc. This implies an indexing of the subsets of C. Any subset F of C has a membership vector $s = (s_1, s_2, \ldots s_{|C|})$, in which s_m is 1 if the member of C with index m participates in F and otherwise 0. Then the index of F is defined as

$$1 + \sum_{m=1}^{|C|} s_m 2^{m-1}.$$

Let F be the subset with index k. Then

$$g_k := \prod_{X_i \in X} Z(X_i \setminus F) \qquad (5.3)$$

and

$$r_k := \prod_{Y_j \in Y} Z(Y_j \setminus F). \qquad (5.4)$$

A holographic reduction is simply the following. Let B be a matrix and g' be a vector such that

$$g'B = g. \qquad (5.5)$$

Then obviously,

$$Z(G) = g'Br = g'(Br). \qquad (5.6)$$

That is, the holographic map of the weighted perfect matchings is the dot product of g' and Br. We are looking for those matrices B for which this dot product has a combinatorial meaning. After heavy algebraic considerations, we will arrive at a picture that the recognizer and generators communicate via the possible configurations (presence or absence of edges) of C and compute a value which is the value of a solution of a problem instance represented by the current configuration.

We are seeking for B as tensor products of two vectors of dimension 2, $n = (n_1, n_2)$ and $p = (p_1, p_2)$. Recall that the tensor product of two vectors, $v = (v_1, v_2, \ldots, v_n)$ and $w = (w_1, w_2, \ldots, w_m)$ is

$$v \otimes w = (v_1 w_1, v_1, w_2, \ldots v_1 w_m, v_2 w_1, \ldots v_2 w_m, \ldots, v_n w_m).$$

Observe that the tensor product is associative but not commutative. If

$$k = 1 + \sum_{m=1}^{|C|} s_m 2^{m-1}$$

then the kth row of B is the tensor product whose m^{th} factor is p if $s_m = 1$ and n if $s_m = 0$.

The consequence of constructing B in this way is that the vectors g' and

Br can be directly computed via the matchgates, and the computation we have to perform on a matchgate is independent from other matchgates.

First, we introduce the computation for recognizer matchgates. Let Y_j be a recognizer matchgate, let $\{e_{i_1}, e_{i_2}, \ldots, e_{i_{l_j}}\}$ be the edges incident to the input vertices of Y_j and let u_j be its standard signature. Let z be a 0-1 vector of dimension $|C|$. Let b^j be the tensor product of l_j vectors whose m^{th} factor is p if the i_m^{th} coordinate of z is 1 and n if the i_m^{th} coordinate of z is 0. Then we define

$$ValR(Y_j, z) := u_j b^j,$$

that is, the inner product of u_j and b^j. We can now state the main theorem for recognizer gates.

Theorem 57. *Let z be a 0-1 vector of dimension $|C|$, and let*

$$k := 1 + \sum_{m=1}^{|C|} z_m 2^{m-1}.$$

Then the kth coordinate of Br is

$$\prod_{Y_j \in Y} ValR(Y_j, z). \tag{5.7}$$

Proof. Observe that the kth coordinate of Br is the inner product of b and r, where b is the tensor product of $|C|$ vectors in which the mth factor is p if the mth coordinate of z is 1 and n if the mth coordinate of z is 0. Therefore it is sufficient to show that

$$br = \prod_{Y_j \in Y} ValR(Y_j, z).$$

On the left-hand side, we have the sum

$$\sum_{i=1}^{2^{|C|}} b_i \prod_{Y_j \in Y} Z(Y_j \setminus F_i) \tag{5.8}$$

where F_i is the subset of C with index i, where index i is defined for F as above. Each factor on the right-hand side is also an inner product, that is, the right-hand side is

$$\prod_{Y_j \in Y} \sum_{t_j=1}^{l_j} Z(Y_j \setminus F_{t_j}) b_{t_j}^j. \tag{5.9}$$

Observe that $\sqcup C_j = C$ and thus $\sum l_j = |C|$. That is, if we factorize the product in Equation (5.9) we get exactly $2^{|C|}$ terms that correspond to the $2^{|C|}$ terms in Equation (5.8). Indeed, for any i, F_i can be unequivocally factorized as $\sqcup F_{t_j}$ such that for each Y_j, $Y_j \setminus F_i = Y_j \setminus F_{t_j}$ and $b_i = \prod b_{t_j}^j$. □

Similar computations can be developed for generators. Let X_i be a generator with outgoing vertices $v_{j_1}, v_{j_2}, \ldots, v_{j_{l_i}}$ corresponding to edges $e_{j_1}, e_{j_2}, \ldots, e_{j_{l_i}}$ incident to these edges. Let u_i be the standard signature of X_i. Let $b_1, b_2, \ldots, b_{2^{l_i}}$ be vectors such that for

$$t = 1 + \sum_{m=1}^{l_i} s_m 2^{m_i}$$

b_t is the tensor product of l_i vectors such that the mth factor is p if $s_m = 1$ and n if $s_m = 0$. Let $w^i = (w_1^i, w_2^i, \ldots, w_{2^{l_i}}^i)$ be the vector satisfying

$$u_i = \sum_{m=1}^{2^{l_i}} w_m^i b_m. \tag{5.10}$$

Then we define

$$ValG(X_i, b_m) := w_m^i. \tag{5.11}$$

For these values the following theorem holds.

Theorem 58. *Let*

$$k = 1 + \sum_{m=1}^{|C|} s_m 2^{m-1}.$$

For each generator X_i, we define $b_{i,k}$ as a tensor product of l_i vectors, such that the tth factor is p if $s_{j_t} = 1$ and p if $s_{j_t} = 0$, where j_t is the index of the edge incident to the output vertex v_{j_t}. Then the kth coordinate of g' is

$$\prod_{X_i \in X} ValG(X_i, b_{i,k}). \tag{5.12}$$

Proof. In an analogous way to constructing B, for each generator X_i, we can define a matrix B^i whose rows are tensor products of p and n. Observe the following.

(a) The relationship between the standard signature and the vector g is that

$$g = u^1 \otimes u^2 \otimes \ldots \otimes u^{|X|}.$$

(b) For each X_i, it holds that

$$u^i = w^i B^i.$$

(c) We can construct B as

$$B = B^1 \otimes B^2 \otimes \ldots \otimes B^{|X|}.$$

Then the remainder of the proof is simply applying the basic properties of the tensor product. Indeed,

$$
\begin{aligned}
g &= u^1 \otimes u^2 \otimes \ldots \otimes u^{|X|} = (w^1 B^1) \otimes (w^2 B^2) \otimes \ldots \times (w^{|X|} B^{|X|}) = \\
&\quad (w^1 \otimes w^2 \otimes \ldots \otimes w^{|X|})(B^1 \otimes B^2 \otimes \ldots \otimes B^{|X|}) = \\
&\quad (w^1 \otimes w^2 \otimes \ldots \otimes w^{|X|})B.
\end{aligned}
\tag{5.13}
$$

That is, we get that

$$
g' = (w^1 \otimes w^2 \otimes \ldots \otimes w^{|X|}).
\tag{5.14}
$$

Observe that this is exactly our claim, that is, the kth coordinate of g' is the appropriate product of $w^i_{m_i}$ terms.

□

For any $F \subseteq C$, we define $z(F)$ to be the membership vector of F, and for any $F \subseteq C$ and $X_i \in X$ whose output vertices are incident to edges $e_{j_1}, e_{j_2}, \ldots, e_{j_{l_i}}$, we define $b_i(F)$ as the tensor product of l_i. The kth factor in the tensor product is p if $e_{j_k} \in F$ and n if $e_{j_k} \notin F$. We get that

$$
Z(G) = g'(Br^T) = \sum_{F \subseteq C} \prod_{X_i \in X} ValG(X_i, b_i(F)) \prod_{Y_i \in Y} ValR(Y_i, z(F)).
\tag{5.15}
$$

Valiant denoted a matchgrid with Ω, and defined the *Holant* of a matchgrid as

$$
Hol(\Omega) := \sum_{F \subseteq C} \prod_{X_i \in X} ValG(X_i, b_i(F)) \prod_{Y_i \in Y} ValR(Y_i, z(F)).
\tag{5.16}
$$

We arrived at the key theorem of the holographic reductions.

Theorem 59. *Let $\Omega = (X, Y, C)$ be a matchgrid, and let G be the edge-weighted graph building Ω. Then for any base,*

$$
Z(G) = Hol(\Omega).
\tag{5.17}
$$

A *holographic algorithm* builds a matchgrid whose Holant in some base is the solution of the problem instance. If the matchgrid is planar, then the partition function of its underlying weighted graph can be computed in polynomial time, thus, the problem can be solved in polynomial time (given that the construction of the matchgrid can be done in polynomial time). In the next section, we give computational problems solvable in polynomial time using holographic algorithms.

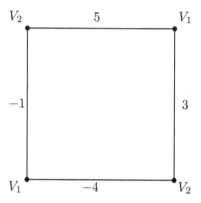

FIGURE 5.1: An edge-weighted bipartite planar graph as an illustrative example of the #X-matchings problem. See text for details.

5.2 Examples

5.2.1 #X-matchings

Recall that counting the not necessarily perfect matchings even in planar graphs is #P-complete, see Subsection 4.2.4. On the other hand, the following problem is in FP.

Problem 12.
Name: #X-MATCHINGS.
Input: a planar bipartite graph, $G = (V_1, V_2, E)$, where each vertex in V_1 have degree 2, and a weight function $w : E \to \mathbb{R}$.
Output: the sum of the weights of matchings, where a weight of a matching consists of the product of the weights of the edges participating in the matching and also the product of -1 times the sum of the edge weights incident to each vertex in V_2 not covered by any edge of the matching.

We give an illustrative example in Figure 5.1. There are two vertices in the class V_1 and also two vertices in V_2. There are two perfect matchings with the scores -3 and -20. There are five imperfect matchings, the empty matching and the matchings containing only one edge. The empty matching has the score $(-(-1 + 5))(-(-4 + 3)) = -4$. The other four matchings have weights $5, -1, -12$ and $+16$. Thus the value to be computed for this problem instance is $-3 - 20 - 4 + 5 - 1 - 12 + 16 = -19$.

The holographic algorithm transforms any edge-weighted, bipartite, planar graph G into a matchgrid. In the matchgrid, edges of G are represented with the edges in C. Each vertex in V_1 is replaced with a generator matchgate. Each vertex in V_2 is replaced with a recognizer matchgate. The weights for the edges in G will appear in the recognizer matchgates. The matchgates are constructed in such a way that for each $F \subseteq C$,

$$\prod_{X_i \in X} ValG(X_i, b_i(F)) \prod_{Y_i \in Y} ValR(Y_i, z(F))$$

is the weight of the corresponding matching in G. We will use the base $n = (-1, 1), p = (1, 0)$.

The generator matchgate contains two output vertices connected with an edge with weight -1. Its standard signature is $(-1, 0, 0, 1)$. It is easy to see that

$$(-1, 0, 0, 1) = n \otimes n + n \otimes p + p \otimes n. \tag{5.18}$$

That is, the signature of this matchgate in the given basis is $(1, 1, 1, 0)$. This is what we would like to get. Indeed, we would like to have a factor 1 for all cases where at most one edge is incident to a vertex in V_1.

Each vertex in $v \in V_2$ is replaced with a recognizer matchgate Y. The recognizer matchgate is a star tree. The number of leaves is the degree of v and the weights of the edges of the matchgate are the weights of the edges incident to v.

The standard signature of such a matchgate is w_i for those coordinates corresponding the cases where all the input vertices are omitted except the vertex incident to the edges with weight w_i. All other coordinates are 0. It is easy to see that the tensor product

$$n \otimes n \otimes \ldots \otimes n$$

is -1 at those coordinates where the standard signature of the matchgate is w_i. Therefore, when z is the all-0 vector, then

$$ValR(Y, z) = -\sum_i w_i.$$

The following two observations are also easy to see. If z is the all-0 vector, except its value is 1 in position i, then

$$ValR(Y, z) = w_i,$$

and if z contains at least two 1s, then

$$ValR(Y, z) = 0.$$

These are exactly the values what we would like to get. Indeed, if a vertex v is not covered by an edge in the matching, then its contribution to the score of the matching is $-\sum_i w_i$. If it is incident to an edge in the matching, and the edge has weight w_i, then its contribution is w_i. We do not consider configurations when v is incident to more than one edge in the matching.

We get that our example problem instance can be solved with the match-grid in Figure 5.2. As an edge-weighted graph G, it is an even cycle, thus

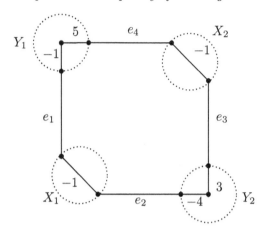

FIGURE 5.2: The matchgrid solving the #X-matching problem for the graph in Figure 5.1. The edges labeled by e_i belong to the set of edges C, and are considered to have weights 1. See the text for details.

it has two perfect matchings with weights -15 and -4. That is, its partition function $Z(G)$ is -19, just the solution of the #X-matchings problem instance. Observe that the relationship between the 7 matchings in the #X-matching problem and the 2 perfect matchings of the underlying matchgrid can no longer be identified.

To give a detailed computation of the holographic reduction, we also compute g, r, g' and Br for this particular problem instance. Both generators have standard signature $(-1, 0, 0, 1)$. Vector g is their tensor product, that is

$$g = (-1, 0, 0, 1) \otimes (-1, 0, 0, 1) =$$
$$(1, 0, 0, -1, 0, 0, 0, 0, 0, 0, 0, 0, -1, 0, 0, 1).$$

The recognizer Y_1 has standard signature $(0, 3, -4, 0)$, the recognizer Y_2 has standard signature $(0, 5, -1, 0)$. However, edges e_1 and e_4 are incident to the input vertices of Y_1 and edges e_2 and e_3 are incident to the input vertices of Y_2. Therefore, the coordinates of the tensor product of the standard signatures have to be permuted to follow the order of different subsets of C. Thus, the recognizer vector is

$$r^T = (0, 0, 0, 15, 0, -20, 0, 0, 0, 0, -3, 0, 4, 0, 0, 0).$$

The scalar product gr is indeed -19. Matrix B representing the base transformation is:

$$\begin{pmatrix}
1 & -1 & -1 & 1 & -1 & 1 & 1 & -1 & -1 & 1 & 1 & -1 & 1 & -1 & -1 & 1 \\
-1 & 1 & 1 & -1 & 1 & -1 & -1 & 1 & 0 & 0 & 0 & 0 & 0 & 0 & 0 & 0 \\
-1 & 1 & 1 & -1 & 0 & 0 & 0 & 0 & 1 & -1 & -1 & 1 & 0 & 0 & 0 & 0 \\
1 & -1 & -1 & 1 & 0 & 0 & 0 & 0 & 0 & 0 & 0 & 0 & 0 & 0 & 0 & 0 \\
-1 & 1 & 0 & 0 & 1 & -1 & 0 & 0 & 1 & -1 & 0 & 0 & -1 & 1 & 0 & 0 \\
1 & -1 & 0 & 0 & -1 & 1 & 0 & 0 & 0 & 0 & 0 & 0 & 0 & 0 & 0 & 0 \\
1 & -1 & 0 & 0 & 0 & 0 & 0 & 0 & -1 & 1 & 0 & 0 & 0 & 0 & 0 & 0 \\
1 & -1 & 0 & 0 & 0 & 0 & 0 & 0 & 0 & 0 & 0 & 0 & 0 & 0 & 0 & 0 \\
-1 & 0 & 1 & 0 & 1 & 0 & -1 & 0 & 1 & 0 & -1 & 0 & -1 & 0 & 1 & 0 \\
1 & 0 & -1 & 0 & -1 & 0 & 1 & 0 & 0 & 0 & 0 & 0 & 0 & 0 & 0 & 0 \\
1 & 0 & -1 & 0 & 0 & 0 & 0 & 0 & -1 & 0 & 1 & 0 & 0 & 0 & 0 & 0 \\
-1 & 0 & 1 & 0 & 0 & 0 & 0 & 0 & 0 & 0 & 0 & 0 & 0 & 0 & 0 & 0 \\
1 & 0 & 0 & 0 & -1 & 0 & 0 & 0 & -1 & 0 & 0 & 0 & 1 & 0 & 0 & 0 \\
-1 & 0 & 0 & 0 & 1 & 0 & 0 & 0 & 0 & 0 & 0 & 0 & 0 & 0 & 0 & 0 \\
-1 & 0 & 0 & 0 & 0 & 0 & 0 & 0 & 1 & 0 & 0 & 0 & 0 & 0 & 0 & 0 \\
1 & 0 & 0 & 0 & 0 & 0 & 0 & 0 & 0 & 0 & 0 & 0 & 0 & 0 & 0 & 0
\end{pmatrix}$$

It is easy to verify that

$$(Br)^T = (-4, 5, -12, 15, 16, -20, 0, 0, -1, 0, -3, 0, 4, 0, 0, 0).$$

The signature of the generators in the new base is $(1, 1, 1, 0)$. The vector g' is their tensor product, that is

$$g' = (1, 1, 1, 0) \otimes (1, 1, 1, 0) =$$
$$(1, 1, 1, 0, 1, 1, 1, 0, 1, 1, 1, 0, 0, 0, 0, 0).$$

It is easy to see that $g'B$ is indeed g. Finally, the non-zero terms in the scalar product of g' and Br are

$$-4, 5, -12, 16, -20, -1, 3,$$

which indeed correspond to the scores of the 7 matchings of our example #X-matching problem instance.

Finally, it should be clear that the holographic reduction is used only to prove that the partition function of the weighted graph corresponding to the matchgrid is exactly the solution of the computing problem that we solve via holographic reduction. The signature vectors g and r have dimension $2^{|C|}$, that is, they clearly grow exponentially with the problem size. Instead of performing the computations directly via the holographic transformation, the holographic algorithm instead computes a Pfaffian orientation of the weighted graph corresponding to the matchgrid, and it computes the Pfaffian of the oriented, weighted adjacency graph using a polynomial running time algorithm, see Theorem 31.

A possible Pfaffian orientation is when e_1 is oriented clockwise, and all other edges anti-clockwise. If we start numbering the vertices anticlockwise starting with the vertex of X_1 incident to e_1, then the Pfaffian of the oriented, weighted adjacency matrix is

$$
Pf
\begin{pmatrix}
0 & -1 & 0 & 0 & 0 & 0 & 0 & 0 & 0 & 1 \\
1 & 0 & 1 & 0 & 0 & 0 & 0 & 0 & 0 & 0 \\
0 & -1 & 0 & -4 & 0 & 0 & 0 & 0 & 0 & 0 \\
0 & 0 & 4 & 0 & 3 & 0 & 0 & 0 & 0 & 0 \\
0 & 0 & 0 & -3 & 0 & 1 & 0 & 0 & 0 & 0 \\
0 & 0 & 0 & 0 & -1 & 0 & -1 & 0 & 0 & 0 \\
0 & 0 & 0 & 0 & 0 & 1 & 0 & 1 & 0 & 0 \\
0 & 0 & 0 & 0 & 0 & 0 & -1 & 0 & 5 & 0 \\
0 & 0 & 0 & 0 & 0 & 0 & 0 & -5 & 0 & -1 \\
-1 & 0 & 0 & 0 & 0 & 0 & 0 & 0 & 1 & 0
\end{pmatrix}
= -19.
$$

5.2.2 #Pl-3-(1,1)-Cyclechain

A *cycle-chain cover* of a graph $G = (V, E)$ is a subgraph $G' = (V, E')$, $E' \subseteq E$, such that all components in G' are cycles or paths. Let CC denote a cycle-chain cover, let $c(CC)$ denote the number of cycles in CC, let $p(CC)$ denote the number of paths in CC, and let $\mathcal{C}(G)$ denote the set of cycle covers of G. Then the (x, y)-cycle-chain sum is defined as

$$
\sum_{CC \in \mathcal{C}(G)} x^{c(CC)} y^{p(CC)}.
$$

It is known that counting the number of Hamiltonian cycles in planar 3-regular graphs is #P-complete [119]. Therefore, computing the $(0, k)$-cycle-chain sum for arbitrary k is also #P-hard, see also Exercise 7. On the other hand, the following cycle-chain sum problem can be solved in polynomial time.

Problem 13.
Name: #PL-3-(1,1)-CYCLECHAIN.
Input: a planar 3-regular graph G.
Output: the number of cycle-chains in G.

We would like to generate a matchgrid such that for its corresponding weighted graph G', it holds that $Z(G')$ is the number of cycle chains in G. We add a vertex to the middle of each edge to transform the planar graph into a planar bipartite graph. We would like to transform these new vertices into generator matchgates such that in some base, they have signature $(1, 0, 0, 1)$. That is, the absence of both new edges in set C incident to the output vertices of the matchgate correspond to the absence of the edge in G, while the presence of both these new edges in C corresponds to the presence of the edge in G.

We would like to transform any vertex with degree 3 into a matchgate with signature $(0, 1, 1, 1, 1, 1, 1, 0)$. That is, one or two of its input nodes are incident to presented edges in C.

We choose the base $n = (1, 1), p = (1, -1)$. A possible generator matchgate is

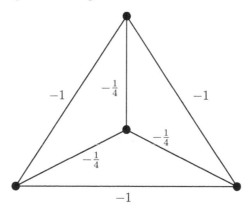

since its standard signature is $(2, 0, 0, 2)$, and indeed,

$$n \otimes n + p \otimes p = (1, 1, 1, 1) + (1, -1, -1, 1) = (2, 0, 0, 2).$$

A possible recognizer matchgate is

where the outer vertices are the 3 input vertices. It is easy to see that its standard signature is $\left(\frac{3}{4}, 0, 0, -\frac{1}{4}, 0, -\frac{1}{4}, -\frac{1}{4}, 0\right)^T$ and its signature in the given base is

$$\begin{pmatrix} n \otimes n \otimes n \\ p \otimes n \otimes n \\ n \otimes p \otimes n \\ p \otimes p \otimes n \\ n \otimes n \otimes p \\ p \otimes n \otimes p \\ n \otimes p \otimes p \\ p \otimes p \otimes p \end{pmatrix} \begin{pmatrix} \frac{3}{4} \\ 0 \\ 0 \\ -\frac{1}{4} \\ 0 \\ -\frac{1}{4} \\ -\frac{1}{4} \\ 0 \end{pmatrix} =$$

$$\begin{pmatrix} 1 & 1 & 1 & 1 & 1 & 1 & 1 & 1 \\ 1 & 1 & 1 & 1 & -1 & -1 & -1 & -1 \\ 1 & 1 & -1 & -1 & 1 & 1 & -1 & -1 \\ 1 & 1 & -1 & -1 & -1 & -1 & 1 & 1 \\ 1 & -1 & 1 & -1 & 1 & -1 & 1 & -1 \\ 1 & -1 & 1 & -1 & -1 & 1 & -1 & 1 \\ 1 & -1 & -1 & 1 & 1 & -1 & -1 & 1 \\ 1 & -1 & -1 & 1 & -1 & 1 & 1 & -1 \end{pmatrix} \begin{pmatrix} \frac{3}{4} \\ 0 \\ 0 \\ -\frac{1}{4} \\ 0 \\ -\frac{1}{4} \\ -\frac{1}{4} \\ 0 \end{pmatrix} = \begin{pmatrix} 0 \\ 1 \\ 1 \\ 1 \\ 1 \\ 1 \\ 1 \\ 0 \end{pmatrix}.$$

To give an example, consider the simplest 3-regular graph, K_4. Its set of cycle-chain covers consists of

(a) 3 perfect matchings,

(b) 3 Hamiltonian cycles,

(c) 12 Hamiltonian paths,

that is, altogether 18 cycle-chain covers.

Its matchgrid consists of 4 recognizer matchgates and 6 generator matchgates, altogether 40 vertices. The recognizers and the generators are connected with 12 edges belonging to set C. When none of the edges in C participates in a perfect matching:

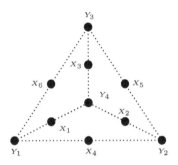

there are 81 perfect matchings: 3 perfect matchings in each recognizer matchgate and 1-1 perfect matchings in the generator matchgates, making $3^4 = 81$ combinations. Each of the global perfect matchings have a score $\frac{1}{4}$. There are perfect matchings when exactly two of the generators have no edges in C incident to both of their output vertices. These 4 missed edges must be in this configuration

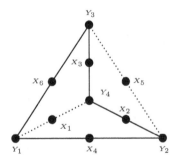

or the other two of them symmetric to it. (We can say that they are related to the perfect matchings of K_4.) Each of them has a score $\frac{1}{4}$.

When exactly 3 of the generators have edges in C incident to both of their output vertices, they must be in this configuration:

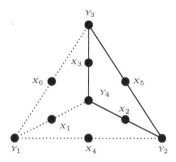

or the other 3 of them symmetric to this to have perfect matchings. (We can say that they are related to the triangles in K_4.) In each of these 4 cases, there are 3 perfect matchings, each of them having a score $-\frac{1}{4}$.

There are no other subsets of the set C for which there are perfect matchings, since any generator matchgate must have either 0 or 2 edges in C incident to its output vertices and any recognizer matchgate must have either 0 or 2 edges in C incident to its input vertices.

To summarize, there are 96 perfect matchings in the matchgrid, 84 having score $\frac{1}{4}$ and 12 having score $-\frac{1}{4}$. Thus, the partition function of the matchgrid is indeed $84 \times \frac{1}{4} + 12 \times \left(-\frac{1}{4}\right) = 18$, the number of cycle-chain covers of K_4.

5.2.3 #Pl-3-NAE-ICE

The so-called "ice" problems are considered in statistical physics. They are orientation problems. That is, the input is an unoriented graph G, and the solutions are assignments of a direction to each of its edges satisfying some constraint. We are interested in the number of such orientations. In its initial work, Linus Pauling proposed to count the orientations of a planar squared lattice, where each vertex has two incoming and two outgoing edges [141]. We know that counting the Eulerian orientations remains #P-complete even restricted to planar graphs [49]. On the other hand, the following problem can be solved in polynomial time using holographic reduction.

Problem 14.
Name: #PL-3-NAE-ICE.
Input: a planar graph G with maximum degree 3.
Output: the number of orientations of the edges of G such that no vertex has all incoming or all outgoing edges.

Here NAE stands for "not all equal". Let G be a planar graph with maximum degree 3. We would like to generate a matchgrid such that for its corresponding weighted graph G', it holds that $Z(G')$ is the number of not-all-equal orientations of G. First, we add a vertex to the middle of each edge of G to get a planar bipartite graph. We would like to transform these new vertices to generator matchgates such that in some base, they have signature $(0, 1, 1, 0)$.

That is, the presence of one of the new edges in the set C will correspond to an orientation of an edge in G.

We would like to replace each original vertex in G to a recognizer matchgate. If the original vertex has degree one, then the signature of the recognizer matchgate in the given base is $(1, 1)$, and the signature must be $(0, 1, 1, 0)$ (must be $(0, 1, 1, 1, 1, 1, 1, 0)$, respectively) for degree 2 (for degree 3, respectively) vertices.

We construct such matchgates in the base $n = (1, 1)$, $p = (1, -1)$. Then

$$n \otimes p + p \otimes n = (1, -1, 1, -1) + (1, 1, -1, -1) = (2, 0, 0, -2).$$

A possible generator is

$$\bullet \overset{1}{\underset{}{\rule{2em}{0.4pt}}} \bullet \overset{-2}{\underset{}{\rule{2em}{0.4pt}}} \bullet \overset{2}{\underset{}{\rule{2em}{0.4pt}}} \bullet$$

where the first and the last vertex are the output vertices. Indeed, this graph has one perfect matching with score 2. If only one of the output vertices are removed, then the number of vertices is odd, therefore, there is no perfect matching in it. Finally, if both output vertices are removed, it has one perfect matching with score -2.

A possible recognizer matchgate representing a degree 1 vertex is

$$\bullet \overset{1}{\underset{}{\rule{2em}{0.4pt}}} \bullet$$

where any of the two vertices is the only input vertex. Its standard signature is $(1, 0)^T$, and indeed

$$\begin{pmatrix} n \\ p \end{pmatrix} \begin{pmatrix} 1 \\ 0 \end{pmatrix} = \begin{pmatrix} 1 & 1 \\ 1 & -1 \end{pmatrix} \begin{pmatrix} 1 \\ 0 \end{pmatrix} = \begin{pmatrix} 1 \\ 1 \end{pmatrix}.$$

A possible recognizer matchgate for degree 2 vertices is

$$\bullet \overset{1}{\underset{}{\rule{2em}{0.4pt}}} \bullet \overset{-0.5}{\underset{}{\rule{2em}{0.4pt}}} \bullet \overset{0.5}{\underset{}{\rule{2em}{0.4pt}}} \bullet$$

where the first and the last vertex are the input vertices. Indeed, it has standard signature $(0.5, 0, 0, -0.5)^T$ and

$$\begin{pmatrix} n \otimes n \\ p \otimes n \\ n \otimes p \\ p \otimes p \end{pmatrix} \begin{pmatrix} 0.5 \\ 0 \\ 0 \\ -0.5 \end{pmatrix} = \begin{pmatrix} 1 & 1 & 1 & 1 \\ 1 & -1 & 1 & -1 \\ 1 & 1 & -1 & -1 \\ 1 & -1 & -1 & 1 \end{pmatrix} \begin{pmatrix} 0.5 \\ 0 \\ 0 \\ -0.5 \end{pmatrix} = \begin{pmatrix} 0 \\ 1 \\ 1 \\ 0 \end{pmatrix}.$$

The recognizer matchgate constructed in the #Pl-3-(1,1)-Cyclechain problem works perfectly for the degree 3 vertices, since each degree 3 vertex must have either one or two edges incident to it.

We again illustrate the problem with a small problem instance. It is in Figure 5.3. The problem instance has 4 not-all-equal orientations. Indeed, the

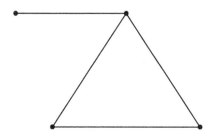

FIGURE 5.3: An example problem instance for the problem #Pl-3-NAE-ICE.

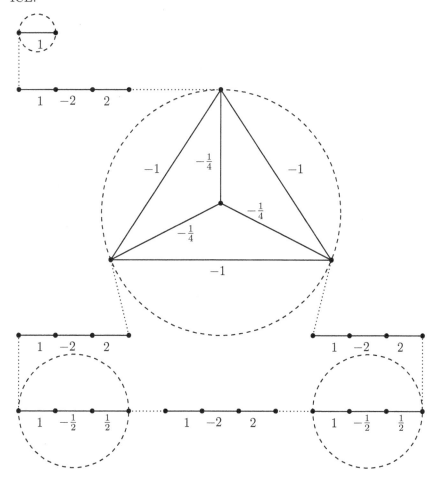

FIGURE 5.4: The matchgrid solving the #Pl-3-NAE-ICE problem for the problem instance in Figure 5.3. The edges belonging to the edge set C are dotted. The recognizer matchgates are put into dashed circles.

triangle can be oriented in two different ways, and an edge incident to the degree 1 vertex can be arbitrary oriented independently from the orientation of the triangle.

The corresponding matchgate that computes the number of orientations is in Figure 5.4. There are three perfect matchings of the arity 3 recognizer matchgate, and each of them can be extended in a single unique way to a perfect matching of the entire graph. Each of these perfect matchings has score 1. There is no perfect matching of the graph containing the edge in C incident to the upper input vertex of the arity 3 matchgates. There is one perfect matching in which the bottom 2 input vertices of the arity 3 matchgate are incident to edges in C. This perfect matching also has score 1. Therefore the partition function of the weighted graph building the matchgate is indeed 4, which is the number of not-all-equal orientations of the graph in Figure 5.3.

Although in this example there are 4 orientations of the input graph and 4 perfect matchings of the matchgate graph, there is no natural one-to-one correspondence between these solutions.

5.2.4 #Pl-3-NAE-SAT

For any logical formula,

$$c_1 \wedge c_2 \wedge \ldots \wedge c_k$$

we can assign a bipartite graph $G = (U, V, E)$, where U represents the clauses $c_1, c_2, \ldots, c_k$, V represents the logical variables $x_1, x_2, \ldots, x_n$, and there is an edge connecting u_i with v_j if x_j participates in clause c_i. A logical formula is called *planar* if G is planar. A clause is a not-all-equal clause if it is TRUE when there are two literals with different values. A not-all-equal formula is a logical formula in which all clauses are not-all-equal clauses.

We know that Pl-3SAT (where the problem instances are planar 3CNFs) is an NP-complete decision problem, and #Pl-3SAT is #P-complete [96]. The existence problem of Pl-Mon-NAE-SAT (where the problem instances are planar, monotone not-all-equal formulae) is reducible to the Four Color Theorem, and therefore, always have a solution [13]. However, counting the 4-colorings of a planar graph is #P-complete [179]. On the other hand, the following problem is solvable in polynomial time.

Problem 15.
Name: #PL-3-NAE-SAT.
Input: a planar, not-all-equal formula Φ in which each clause has 2 or 3 literals.
Output: the number of satisfying assignments of Φ.

We are going to construct a matchgrid using again the base $n = (1, 1), p = (1, -1)$, which appeared to be a very useful base in designing holographic reductions.

Let Φ be a planar not-all-equal formula, and let $G = (U, V, E)$ be its corresponding planar graph. Each clause vertex in U will be represented with a recognizer matchgate, each edge will be represented with a generator matchgate, and each variable vertex in V will be represented with a subgraph possibly containing several recognizer and generator matchgates.

We already introduced the arity 2 and arity 3 not-all-equal recognizer matchgates in the #Pl-3-NAE-ICE problem. We give here the arity 2 all-equal recognizer gate:

$$\bullet \overset{1}{\rule{2cm}{0.4pt}} \bullet \overset{0.5}{\rule{2cm}{0.4pt}} \bullet \overset{0.5}{\rule{2cm}{0.4pt}} \bullet$$

where the first and the last vertices are the input vertices. Indeed, it has standard signature $(0.5, 0, 0, 0.5)^T$ and

$$
\begin{pmatrix} n \otimes n \\ p \otimes n \\ n \otimes p \\ p \otimes p \end{pmatrix}
\begin{pmatrix} 0.5 \\ 0 \\ 0 \\ 0.5 \end{pmatrix}
=
\begin{pmatrix} 1 & 1 & 1 & 1 \\ 1 & -1 & 1 & -1 \\ 1 & 1 & -1 & -1 \\ 1 & -1 & -1 & 1 \end{pmatrix}
\begin{pmatrix} 0.5 \\ 0 \\ 0 \\ 0.5 \end{pmatrix}
=
\begin{pmatrix} 1 \\ 0 \\ 0 \\ 1 \end{pmatrix}.
$$

A possible all-equal arity 3 recognizer matchgate is

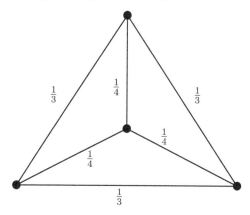

where the outer vertices are the 3 input vertices. It is easy to see that its standard signature is $\left(\frac{1}{4}, 0, 0, \frac{1}{4}, 0, \frac{1}{4}, \frac{1}{4}, 0\right)^T$ and its signature in the given base is

$$
\begin{pmatrix}
n \otimes n \otimes n \\
p \otimes n \otimes n \\
n \otimes p \otimes n \\
p \otimes p \otimes n \\
n \otimes n \otimes p \\
p \otimes n \otimes p \\
n \otimes p \otimes p \\
p \otimes p \otimes p
\end{pmatrix}
\begin{pmatrix}
\frac{1}{4} \\ 0 \\ 0 \\ \frac{1}{4} \\ 0 \\ \frac{1}{4} \\ \frac{1}{4} \\ 0
\end{pmatrix}
=
$$

$$
\begin{pmatrix}
1 & 1 & 1 & 1 & 1 & 1 & 1 & 1 \\
1 & 1 & 1 & 1 & -1 & -1 & -1 & -1 \\
1 & 1 & -1 & -1 & 1 & 1 & -1 & -1 \\
1 & 1 & -1 & -1 & -1 & -1 & 1 & 1 \\
1 & -1 & 1 & -1 & 1 & -1 & 1 & -1 \\
1 & -1 & 1 & -1 & -1 & 1 & -1 & 1 \\
1 & -1 & -1 & 1 & 1 & -1 & -1 & 1 \\
1 & -1 & -1 & 1 & -1 & 1 & 1 & -1
\end{pmatrix}
\begin{pmatrix}
\frac{1}{4} \\ 0 \\ 0 \\ \frac{1}{4} \\ 0 \\ \frac{1}{4} \\ \frac{1}{4} \\ 0
\end{pmatrix}
=
\begin{pmatrix}
1 \\ 0 \\ 0 \\ 0 \\ 0 \\ 0 \\ 0 \\ 1
\end{pmatrix} .
$$

We also already introduced the all-equal generator matchgates in the #Pl-3-(1,1)-Cyclechain problem, and the not-all-equal generator matchgates in the #Pl-3-NAE-ICE problem. We can use these matchgates to build a matchgrid in the following way.

(a) Replace all vertices in U with the appropriate arity (2 or 3) not-all-equal recognizer matchgate.

(b) Replace all degree 2 and degree 3 vertices in V with an arity 2 or arity 3 all-equal recognizer matchgate.

(c) Replace each degree k, $k > 3$ vertex in V with a chain of $k - 2$ arity 3 all-equal recognizer matchgates, connected with $k - 1$ (arity 2) all-equal generator matchgates. The first and the last recognizer matchgates have 2 free input vertices, and all other recognizer matchgates have 1 free input vertex. Therefore, this component has k free input vertices.

(d) Replace and edge (u_i, v_j) in G with a not-all-equal generator matchgate if variable x_j is a negated literal in clause c_i, and replace it with an all-equal generator matchgate if variable x_j is not negated in clause x_i.

5.2.5   #$_7$Pl-Rtw-Mon-3SAT

Our last example is one of the most curious problems solvable in polynomial time using holographic reduction. A logical formula is called *read-twice* if each variable appears in exactly 2 clauses. The Pl-Rtw-Mon-3SAT problem asks the satisfiability of a planar, read-twice, monotone 3CNF. It is trivially in P, since the all TRUE assignments naturally satisfies it. Surprisingly, #Pl-Rtw-Mon-3SAT is still a #P-complete problem, furthermore, deciding if there are an even number of satisfying assignments is $\oplus$P-complete, and thus, NP-hard [174] (problems in $\oplus$P ask the parity of the number of witnesses of problems in NP; the notation $\oplus$P is pronounced "parity-p" and also denoted by PP). On the other hand, the following problem can be solved in polynomial time.

Problem 16.
Name: #$_7$PL-RTW-MON-3SAT.
Input: a planar, read twice, monotone 3CNF.
Output: the number of satisfying assignments modulo 7.

We would like to design a matchgate in which each clause is replaced with an arity 3 recognizer matchgate with a signature

$$Br = (0, 1, 1, 1, 1, 1, 1, 1)^T$$

in some base (modulo 7), and each variable is replaced with an arity 2 generator matchgate with a signature

$$g' = (1, 0, 0, 1)$$

in some base, also modulo 7. That is, in the matchgrid, each $F \subseteq C$ having a value 1 represents a satisfying assignment.

We will work in the base $n = (5, 4)$, $p = (1, 1)$, and all computations are in the finite field $\mathbb{F}_7$. Each clause is replaced with a recognizer matchgate

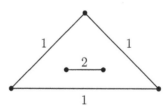

where the 3 vertices of the triangle are the input vertices. It is easy to see that it has standard signature

$$r = (0, 2, 2, 0, 2, 0, 0, 2)^T$$

and thus its signature in the given base is

$$
\begin{pmatrix}
n \otimes n \otimes n \\
p \otimes n \otimes n \\
n \otimes p \otimes n \\
p \otimes p \otimes n \\
n \otimes n \otimes p \\
p \otimes n \otimes p \\
n \otimes p \otimes p \\
p \otimes p \otimes p
\end{pmatrix}
\begin{pmatrix}
0 \\ 2 \\ 2 \\ 0 \\ 2 \\ 0 \\ 0 \\ 2
\end{pmatrix}
=
$$

$$
\begin{pmatrix}
6 & 2 & 2 & 3 & 2 & 3 & 3 & 1 \\
4 & 6 & 6 & 2 & 4 & 6 & 6 & 2 \\
4 & 6 & 4 & 6 & 6 & 2 & 6 & 2 \\
5 & 4 & 5 & 4 & 5 & 4 & 5 & 4 \\
4 & 4 & 6 & 6 & 6 & 6 & 2 & 2 \\
5 & 5 & 4 & 4 & 5 & 5 & 4 & 4 \\
5 & 5 & 5 & 5 & 4 & 4 & 4 & 4 \\
1 & 1 & 1 & 1 & 1 & 1 & 1 & 1
\end{pmatrix}
\begin{pmatrix}
0 \\ 2 \\ 2 \\ 0 \\ 2 \\ 0 \\ 0 \\ 2
\end{pmatrix}
=
\begin{pmatrix}
0 \\ 1 \\ 1 \\ 1 \\ 1 \\ 1 \\ 1 \\ 1
\end{pmatrix}.
$$

(Recall that all computations are modulo 7.)

Each variable is replaced with a generator matchgate

where the first and the last vertices are the output vertices. It has standard signature

$$g = (5, 0, 0, 3)$$

therefore, its signature in the given base is indeed $g' = (1, 0, 0, 1)$ as

$$(5, 0, 0, 3) = n \otimes n + p \otimes p = (4, 6, 6, 2) + (1, 1, 1, 1).$$

To give an illustrative example, consider the planar, monotone, read twice 3CNF

$$\Phi := (x_1 \vee x_2 \vee x_3) \wedge (x_1 \vee x_4 \vee x_6) \wedge (x_2 \vee x_4 \vee x_5) \wedge (x_3 \vee x_5 \vee x_6).$$

Its planar drawing can be obtained from K_4 by adding a vertex in the middle of each edge. The vertices of K_4 are the clause vertices, and the additional vertices are the variable vertices. The number of satisfying assignments of Φ are those subgraphs of (the vertex-labeled) K_4 that do not contain an isolated vertex. It is easy to see that there are 42 such subgraphs of K_4, thus there are 42 satisfying assignments of Φ. Indeed, these subgraphs of K_4 are the following:

1. There are 4 perfect matchings.

2. There are 16 subgraphs with 3 vertices and without an isolated vertex, the $\binom{6}{3}$ subgraphs with 3 vertices minus the 4 subgraphs consisting of a triangle and an isolated vertex.

3. When 2, 1, or 0 edges are omitted from K_4, and there cannot be isolated vertices; the number of such graphs is $\binom{6}{2} = 15$, $\binom{6}{1} = 6$, $\binom{6}{0} = 1$, respectively.

That is, the number of satisfying assignments of Φ is 0 modulo 7.

The perfect matchings in the matchgrid correspond to those subgraphs of K_4 in which every degree is either 1 or 3. Indeed, each generator matchgate must have either 0 or 2 edges in C incident to its output edges to have a perfect matching. The two cases correspond to the presence or absence of the corresponding edge in K_4. Similarly, each recognizer matchgate must have either 1 or 3 edges in C incident to its input edges to have a perfect matching. These two cases correspond to having degree 1 or degree 3 on the corresponding vertex of K_4. It is easy to see that each subgraph of K_4 with the prescribed constraint corresponds to exactly one prefect matching of the matchgrid, and the score of that perfect matching is the product of the score of the perfect matchings of the matchgates. The subgraphs with all degree 1 or 3 are the following:

1. There are 4 perfect matchings in K_4. It is easy to see that each corresponding perfect matching has score $2^6 3^2 5^4 \pmod 7 = 4$.

2. There are 4 star trees. Each of the corresponding perfect matchings has a score $2^6 3^3 5^3 \pmod 7 = 1$.

3. The K_4 itself. Its corresponding perfect matching has score $2^6 3^6 \pmod 7 = 1$.

Therefore, the partition function of the matchgrid is indeed

$$4 \times 4 + 4 \times 1 + 1 \times 1 \pmod 7 = 0.$$

Valiant found a holographic reduction for the $\#_7$Pl-Rtw-Mon-3SAT problem using higher-order tensor algebra (not described in this book) [174]. Jin-Yi Cai and Pinyan Lu showed such higher-order tensor algebras are not needed in holographic reductions, any higher-order holographic reduction can be efficiently transformed into such a holographic reduction that has been introduced in this book [36, 34]. They gave a holographic reduction in base $n = (1, 6), p = (5, 3)$ [35]. They also proved that characteristic 7 is the unique characteristic of a field for which there is a common basis in which both an arity 2 generator matchgate with signature $(1, 0, 0, 1)$ and an arity 3 recognizer matchgate with signature $(0, 1, 1, 1, 1, 1, 1, 1)$ exist. That is, the $\#_k$Pl-Rtw-Mon-3SAT problem can be solved with a holographic reduction only for $k = 7$. In this book, we gave another holographic reduction, see Exercises 9 and 10 for obtaining the holographic reduction introduced here.

5.3   Further results and open problems

5.3.1   Further results

- Jin-Yi Cai and Pinyan Lu studied the matchgates with symmetric signature vectors [35]. A signature vector is symmetric if the value of each coordinate depends on how many input or output vertices are removed, and does not depend on which vertices are removed. Symmetric signatures are represented in square brackets ([ ]), listing the values for each number of omitted input or output vertices. For example, (a, b, b, c, b, c, c, d) is represented as $[a, b, c, d]$. This research led to a polynomial running time algorithm whose input is a set of symmetric generator and recognizer signature vectors, and the output is a base and set of matchgates such that their signature vectors are exactly the given vectors in the obtained base or the algorithm reports that no such base and matchgates exist [31]. Note that it is sufficient to give the $k + 1$ possible values for an arity k signature, since the signature vectors are symmetric.

- Constraint Satisfaction Problems (CSP) are generalizations of the satisfiability problems. The input of a #CSP is a set of variables $\mathcal{X} = \{x_1, x_2, \ldots, x_n\}$ and a set of functions $\mathcal{F} = \{f_1, f_2, \ldots, f_m\}$, and each function takes a subset of $\mathcal{X}$ as its arguments. Each variable x_i has a domain D_i. The problem asks to compute

$$\sum_{(x_1, x_2, \ldots, x_n) \in D_1 \times D_2 \times \ldots \times D_n} \prod_{f \in F} f(x_{i_1}, x_{i_2}, \ldots, x_{i_{a(f)}})$$

where $a(f)$ is the arity of function f. A #CSP problem is planar, if the graph $G = (U, V, E)$ is planar, where $U = \mathcal{X}$, $V = \mathcal{F}$, and there is an edge between x_i and f_j if x_i is an argument of f_j. Jin-Yi Cai, Pinyan Lu and Mingji Xia considered the #CSP problems when each domain is $\{0, 1\}$, and they proved dichotomy theories. They showed that the tractable planar #CSP problems are exactly those which can be computed with holographic algorithms [38].

- Jin-Yi Cai, Pinyan Lu and Mingji Xia introduced the Fibonacci gates [37]. Fibonacci gates have symmetric signatures $[f_0, f_1, \ldots, f_k]$ where for each f_i it holds that $f_i = f_{i-1} + f_{i-2}$. The authors showed that the Holant problems with Fibonacci gates are solvable for arbitrary graphs in polynomial time not only for planar graphs.

- Mingji Xia, Peng Zhang and Wenbo Zhao used holographic reductions to prove #P-completeness of several counting problems [184]. First they used polynomial interpolation (see Subsection 4.2.4) to prove that counting the vertex covers in planar, bipartite 3-regular graphs is in #P-complete. Then they used holographic reduction to prove that counting the (not necessarily perfect) matchings in 2-3 regular, planar, and bipartite graphs is also in #P-complete. Then they consider ternary symmetric Boolean functions in the form

$$f_S(l_1, l_2, l_3) = \begin{cases} 1 & \text{if } l_1 + l_2 + l_3 \in S \\ 0 & \text{otherwise} \end{cases}$$

where $l_i \in \{0, 1\}$ are the literals and $S \subseteq \{0, 1, 2, 3\}$. A literal l_i might be x_i or $\overline{x_i}$, where x_i is a logical variable. When only positive literals are allowed, we denote it with the "-Mon-" tag in the description of the problem. The 3Pl-Rtw-f_S-SAT problem asks if a read-twice, planar formula in which each clause is an f function that is satisfiable. Using holographic reductions, the authors showed that the #3Pl-Rtw-Mon-$f_{\{0,1\}}$-SAT, the #3Pl-Rtw-Mon-$f_{\{0,1,2\}}$-SAT and #3Pl-Rtw-$f_{\{0,1,2\}}$-SAT problems are all in #P-complete.

- Also using holographic reductions, Jin-Yi Cai, Heng Guo and Tyson Williams proved that edge coloring of an r-regular planar graph with k colors is in #P-complete for all $k \geq r \geq 3$ [33].

5.3.2 Open problems

The ultimate goal of the research on holographic reduction is to find the border of P and NP-complete (the border of FP and #P-complete) or find an "accidental" algorithm that solves an NP-complete (#P-complete) problem in polynomial time. The following list of open problems highlights that possibly we are very far from this ultimate goal.

- Is there a combinatorial explanation of holographic reductions? The holographic reductions are essentially "carefully designed cancellations in tensor spaces" [37]. We saw that the cancellations appearing in division-free computations of the determinant have a nice combinatorial description (clow sequences, see Section 3.1). So far, nobody has been able to find such a combinatorial explanation/description; on the other hand, there is no proof that such combinatorial explanation does not exist. Also, we do not know if any of the introduced problems solvable with holographic algorithms has a polynomial running time algorithm without holographic reduction.

- Are there other computational paradigms leading to efficient algorithms yet to be discovered? There are still many counting problems with unknown computational complexity, and some of them were mentioned at the end of Chapter 4. We cannot exclude the possibility that some of these seemingly #P-complete counting problems are actually in FP, just the efficient algorithms solving them are not discovered yet.

5.4   Exercises

1. ○ Show that for any positive integer k, there exists a planar, 3-regular graph with $2k + 2$ vertices.

2. * Professor F. Lake tells his class that matchgates having weights on the edges in the set C have more computational power. Should they believe him?

3. Show that for any row vectors u_1 and u_2 and matrices A_1 and A_2, it holds that
$$(u_1 A_1) \otimes (u_2 A_2) = (u_1 \otimes u_2)(A_1 \otimes A_2).$$

4. Show that for $n = (-1, 1)$ and $p = (1, 0)$, it indeed holds that
$$(1, 1, 1, 0) = n \otimes n + n \otimes p + p \otimes n.$$

5. Let G be the planar bipartite graph constructed from the octahedron by adding a vertex to the middle of each edge. Compute the X-matching of G when each edge has weight 1.

6. ○ Show that it is possible to compute the matchings of a 2-4 regular, planar bipartite graph modulo 5 in polynomial time.

7. * Show that computing the $(0, k)$-cycle-chain sum in planar 3-regular graphs is #P-hard by reducing the counting of the Hamiltonian cycles in planar 3-regular graphs to it.

8. Compute the number of cycle-chain covers of the 3-dimensional cube.

9. * Show that there is an arity 3 recognizer matchgate with signature $(0, 1, 1, 1, 1, 1, 1, 1)$ in base $n = (1 + \omega, 1 - \omega)$, $p = (1, 1)$, where ω is the complex number $-\frac{1}{2} + \frac{\sqrt{3}}{2} i$. Use the fact that
$$\omega^2 = -\omega - 1.$$

10. * Show that $\omega = 4$ satisfies the following equalities in field $\mathbb{F}_7$:
$$\omega^3 = 1$$
$$\omega^2 = -\omega - 1,$$
thus, it is possible to construct an arity 3 recognizer matchgate with signature $(0, 1, 1, 1, 1, 1, 1, 1)$ in base $n = (1 + \omega, 1 - \omega) = (5, -3) = (5, 4) \pmod 7$, $p = (1, 1)$ over field $\mathbb{F}_7$ by simply copying the solution of Exercise 9.

11. ∘ Show that it is #P-complete to compute the number of subgraphs of a 3-regular planar graph that do not contain any isolated vertex.

12. Show that it is #P-complete to compute the number of subgraphs of a 3-regular planar graph that contain at least one isolated vertex.

5.5 Solutions

Exercise 1. Construct recursively an infinite series starting with K_4.

Exercise 2. No, he is not right. Any such matchgrid might be mimicked by inserting a path of length 2 between the edge in C with weight w and (for example) the output node of the incident recognizer matchgate. The path will belong to the recognizer matchgate, and the two edges will have weights 1 and w.

Exercise 6. Observe that the number to be computed is the score of the X-matching when each weight is 1.

Exercise 7. The reduction is based on polynomial interpolation. Let G be a planar, 3-regular graph, and consider the polynomial

$$f_G(x) = \sum_{i=1}^{\lfloor \frac{n}{3} \rfloor} a_i x^i$$

where n is the number of vertices of G, and a_i is the number of cycle covers of G with exactly i cycles. Clearly, a_1 is the number of Hamiltonian cycles of G. This coefficient can be obtained by evaluating the polynomial $f_G(x)$ at $\lfloor \frac{n}{3} \rfloor$ different points. Observe that $f_G(k)$ is the $(0, k)$-cycle-chain sum of G.

Exercise 9. The following recognizer matchgate works:

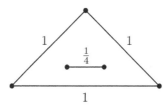

since its standard signature is

$$r = \left(0, \frac{1}{4}, \frac{1}{4}, 0, \frac{1}{4}, 0, 0, \frac{1}{4}\right)^T$$

and thus its signature in the given base is

$$
\begin{pmatrix}
n \otimes n \otimes n \\
p \otimes n \otimes n \\
n \otimes p \otimes n \\
p \otimes p \otimes n \\
n \otimes n \otimes p \\
p \otimes n \otimes p \\
n \otimes p \otimes p \\
p \otimes p \otimes p
\end{pmatrix}
\begin{pmatrix}
0 \\ \frac{1}{4} \\ \frac{1}{4} \\ 0 \\ \frac{1}{4} \\ 0 \\ 0 \\ \frac{1}{4}
\end{pmatrix}
=
$$

$$
\begin{pmatrix}
-1 & 1+2\omega & 1+2\omega & 3 & 1+2\omega & 3 & 3 & -3-6\omega \\
\omega & 2+\omega & 2+\omega & -3\omega & \omega & 2+\omega & 2+\omega & -3\omega \\
\omega & 2+\omega & \omega & 2+\omega & 2+\omega & -3\omega & 2+\omega & -3\omega \\
1+\omega & 1-\omega & 1+\omega & 1-\omega & 1+\omega & 1-\omega & 1+\omega & 1-\omega \\
\omega & \omega & 2+\omega & 2+\omega & 2+\omega & 2+\omega & -3\omega & -3\omega \\
1+\omega & 1+\omega & 1-\omega & 1-\omega & 1+\omega & 1+\omega & 1-\omega & 1-\omega \\
1+\omega & 1+\omega & 1+\omega & 1+\omega & 1-\omega & 1-\omega & 1-\omega & 1-\omega \\
1 & 1 & 1 & 1 & 1 & 1 & 1 & 1
\end{pmatrix}
\begin{pmatrix}
0 \\ \frac{1}{4} \\ \frac{1}{4} \\ 0 \\ \frac{1}{4} \\ 0 \\ 0 \\ \frac{1}{4}
\end{pmatrix}
=
\begin{pmatrix}
0 \\ 1 \\ 1 \\ 1 \\ 1 \\ 1 \\ 1 \\ 1
\end{pmatrix}.
$$

Exercise 10. It is trivial to check the inequalities. The weight $\frac{1}{4}$ should be replaced with 2 since $2 \times 4 = 1 \pmod 7$.

Part II

Computational Complexity
of Sampling

Chapter 6

Methods of random generations

The idea to use random numbers in scientific computation dates back to the 18th century. The French mathematician, naturalist. and encyclopédist, Georges-Louis Leclerc, Comte de Buffon introduced a problem that we know today as Buffon's needle problem [30]. It asks to find the probability that a randomly thrown needle of length l hits one of the parallel lines equally spaced on the floor a distance $t > l$ apart. We can assume that the distance of the center of the needle from the closest line is uniformly distributed between 0 and $\frac{t}{2}$. We can also assume that the angle the needle closes with any of the parallel lines is also uniformly distributed between 0 and π. The needle hits one of the lines if

$$x < \sin(\alpha)\frac{l}{2} \tag{6.1}$$

where x is the distance of the center from the closest line and α is the angle of the needle. Using geometric probability, the probability that the needle crosses a line is

$$\frac{\int_{\alpha=0}^{\pi} \sin(\alpha)\frac{l}{2}d\alpha}{\frac{t}{2}\pi} = \frac{2l}{t\pi}. \tag{6.2}$$

From this theoretical result, the value of π can be experimentally estimated by throwing the needle several times. If the measured frequency of the needle hits is f, then the estimation for π is

$$\hat{\pi} = \frac{2l}{tf}. \tag{6.3}$$

6.1   Generating random numbers

Random algorithms use random number generators. Typical computers cannot generate truly random numbers, however, they generate *pseudorandom* numbers. Pseudorandom number generators generate integer numbers from a given interval $[a, b]$ mimicking the uniform distribution. Then these numbers can be used to generate (pseudo)random numbers uniformly distributed on the $[0, 1]$ interval by projecting $[a, b]$ to $[0, 1]$. Of course, not all real numbers can be achieved in this way, only a finite subset of them. However, they well approximate the uniform distribution on the $[0, 1]$ interval. Once we have random numbers uniformly distributed on $[0, 1]$, then we can easily generate random numbers following other distributions.

Assume that π is a distribution on the integer numbers $1, \ldots, n$. The following procedure will generate a random number following π. Set x to 1, p to 0, and iterate the following until a value is returned. Draw u uniformly on $[0, 1]$. If $u \leq \frac{\pi(x)}{1-p}$, return x. Otherwise add $\pi(x)$ to p and increment x. The so-generated random ξ indeed follows distribution π, since

$$P(\xi = x) = (1 - \pi(0)) \left(1 - \frac{\pi(1)}{1 - \pi(0)}\right) \cdots \frac{\pi(x)}{1 - \sum_{i=1}^{x-1} \pi(i)} = \pi(x). \quad (6.4)$$

Assume that π is a continuous distribution with a cumulative distribution function F. Furthermore, assume that $F^{-1}(x)$ can be easily computed for any x. Then the following method, called the *inversion method*, generates a random variable following π. Generate u from the uniform distribution on $[0, 1]$. Return $F^{-1}(u)$. Then the so-generated random variable ξ indeed follows the distribution π, since

$$P(\xi \leq x) = F(x), \quad (6.5)$$

and the cumulative distribution function unequivocally defines the distribution. Below we give an example of how to use the inversion method.

Example 12. *Generate a random variable following the exponential distribution with parameter λ.*

Solution. The cumulative density function is

$$y = 1 - e^{-\lambda x}. \quad (6.6)$$

From this we can easily compute the inverse of the c.d.f.:

$$x = \frac{\ln(1 - y)}{-\lambda}. \quad (6.7)$$

It is easy to see that $1 - u$ is also uniform on $[0, 1]$, if u is uniform in $[0, 1]$. Thus the following procedure generates a random variable following the exponential distribution with parameter λ. Draw a random u following the uniform distribution on $[0, 1]$. Return $\frac{\ln(u)}{-\lambda}$. ■

Sometimes the cumulative density function does not have an analytic, easy-to-compute form, even its inverse. Interestingly, there is still a simple method that generates normally distributed random variables first proved by George Edward Pelham Box and Mervin Edgar Muller [22].

Theorem 60. *Let u_1 and u_2 be uniformly distributed values on $[0,1]$. Then*

$$z_1 := \sqrt{-2\ln(u_1)}\cos(2\pi u_2) \tag{6.8}$$

and

$$z_2 := \sqrt{-2\ln(u_1)}\sin(2\pi u_2) \tag{6.9}$$

are independent random variables, both of them following the standard normal distribution.

In many cases, more sophisticated methods are needed, see also the next section.

6.2 Rejection sampling

Rejection sampling was introduced by von Neumann [181]. Let π and p be two distributions over the same domain X. Assume that π is a distribution we would like to sample from, and p is a distribution we can sample from. We will call p the *sampling distribution* and π is called the *target distribution*. We are not required to be able to calculate the exact densities (probabilities), but we have to be able to calculate functions f and g that are proportional to π and p, respectively. Let the normalizing constants be n_1 and n_2, namely, for any $x \in X$, the equations

$$\pi(x) = n_1 f(x) \tag{6.10}$$

and

$$p(x) = n_2 p(x) \tag{6.11}$$

hold. Furthermore, there is a so-called *enveloping constant c*, that satisfies that for any $x \in X$,

$$cg(x) \geq f(x). \tag{6.12}$$

If these conditions are given, we can apply the rejection sampling method. It consists of two steps.

1. Draw a random x following the distribution p.

2. Draw a random number u following the uniform distribution on the interval $[0,1]$. Accept the sample x if

$$ucg(x) < f(x). \tag{6.13}$$

Otherwise, reject sample x, and go back to the first step.

The following theorem holds for the rejection sampling.

Theorem 61. *The accepted samples in the rejection sampling follow the distribution π.*

Proof. We apply the Bayes theorem on conditional probabilities. Let A denote the event that a sample is generated. First, we calculate $P(A)$. If X is a discrete space, then the probability that A happens is the sum of the probability that x is drawn multiplied by the probability that x is accepted, summing over all members $x \in X$. That is,

$$P(A) = \sum_{x \in X} p(x) \frac{f(x)}{cg(x)} = \sum_{x \in X} p(x) \frac{n_1 \pi(x)}{cn_2 p(x)} = \frac{n_1}{cn_2} \sum_{x \in X} \pi(x) = \frac{n_1}{cn_2}. \quad (6.14)$$

(For a continuous space, similar calculation holds, just the summation can be replaced with integration.) Now we can use the Bayes theorem. For an arbitrary x, the equation

$$P(x|A) = \frac{P(A|x)P(x)}{P(A)} = \frac{\frac{f(x)}{cg(x)}p(x)}{\frac{n_1}{cn_2}} = \frac{\frac{n_1 \pi(x)}{cn_2 p(x)}p(x)}{\frac{n_1}{cn_2}} = \pi(x) \quad (6.15)$$

holds, namely, the accepted samples indeed follow the distribution π. $\qquad\square$

We show an example of how to sample satisfying assignments of a disjunctive normal form. Surprisingly, we can sample uniformly satisfying assignments of a disjunctive normal form in *expected* polynomial running time using rejection sampling. This rejection sampling can be easily turned to an FPAUS.

Example 13. *Let a DNF Φ be given with n variables. Sample satisfying assignments of Φ using the rejection method. Show that #DNF is in FPAUS.*

Solution. A very naïve approach for a rejection sampling is to choose p as the uniform distribution of all possible assignments. That is, generate random assignments uniformly, and accept those which satisfy Φ. We know that for any assignment γ, $p(\gamma) = \frac{1}{2^n}$ and it is easy to sample from p. Let S denote the set of satisfying assignments. Then the target distribution π is such that for an $x \in S$, $\pi(x) = \frac{1}{|S|}$ and for $x \in \overline{S}$, $\pi(x) = 0$. The function g can be defined as

$$g(x) := \begin{cases} \frac{1}{2^n} & \text{if } x \in S \\ 0 & \text{if } x \in \overline{S} \end{cases} \quad (6.16)$$

and then the enveloping constant might also be 1. Let l denote the number of literals in the shortest conjunctive clause in Φ. Then we know that

$$|S| \geq 2^{n-l}. \quad (6.17)$$

Therefore, the normalizing constant n_2 cannot be larger than 2^l. What follows is that the acceptance probability cannot be smaller than 2^{-l}, since $n_1 = c = 1$, and the acceptance probability calculated in Equation (6.14) is $\frac{n_1}{n_2 c} = \frac{1}{n_2}$. If Φ contains a short clause (that is, $l = O(\log(n))$), this naïve method already provides an efficient sampling of satisfying assignments, since the expected number of trials to generate one satisfying assignment has an upper bound polynomial in n. On the other hand, if all clauses in Φ are long, then this approach might be inefficient.

However, satisfying assignments can be generated efficiently even if Φ contains only long clauses. Let $\Phi = C_1 \vee C_2 \vee \ldots \vee C_k$, where C_i, $i = 1, \ldots, k$ are the clauses. Let S_i denote the set of assignments that satisfy clause C_i. We know that $|S_i| = 2^{n-l_i}$, where l_i is the number of literals in C_i. Clearly,

$$S = \cup_{i=1}^k S_i. \tag{6.18}$$

Define

$$M := \sum_{i=1}^k |S_i|. \tag{6.19}$$

Clearly, $M \geq |S|$. Also observe that any satisfying assignment is counted at most k times in M, therefore it also holds that

$$|S| \geq \frac{M}{k}. \tag{6.20}$$

We define distribution p via the following method. Select a random clause C_i with probability $\frac{2^{n-l_i}}{M}$, then select a random satisfying assignment γ uniformly from S_i. A random clause γ has probability $\frac{t(\gamma)}{M}$, where $t(\gamma)$ is the number of clauses that γ satisfies.

What follows is that we can calculate exactly p for any γ, therefore g might be p and then $n_2 = 1$. We can set f to be the constant $\frac{k}{M}$. Since for any γ, $\pi(\gamma) = \frac{1}{|S|}$, we get that the normalizing constant $n_1 \geq 1$. Since the minimum probability in p is $\frac{1}{M}$, k is an appropriate number for the enveloping constant. We get that the acceptance probability is

$$P(A) = \frac{n_1}{cn_2} = \frac{n_1}{k} \geq \frac{1}{k} \tag{6.21}$$

since $n_1 \geq 1$. Hence the expected number of trials for a satisfying assignment is upper bounded by k, the number of clauses in Φ, which is less than the length of Φ.

Next, we show that this rejection sampler can be the core of an FPAUS. Let the DNF Φ and $\varepsilon > 0$ be the input of the FPAUS, and let k denote the number of clauses in Φ. Then do the rejection sampling till the first acceptance but at most $2k \log\left(\frac{1}{\varepsilon}\right)$ times. The probability that all proposed satisfying assignments are rejected is at most

$$\left(1 - \frac{1}{k}\right)^{2k \log\left(\frac{1}{\varepsilon}\right)} \approx \varepsilon^2 \ll \varepsilon. \tag{6.22}$$

If all proposals are rejected, then return with the last proposal. Then the generated satisfying assignment follows the distribution

$$\pi = (1 - \alpha)U + \alpha p, \tag{6.23}$$

where U is the uniform distribution, p is an unknown distribution, and $0 < \alpha < \varepsilon$. It is easy to show that the total variation distance between π and the uniform one is less than ε, since

$$d_{TV}(U, \pi) = \frac{1}{2} \sum_{x \in X} |U(x) - \pi(x)| =$$

$$\frac{1}{2} \sum_{x \in X} |U(x) - (1 - \alpha)U(x) - \alpha p(x)| = \frac{1}{2} \sum_{x \in X} \alpha |U(x) - p(x)| =$$

$$\alpha d_{TV}(U, p) < \alpha < \varepsilon. \tag{6.24}$$

The inequality $\alpha d_{TV}(U, p) < \alpha$ comes from the fact that the total variation distance between any two distributions is at most 1.

Since one rejection sampling step can be done in polynomial time, the running time of the algorithm is clearly polynomial in both the size of the problem and $-\log(\varepsilon)$, and thus, the described procedure is indeed an FPAUS. ∎

Generating uniform satisfying assignments might also be used to estimate the number of satisfying assignments. From a set of samples, we can estimate the probability that the logical variable x_1 is TRUE in satisfying assignments, and its complement probability, the probability that x_1 is FALSE in satisfying assignments. One of them is greater than or equal to 0.5. Let $f(x_1)$ be the frequency of samples (satisfying assignments) in which x_1 is TRUE. If $f(x_1) \geq 0.5$, then let Φ_1 be the DNF obtained from Φ such that all clauses are removed in which $\overline{x_1}$ is a literal, and each x_1 is removed from all clauses in which x_1 is a literal. The number of satisfying assignments of Φ_1 is the number of satisfying assignments of Φ in which x_1 is TRUE. Particularly,

$$|S_{\Phi_1}| = P(x_1 = TRUE)|S_\Phi| \tag{6.25}$$

where S_{Φ_1} denotes the set of satisfying assignments of Φ_1 and S_Φ denotes the set of satisfying assignments of Φ.

If $f(x_1) < 0.5$, then let Φ_1 be the DNF obtained from Φ such that all clauses are removed in which x_1 is a literal, and each $\overline{x_1}$ is removed from all clauses in which $\overline{x_1}$ is a literal. The number of satisfying assignments of Φ_1 is the number of satisfying assignments of Φ in which x_1 is FALSE. Particularly,

$$|S_{\Phi_1}| = P(x_1 = FALSE)|S_\Phi| \tag{6.26}$$

where S_{Φ_1} denotes the set of satisfying assignments of Φ_1 and S_Φ denotes the set of satisfying assignments of Φ.

We can generate random satisfying assignments of Φ_1 and can estimate the fraction of satisfying assignments in which x_2 is TRUE and the fraction of satisfying assignments in which x_2 is FALSE. Similar to the previous case, we can generate Φ_2, which is a DNF of random variables $x_3, \ldots, x_n$, and whose satisfying assignments have cardinality as the number of assignments of Φ_1 with a prescribed value for x_2.

Eventually, we can generate a DNF Φ_{n-1} that contains only literals x_n and $\overline{x_n}$ and has either 1 or 2 satisfying assignments depending on whether only one of the literals or both of them appears in it. Therefore we know $|S_{\Phi_{n-1}}|$, and we have estimations for all fractions $\frac{|S_{\Phi_{i-1}}|}{|S_{\Phi_i}|}$. We know that

$$|S_\Phi| = \frac{|S_\Phi|}{|S_{\Phi_1}|} \times \frac{|S_{\Phi_1}|}{|S_{\Phi_2}|} \times \ldots \times \frac{|S_{\Phi_{n-2}}|}{|S_{\Phi_{n-1}}|} \times |S_{\Phi_{n-1}}|, \qquad (6.27)$$

therefore the product of the estimated fractions and $|S_{\Phi_{n-1}}|$ is an estimation for $|S_\Phi|$. Although the errors in the estimations are multiplied, it turns out that very good estimation can be obtained in polynomial running time. This will be proved in Subsection 7.3.1.

When the aim is to estimate the number of satisfying assignments, we also can keep each generated sample, and can appropriately weight it instead of rejecting. This is prescribed in the next section.

6.3   Importance sampling

Let π and p be two distributions over the same domain X. Just like in the previous section, p is the sampling distribution and π is the target distribution. Let $f : X \to \mathbb{R}$ be a function, and assume that we would like to estimate the expected value of f under the distribution π, that is

$$E_\pi[f] := \sum_{x \in X} f(x)\pi(x). \qquad (6.28)$$

Assume that we can sample only from the distribution p. We would like to find function g satisfying that

$$E_p[g] = E_\pi[f] \qquad (6.29)$$

namely, the expected value of g under the distribution p is the expected number of f under the distribution p. It is easy to see that the following theorem holds.

Theorem 62. *If*

$$g(x) := \frac{\pi(x)}{p(x)} f(x) \qquad (6.30)$$

then

$$E_p[g] = E_\pi[f]. \tag{6.31}$$

Proof. Indeed,

$$E_p[g] = \sum_{x \in X} g(x)p(x) = \sum_{x \in X} \frac{\pi(x)}{p(x)} f(x)p(x) = \sum_{x \in X} f(x)\pi(x) = E_\pi[f]. \tag{6.32}$$

$\square$

It is frequently the case that we do not know the probabilities (densities) π and p, but we can calculate $\pi(x)$ and $p(x)$ up to an unknown normalizing constant. Then we can calculate g also up to an unknown normalizing constant. However, we can select an f for which we know the expectation, and can estimate the normalizing constant. The trivial choice for such f is the constant 1 function. The constant 1 function has expectation 1 under any distribution. Then

$$E_\pi[f] = 1 = E_p[g] = nE_p[\tilde{g}] \tag{6.33}$$

where n is the unknown normalizing constant and $\tilde{g}$ is the function we can calculate. If we sample from distribution p, the average $\tilde{g}$ value of the samples is an estimation for $\frac{1}{n}$. In the following example we show how to use this method to estimate the number of satisfying assignments of a disjunctive normal form.

Example 14. *Let Φ be a disjunctive normal form containing k clauses. Estimate the number of satisfying assignments of Φ using importance sampling.*

Solution. Let p be the distribution defined in the solution of Example 13. Recall that we can calculate the probability of any satisfying assignment γ in the distribution p:

$$p(\gamma) = \frac{t(\gamma)}{M}. \tag{6.34}$$

We cannot calculate $\pi(\gamma)$, where π is the uniform distribution of satisfying assignments. However, we can calculate π up to an unknown normalizing constant, that is constant 1, and then the normalizing constant is $\frac{1}{|S|}$, where S is the set of satisfying assignments. Let f be the constant 1 function, then

$$g(\gamma) = \frac{\pi(\gamma)}{p(\gamma)} f(\gamma) = \frac{\pi(\gamma)}{\frac{t(\gamma)}{M}} 1 = \frac{1}{|S|} \frac{M}{t(\gamma)} = \frac{1}{|S|} \tilde{g}(\gamma). \tag{6.35}$$

Namely, if we generate satisfying assignments following the distribution p, and calculate the average $\tilde{g}$ value (that is, $\tilde{g}(\gamma) = \frac{M}{t(\gamma)}$) that will be an estimation for the inverse of the normalizing constant, that is $|S|$. Furthermore, we know that the average of the samples is an unbiased estimator for the expectation. Therefore if $\gamma_1, \gamma_2, \ldots, \gamma_N$ are samples following the distribution p, then

$$\frac{\sum_{i=1}^{N} \frac{M}{t(\gamma_i)}}{N} \tag{6.36}$$

is an unbiased estimator for the number of satisfying assignments of Φ. ■

There is one more important property of function $\tilde{g}$ in the previous example. The smallest value of $\tilde{g}$ cannot be smaller than $\frac{M}{k}$, where k is the number of clauses and the largest value of $\tilde{g}$ cannot be larger than M. What follows is that the variance

$$V_p[\tilde{g}] := \sum_{x \in X} (\tilde{g}(x) - E_p[\tilde{g}])^2 p(x) \qquad (6.37)$$

cannot be larger than M^2. Indeed, the variance is the expectation of the squared values minus the squared expectation. Then an upper bound on the variance is the expectation of the squared values, and then an upper bound for this latter is the maximum squared value. That is, M^2. On the other hand, we know that the expectation cannot be smaller than the smallest value, that is $\frac{M}{k}$. Therefore the standard deviation of the $\tilde{g}$ values under the distribution p is at most k times more than the expected value. What follows is that a small number of samples, say $O(k^3)$ samples, is sufficient for the standard error to be much smaller than the expectation. This means that it is computationally efficient to estimate the number of satisfying assignments of a disjunctive normal form using the introduced importance sampling.

It is quite exceptional that a rejection sampling or an importance sampling provides a good estimation of the size of the space of interest, although we can generate artificial examples where the importance sampling has a much smaller standard deviation (actually, 0 variance and thus 0 deviation) than sampling from the correct distribution.

Example 15. *Consider the biased cube C with which throwing value k has probability proportional to k. Work out the importance sampling method to estimate the expected value of the unbiased cube using samples from the biased cube C.*

Solution. The sum of the possible values is 21, therefore if ξ denotes the thrown value, then

$$P(\xi = k) = \frac{k}{21}.$$

This is the sampling distribution p. The desired distribution π is the uniform distribution, where each outcome has probability $\frac{1}{6}$. The function f assigns a value k to the event "a value k has been thrown". Then the g function we are looking for is

$$g(k) = \frac{\pi(k)}{p(k)} k = \frac{\frac{1}{6}}{\frac{k}{21}} k = \frac{21}{6} = 3.5.$$

Namely, g is the constant 3.5 function, which is the expectation of the thrown value on an unbiased die. What follows is that this importance sampling has 0 variance! Namely, one sample from the sampling distribution provides an exact estimation of the expectation in the target distribution. ■

Example 15 is artificial since there, p is proportional to f and π is the uniform distribution, therefore for any x, $p(x)$ is proportional to $f(x)\pi(x)$. The normalizing constant is exactly the expectation of f under distribution π that we are looking for. However, when we do not know this expectation, it is very unlikely that the sampling distribution is close to the product of the function f and the target probability distribution π.

Rather, in many cases, the sampling distribution is built on sequentially, making a random choice in each step. This is especially the case when the space from which we would like to sample contains modular objects like trees, sequences, etc. Then in each step there is some deviation from the target distribution, and these errors are multiplied along the random generations. What follows is that a so-called *sequential importance sampling* might have some extremely large values in its g function for those x, for which

$$p(x) \ll \pi(x). \tag{6.38}$$

Therefore, the variance of the g function might be extremely big, making the method computationally intractable. In many of those cases, Markov chain Monte Carlo methods can help, see Section 6.6.

6.4 Sampling with algebraic dynamic programming

In this section, we prove that if an ensemble of combinatorial objects can be counted with algebraic dynamic programming in polynomial time, then it can be uniformly sampled also in polynomial time. We prove slightly more: if an algebraic dynamic programming computes the sum of the (non-negative) weights of some combinatorial objects in polynomial time, then it is possible to sample those combinatorial objects from the distribution proportional to their weights also in polynomial time.

Theorem 63. *Let $E = (Y, R, f, \mathcal{T})$ be an evaluation algebra with the following properties.*

(a) R is $\mathbb{R}^+ \cup \{0\}$, the non-negative real numbers semiring.

(b) The evaluation algebra solves a counting problem in polynomial time in the sense of Theorem 21.

(c) For any $T_i \in \mathcal{T}$, if $\circ_i$ is an m-ary operation, then

$$T_i(f(a_1), \ldots, f(a_m); p(a_1), \ldots, p(a_m)) := c_{p(a_1),\ldots,p(a_m)} \prod_{j=1}^{m} f(a_j).$$

For any problem instance x, let θ denote the parameter for which the solution of x is $F(S(\theta))$. We further assume that for any $\theta' \in B \cap \theta_\downarrow$, sampling from the distribution

$$\pi(a) := \frac{f(a)}{F(S(\theta'))} \tag{6.39}$$

can be done in polynomial time. Then it is possible to sample from $S(\theta)$ following the distribution

$$\pi(a) := \frac{f(a)}{F(S(\theta))} \tag{6.40}$$

in polynomial time.

Proof. We exhibit a recursive algorithm that generates samples from the prescribed distribution. Let i denote the indexes in the computation

$$F(S(\theta)) = \sum_i T_i(F(S(\theta_{i,1}, \ldots, \theta_{i,m_i}; \theta_{i,1}, \ldots, \theta_{i,m_i}). \tag{6.41}$$

We claim that the following algorithm samples from π:

1. Sample a random i following the distribution

$$p(i) := \frac{T_i(F(S(\theta_1)), \ldots, F(S(\theta_{m_i})); \theta_1, \ldots, \theta_{m_i})}{F(S(\theta))}.$$

2. For each $\theta_j \in \{\theta_1, \ldots, \theta_{m_i}\}$, sample a random a_j following the distribution

$$\pi_j(a_j) := \frac{f(a_j)}{F(S(\theta_j))}.$$

3. Return with $\circ_{i\,j=1}^{m_i} a_j$.

Indeed, the probability for sampling $\circ_{i\,j=1}^{m_i} a_j$ is

$$p(i) \prod_{j=1}^{m_i} \frac{f(a_j)}{F(S(\theta_j))} =$$

$$\frac{c_{\theta_1,\ldots,\theta_{m_i}} \prod_{j=1}^{m_i} F(S(\theta_j))}{F(S(\theta))} \prod_{j=1}^{m_i} \frac{f(a_j)}{F(S(\theta_j))} =$$

$$\frac{c_{\theta_1,\ldots,\theta_{m_i}} \prod_{j=1}^{m_i} f(a_j)}{F(S(\theta_j))} = \pi(\circ_{i\,j=1}^{m_i} a_j). \tag{6.42}$$

Sampling from each π_j can be done in the same way. Therefore, the following recursive algorithm samples a random a from the prescribed distribution:

sampler(θ)
    if $\theta \notin B$
        Generate a random i following the distribution

$$p(i) := \frac{T_i(F(S(\theta_1)), \ldots, F(S(\theta_{m_i})); \theta_1, \ldots, \theta_{m_i})}{F(S(\theta))}.$$

    for each $j \in \{1, 2, \ldots, m_i\}$
        $a_j :=$ sampler(θ_j)
    return $\circ_i{}_{j=1}^{m_i} a_j$
else
    return a following distribution $\frac{f(a)}{F(S(\theta))}$.

We have to prove that the presented recursive algorithm runs in polynomial time. Since $F(S(\theta))$ can be computed in polynomial time, so can $F(S(\theta'))$ for any $\theta' \in \theta_{\downarrow}$. Furthermore, the number of parameters covered by a particular parameter are also polynomially upper bounded as well as the range of the indexes in Equation (6.41). Therefore, sampling from the distribution

$$p(i) := \frac{T_i(F(S(\theta_1)), \ldots, F(S(\theta_{m_i})); \theta_1, \ldots, \theta_{m_i})}{F(S(\theta))}$$

can be done in polynomial time. Also, the number of times the recursive function calls itself is polynomial bounded. Finally, the base cases $\theta \in B$ also can be handled in polynomial time due to the conditions of the theorem. Therefore, the overall running time grows polynomially. $\qquad\square$

We show two examples of how Theorem 63 can be applied.

Example 16. *Let $G = (T, N, \mathfrak{S}, R, \pi)$ be a stochastic regular grammar, and let X be a sequence of length n from alphabet T. Give a random sampling method that samples generations of X following the distribution*

$$p(g) := \frac{\prod_{r \in g} \pi(r)^{m(g,r)}}{P(X)}$$

where $m(g, r)$ is the multiplicity of the rule r in the generation g generating X, and $P(X)$ is the probability that the regular grammar generates X.

Solution. Recall that $P(X)$, that is, the sum of the probabilities of the generations of X can be computed with algebraic dynamic programming, see Subsection 2.3.1. Applying Theorem 63 on that algebraic dynamic programming leads to the following two-phase method.
Phase I. Fill in a dynamic programming table $d(i, W)$ for all $i = 0, 1 \ldots, n$ and $W \in N$ with the initial conditions

$$d(0, W) = \begin{cases} 1 & \text{if } W = \mathfrak{S} \\ 0 & \text{otherwise} \end{cases}$$

and recursions

$$d(i, W) = \sum_{W' \in N} d(i - 1, W') \pi(W' \to x_i W)$$

$$d(i, \varepsilon) = \sum_{W \in N} d(i-1, W)\pi(W \to x_i) + \sum_{W \in N} d(i, W)\pi(W \to \varepsilon).$$

Phase II. Draw a random rewriting rule r in the form $W \to x_n$ or $W \to \varepsilon$ with probability

$$\frac{d(n-1, W)\pi(W \to x_n)}{d(n, \varepsilon)}$$

or

$$\frac{d(n, W)\pi(W \to \varepsilon)}{d(n, \varepsilon)}.$$

Set $g := r$ and set i to $n-1$ if the selected rule is in the form $W \to x_n$ and set i to n if the selected rule is in the form $W \to \varepsilon$. Set W' to the non-terminal in the selected rule. Then do the following iteration. While i is not 0, select a random rewriting rule $W \to x_i W'$ with probability

$$\frac{d(i-1, W)\pi(W \to x_i W')}{d(i, W')}.$$

Let $g := W \to x_i W', g$. Set i to $i-1$ and set W' to W. ∎

Example 17. *Let $G = (T, N, \mathfrak{S}, R, \pi)$ be a stochastic context-free grammar in Chomsky Normal Form, and let X be a sequence of length n from alphabet T. Give a random sampling method that generates a random parse tree $\mathcal{T}$ following the distribution*

$$p(\mathcal{T}) \propto \prod_{r \in \mathcal{T}} \pi(r)^{m(\mathcal{T}, r)},$$

where $m(\mathcal{T}, r)$ is the multiplicity of the rewriting rule r in the parse tree $\mathcal{T}$ generating X, and $\propto$ stands to "proportional to".

Solution. Recall that the sum

$$\sum_{\mathcal{T} \mid \mathcal{T} \text{ generates } X} \prod_{r \in \mathcal{T}} \pi(r)^{m(\mathcal{T}, r)}$$

can be computed with algebraic dynamic programming, see Subsection 2.3.3. Applying Theorem 63 on that algebraic dynamic programming leads to the following two-phase method.

Phase I. Fill in a dynamic programming table $d(i, j, W)$ for all $1 \geq i \geq j \geq n$ and $W \in N$ with the initial condition

$$d(i, i, W) = \pi(W \to x_i)$$

and recursion

$$d(i, j, W) = \sum_{i \geq k < j} \sum_{W_1 \in N} \sum_{W_2 \in N} d(i, k, W_1) d(k+1, j, W_2)\pi(W \to W_1 W_2).$$

Phase II. The random tree is obtained by the following recursive function calling it with parameters $(1, n, \mathfrak{S})$.

TreeSampler(i, j, W)

    if $i < j$

        Generate a random k, W_1 and W_2 following the distribution

$$\frac{d(i, k, W_1)d(k+1, j, W_2)\pi(W \rightarrow W_1 W_2)}{d(i, j, W)}$$

        Let $\mathcal{T}_1 := $**TreeSampler**$(i, k, W_1)$

        Let $\mathcal{T}_2 := $**TreeSampler**$(k+1, j, W_2)$

        Generate a tree $\mathcal{T}$ by merging $\mathcal{T}_\infty$ and $\mathcal{T}_\in$ with rule $W \rightarrow W_1 W_2$

        return $\mathcal{T}$

    else

        return $\mathcal{T} = W \rightarrow x_i$

                                                                     ■

6.5   Sampling self-reducible objects

In this section, we discuss a very important class of counting problems, the self-reducible counting problems. Roughly speaking, a counting problem is self-reducible if any beginning of any solution of a problem instance can be extended as the solutions of another problem instance. This other problem instance must have a size comparable with the original problem instance. Furthermore, the solutions can multifurcate only polynomially at any step. Many natural counting problems are self-reducible. For example, if we already selected a few edges participating in a perfect matching in a graph, then the possible extensions of this subset of edges are the perfect matchings of the remaining graph. Formally, the self-reducible counting problems are defined in the following way.

Definition 50. *A counting problem is* self-reducible *if the following holds:*

1. *There exists a relation $R \subseteq \Sigma^* \times \Sigma^*$ such that whenever xRy, then x describes a problem instance and y describes a solution.*

2. *There exists a polynomial time computable function $g : \Sigma^* \rightarrow \mathbb{N}$ such that $xRy \Rightarrow |y| = g(x)$. Furthermore, $g(x) = poly(|x|)$.*

3. *There exist polynomial time computable functions $\phi : \Sigma^* \times \Sigma^* \rightarrow \Sigma^*$ and $\sigma : \Sigma^* \rightarrow \mathbb{N}$ satisfying*

    (a) $\sigma(x) = O(\log(|x|))$,

(b) $g(x) > 0 \Rightarrow \sigma(x) > 0$,

(c) $|\phi(x, w)| \le |x|$,

(d) $xRy_1y_2 \ldots y_n \Leftrightarrow \phi(x, y_1y_2 \ldots y_{\sigma(x)})Ry_{\sigma(x)+1} \cdots y_n$.

The solution space of a problem instance x of a reducible problem can be represented by a rooted tree called a *count tree* in the following way. The root of the tree is labeled with x. For any node v of the tree, if the node is labeled by x' and $g(x') > 0$, then v is an internal node, and the number of outgoing edges of v is the number of sequences w of length $\sigma(x')$ that there exists a sequence z such that $x'Rwz$. The children of v are labeled with $\phi(x', w)$, and the edge (v, u) is labeled by the sequence w if u is the child of v and is labeled by $\phi(x', w)$. The number of solutions of x is the number of leaves of this tree, and for any x' labeling v, the number of solutions of x' is the number of leaves of the subtree rooted in v. The function σ is called the granulation function. We cannot expect that we can assign a meaning to an arbitrary suffix of a sequence representing a solution, but we require that we can "granulate" sequences representing the solutions such that each appearing suffix is meaningful. For example, the sequence

$$z = \text{``}(v_3, v_4), (v_5, v_6)\text{''}$$

represents a pair of edges, and they form a suffix of the string

$$y = \text{``}(v_1, v_2), (v_3, v_4), (v_5, v_6)\text{''}$$

that represents a perfect matching in K_6, and the two edges described in z form a perfect matching of the complete graph on vertices $\{v_3, v_4, v_5, v_6\}$. On the other hand, the sequence

$$\text{``}_4), (v_5, v_6)\text{''}$$

is meaningless from the same point of view, although it is still a suffix of y. It is natural to allow σ to grow as the logarithm of the size of the problem instance. Indeed, in many cases, both the problem instance and the solution contain indexes, the size of them grows polynomially with the input size, thus the number of characters necessary to write down these indexes grows logarithmically. On the other hand, we do not want to allow larger granulation functions, since we want self-reducible problems to be locally explorable in polynomial time. That is, the number of children of any internal node in the tree representing the solution space can grow only polynomially with the size of the problem instance.

Self-reducible objects are easy to sample if they are easy to count as the following theorem states.

Theorem 64. *If a self-reducible counting problem is in FP, then there exists a sampling algorithm that runs in polynomial time and generates solutions from the uniform distribution.*

Proof. Let x be a problem instance of a self-reducible counting problem, and let $l(v)$ denote the problem instance labeling vertex v in the count tree of x. Further, let $f(x)$ denote the number of solutions of a problem instance x. A random solution following the uniform distribution can be generated with the following recursion.

Set v to the root of its count tree, and set y to the empty sequence. While $g(l(v)) > 0$, select a random node u from the children of v following the distribution

$$\frac{f(l(u))}{f(l(v))}. \tag{6.43}$$

Extend y with the label of the edge (v, u) and set v to u.

It is trivial to see that this recursion generates uniformly a random solution. It is also easy to see that the recursion runs in polynomial time. Indeed, a vertex in the count tree has a polynomial number of children that can be obtained in polynomial time. These vertices are labeled with problem instances that cannot be larger than x, therefore the probabilities in Equation (6.43) can be computed in polynomial time, and thus, a random child can be drawn in polynomial time. Since the depth of the count tree is also a polynomial function of $|x|$, the recursion ends in polynomial number of steps, thus, the overall running time is also a polynomial function of $|x|$. □

If a self-reducible counting problem is in FP, it does not necessarily mean that it has an algebraic dynamic programming computing the number of solutions in polynomial time. Indeed, according to the work of Jerrum and Snir, there is no polynomial monotone circuit that computes the spanning tree polynomial. On the other hand, it is easy to show that the number of spanning trees is a self-reducible counting problem, and the number of spanning trees of a graph can be computed in polynomial time. What follows is that it is possible to generate uniformly a random spanning tree of a graph in polynomial time as the following example shows.

Example 18. *Let $G = (V, E)$ be an arbitrary graph. Generate uniformly a random spanning tree of G.*

Solution. Fix an arbitrary total ordering of the edges of G. Any spanning tree is a subset of edges of G that can be described by a 0-1 vector of length $|E|$, where 0 at position i means the absence of the ith edge and 1 means the presence of the ith edge in the spanning tree.

Let $e = (u, v)$ be an edge in G. We will denote by G^{-e} the (possibly multi)graph that is obtained by contracting the edge e, and having multiple edges between w and x if both u and v are adjacent to w, where x is the new vertex appearing in the edge contraction. If G itself is a multigraph, then the number of edges between w and x is the sum of the number of edges between w and u and the number of edges between w and v. Observe the following. The number of spanning trees in G that contain edge e is the number of spanning trees in G^{-e} in which two spanning trees are distinguished if they contain different edges between w and x.

Also observe that the number of spanning trees in G that do not contain edge e is the number of spanning trees in $G \setminus \{e\}$. Therefore the spanning trees of G can be described with the following count tree. Each vertex v is labeled by a (possibly multi)graph G'. If G' has edges, then v has two children labeled by $G'^{-e'}$ and $G' \setminus \{e'\}$, where e' is the smallest edge in G'. Further, the two edges connecting v to its children are labeled by 1 and 0.

Since the number of spanning trees of multigraphs can be computed in polynomial time, this count tree can be used to uniformly generate random spanning trees of G in polynomial time. ∎

6.6 Markov chain Monte Carlo

Although Buffon already invented random computations in the 18th century, random computations became widespread only after computers were invented. Soon after building the first electronic computer, Metropolis and his coworkers invented Markov chain Monte Carlo methods [128]. Their algorithm was widely used by physicists and chemists. Hastings generalized the Metropolis algorithm in 1970 [91], what we know today as the Metropolis-Hastings algorithm.

Definition 51. *A homogeneous, discrete time, finite space Markov chain is a random process described with the pair (X, T), where X is a finite set called* state space *and T is a function mapping from $X \times X$ to the non-negative real numbers. The members of X are called* states. *The random process is a random walk on the states starting at time $t = 0$. The random walk makes one step in each unit of time. If $x, y \in X$, then $T(y|x)$ is the conditional probability that the process jumps to state y given that it is now in state x. The random walk might jump from a state x back to state x, namely, it is allowed that $T(x|x)$ is greater than 0. However, $T(\cdot|x)$ must be a probability distribution, so we require that for any x,*

$$\sum_{y \in X} T(y|x) = 1 \qquad (6.44)$$

hold. T is also called transition probabilities. *The Markov chain can be described with a directed graph called a* Markov graph. *$\vec{G} = (V, E)$ is the Markov graph of the Markov chain $M = (X, T)$, if V represents the states of the Markov chain, and there is an edge from v_x to v_y if the corresponding transition probability $T(y|x)$ is greater than 0.*

The Markov chain is irreducible *if its Markov graph is strongly connected.*

A Markov chain is aperiodic *if the greatest common divisor of the directed cycle lengths of the Markov graph is* 1. *A Markov chain is* reversible *with respect to a distribution* $\boldsymbol{\pi}$, *if the* detailed balance *is satisfied, that is, for any* $x, y \in X$, *equation*

$$\pi(x)T(y|x) = \pi(y)T(x|y) \tag{6.45}$$

holds.

Throughout the book, all introduced Markov chains will be homogeneous, discrete time and finite space Markov chains, so we simply will refer to them as Markov chains.

The process might start in a deterministic state or in a random state described with a distribution $\mathbf{x}_0$. A deterministic state might also be considered as a distribution in which the distinguished state has probability 1 and all other states have probability 0. We can write the transition probabilities into a matrix $\mathbf{T} = \{t_{i,j}\}$, where $t_{i,j} = T(v_i|v_j)$. Then the distribution of the states after one step in the Markov chain is

$$\mathbf{x}_1 = \mathbf{x}_0\mathbf{T} \tag{6.46}$$

and generally, after t steps,

$$\mathbf{x}_t = \mathbf{x}_0\mathbf{T}^t. \tag{6.47}$$

The central question in the theory of Markov chains is whether the random process converges and if it converges, what is its limit? There are several ways to define convergence of distributions; in the theory of Markov chains, measuring the distance of distributions in total variation distance is the most common way.

Definition 52. *A distribution* $\boldsymbol{\pi}$ *is an* equilibrium distribution *of the Markov chain* $M = (X, T)$ *if for all* $x \in X$, *equation*

$$\pi(x) = \sum_{y \in X} \pi(y)T(x|y) \tag{6.48}$$

holds.

There are two types of convergence. The local convergence means converence from some given starting distribution. Global convergence means convergence from an arbitrary starting distribution.

Definition 53. *A Markov chain converges to distribution* $\boldsymbol{\pi}$ *from a distribution* $\mathbf{x}_0$ *if*

$$\lim_{t \to \infty} d_{TV}(\mathbf{x}_0\mathbf{T}^t, \boldsymbol{\pi}) = 0. \tag{6.49}$$

A stationary distribution $\boldsymbol{\pi}$ *is* globally stable *if the Markov chain converges to* $\boldsymbol{\pi}$ *from an arbitrary distribution.*

The following theorem is a central theorem providing globally stable stationary distributions.

Theorem 65. *Let $M = (X, T)$ be an irreducible and aperiodic Markov chain, and also reversible with respect to the distribution π. Then π is a globally stable stationary distribution.*

We are not going to prove this theorem here, but the proof can be found in many standard textbooks on Markov chains [25, 121]. Below we provide an example for a reversible, irreducible, and aperiodic Markov chain.

Example 19. *Fix a positive integer n. Let X be the set of Dyck words of length $2n$. Provide an irreducible, aperiodic Markov chain which is reversible with respect to the uniform distribution over X, thus, the uniform distribution is the globally stable stationary distribution.*

Solution. Consider the following random perturbation. Let $D \in X$ be an arbitrary Dyck word, which is the current state of the Markov chain. Generate a random i uniformly from $[1, 2n - 1]$. Let D' be the word that we get by swapping the characters in position i and $i + 1$. If D' is also a Dyck word, then the next state in the Markov chain will be D', otherwise, it will be D.

It is indeed a Markov chain, since the Markov property holds, that is, what the next state is depends on only the current state and not on previous states. We claim that this chain is irreducible, aperiodic, and reversible with respect to the uniform distribution.

To prove that the Markov chain is irreducible, first observe that if there is a random perturbation perturbing D_1 to D_2, then there is also a random perturbation from D_2 to D_1. Then it is sufficient to show that any Dyck word can be transformed to a reference Dyck word D_0. Let D_0 be the Dyck word

$$\underbrace{x \ldots x}_{n} \ \underbrace{y \ldots y}_{n}.$$

Any Dyck word can be easily transformed into D_0. Indeed, let D be a Dyck word, and let i be the smallest index such that D contains a y in position i. If $i = n + 1$, then $D = D_0$, and we are ready. Otherwise there is a position i' such that the character in position i' is x, and all characters in position $i, i + 1, \ldots, i' - 1$ are y. Then we can swap characters in positions $i' - 1$ and i', then in positions $i' - 2$ and $i' - 1$, and so on, finally in positions i and $i + 1$. Then now the smallest index where there is a y in the Dyck word is $i + 1$. Therefore in a finite number of steps, the smallest index where there is a y will be $n + 1$, and then we transform the Dyck word into D_0. Then any Dyck word D_1 can be transformed into D_2, since both D_1 and D_2 can be transformed into D_0, and the reverse way of transformations from D_0 to D_2 are also possible transformation steps in the Markov chain. Then we can transform D_1 into D_0 and then D_0 into D_2.

To see that the Markov chain is aperiodic, it is sufficient to show that there are loops in the Markov graph. Indeed, any Dyck word ends with a character y, therefore whenever the random i is $2n - 1$, the Markov chain remains in the same state. Therefore, in the Markov graph, there is a loop on each vertex, that is, there are cycles of length 1. Then the greatest common divisor of the cycle lengths is 1.

We are going to prove that the Markov chain is reversible with respect to the uniform distribution. We only have to show that for any Dyck words D_1 and D_2,

$$T(D_2|D_1) = T(D_1|D_2) \tag{6.50}$$

since in the uniform distribution

$$\pi(D_1) = \pi(D_2) \tag{6.51}$$

naturally holds. It is easy to see that Equation (6.50) holds. If D_1 cannot be transformed into D_2 in a single step, and in this case D_2 cannot be transformed into D_1 in a single step, then both transition probabilities are 0. If D_1 can be transformed into D_2 by swapping the characters in positions i and $i+1$ for some i, then D_2 can be transformed back to D_1 by also swapping the characters in positions i and $i+1$. Both transformations have the same probability, $\frac{1}{2n-1}$, therefore they are equal. ∎

It turns out that essentially any Markov chain can be transformed into a Markov chain that converges to a prescribed distribution π. The technique is the Metropolis-Hastings algorithm described in the following theorem.

Theorem 66. *Let $M = (X, T)$ be an irreducible and aperiodic Markov chain. Furthermore, we require that for any $x, y \in X$, the property*

$$T(y|x) \neq 0 \Longrightarrow T(x|y) \neq 0 \tag{6.52}$$

holds. Let π be a non-vanishing distribution on X. We require that for any x, $\pi(x)$ can be calculated, possibly up to an unknown normalizing constant. In other words, for any $x, y \in X$, the ratio $\frac{\pi(y)}{\pi(x)}$ can be calculated. Then the following algorithm, called the Metropolis-Hastings algorithm, *defines a Markov chain, which is irreducible, aperiodic, and reversible with respect to the distribution π.*

1. *Let the current state be x_t. Generate a random y following the conditional distribution $T(\cdot|x)$. That is, generate a random next state in the Markov chain M assuming that the current state is x.*

2. *Generate a random real number u following the uniform distribution on the interval $[0, 1]$. The next state of the Markov chain, x_{t+1}, is y if*

$$u \leq \frac{\pi(y)T(x|y)}{\pi(x)T(y|x)} \tag{6.53}$$

and it is x_t otherwise.

Proof. The algorithm generates a Markov chain, since the Markov property holds, that is, the next state depends on only the current state and not on the previous states. Let this defined Markov chain be denoted by M'. The ratio in Equation (6.53) is positive, since the distribution π is non-vanishing and $T(x|y)$ cannot be 0 when y is proposed from x due to the required property in Equation (6.52). What follows is that M and M' have the same Markov graph. However, in that case, M' is irreducible and aperiodic, since M was also irreducible and aperiodic. We have to show that M' is reversible with respect to the distribution π. First we calculate the transition probabilities in M'. The way to jump from state x to y in M' is to first propose y when the current state is x. This has probability $T(y|x)$. Then the proposed state y has to be accepted. The acceptance probability is 1 if the ratio in Equation (6.53) is greater than or equal to 1, and the acceptance probability is the ratio itself if it is smaller than 1. Therefore, we get that

$$T'(y|x) = T(y|x) \min \left\{ 1, \frac{\pi(y)T(x|y)}{\pi(x)T(y|x)} \right\}. \tag{6.54}$$

It immediately follows that M' is reversible with respect to π. Indeed,

$$\pi(x)T'(y|x) = \pi(x)T(y|x) \min \left\{ 1, \frac{\pi(y)T(x|y)}{\pi(x)T(y|x)} \right\} =$$
$$\min \left\{ \pi(x)T(y|x), \pi(y)T(x|y) \right\}. \tag{6.55}$$

Namely, $\pi(x)T'(y|x)$ is symmetric to x and y, therefore,

$$\pi(x)T'(y|x) = \pi(y)T'(x|y) \tag{6.56}$$

that is, the detailed balance holds. □

6.7 Exercises

1. Generate a random variable following the normal distribution with mean μ and variance σ^2.

2. Generate a random variable following the Pareto distribution with parameters x_m and α. The support of the Pareto distribution is $[x_m, \infty)$ and its cumulative density function is

$$1 - \left(\frac{x_m}{x} \right)^\alpha.$$

3. * Generate uniformly a random triangulation of a convex n-gon.

4. Generate uniformly a random alignment of two sequences.

5. ○ Let $A, B \in \Sigma^*$ be given together with a similarity function s that maps from $(\Sigma \cup \{-\}) \times (\Sigma \cup \{-\}) \setminus \{-, -\}$ to the real numbers. Generate uniformly an alignment from the set of maximum similarity alignments.

6. ○ Generate uniformly a random perfect matching of a planar graph.

7. Generate uniformly a random Dyck word of length $2n$.

8. Let $G = (V, E)$ be a planar graph, and let $w : E \to \mathbb{R}^+$ be the edge weights. Generate a random perfect matching of G following the distribution that is proportional to the product of the edge weights.

9. Let Σ be a finite alphabet, and let $m : \Sigma \times \Sigma \to \mathbb{Z}^+ \cup \{0\}$ be an arbitrary function. Generate uniformly a random sequence that contains the $\sigma_1 \sigma_2$ substring $m(\sigma_1, \sigma_2)$ times. The running time must be polynomial with the length of the generated sequence.

10. * Apply the rejection method to generate random variables from the tail of a normal distribution. That is, the target distribution has probability density function

$$\frac{\frac{1}{\sqrt{2\pi}} e^{-\frac{x^2}{2}}}{\Phi(-a)}$$

on the domain $[a, \infty)$, $a > 0$. Use the shifted exponential distribution for the auxiliary distribution and find the best enveloping constant.

11. Show that for any two distributions π and p, the inequality

$$d_{TV}(\pi, p) \leq 1$$

holds.

12. * Show that for any two distributions π and p, the equality

$$d_{TV}(\pi, p) = \max_{A \subset X} \sum_{x \in A} (\pi(x) - p(x))$$

holds.

13. Show that the total variation distance is indeed a distance.

14. * Develop a Markov chain Monte Carlo method that converges to the uniform distribution of the possible (not necessarily perfect) matchings of a graph.

15. Develop a Markov chain Monte Carlo method that converges to the uniform distribution of the spanning trees of a graph.

16. ○ Develop a Markov chain Monte Carlo method that converges to the uniform distribution of permutations of length n.

6.8 Solutions

Exercise 3. Let $v_1, v_2, \ldots v_n$ denote the vertices of the polygon. A triangulation can be expressed with the list of edges participating in the triangulation.

We give a recursive function that generates a random triangulation of a polygon defined with a set of edges. For this, we need to define the following family of distributions. The domain of the distribution p_i is $\{0, 1, \ldots, i-1\}$ and

$$p_i(j) := \frac{C_j C_{i-j}}{C_i},$$

where C_i is the ith Catalan number. (Since $C_i = \sum_{j=0}^{i-1} C_j C_{i-j}$, this is indeed a distribution). The recursive function is the following:

    triangulator$(V = \{v_1, v_2, \ldots, v_n\})$
        if $|V| \geq 4$
            Generate a random j from the distribution $p_{|V|-3}$
            $E_0 := \emptyset$
            if $j \neq n$
                $E_0 := E_0 \cup \{(v_1, v_j)\}$
            if $j \neq 3$
                $E_0 := E_0 \cup \{(v_2, v_j)\}$
            $E_1 :=$**triangulator**$(v_1, v_j, v_{j+1}, \ldots, v_n)$
            $E_2 :=$**triangulator**$(v_2, v_3, \ldots, v_j)$
            return $E_0 \cup E_1 \cup E_2$
        else
            return $\emptyset$.

Exercise 5. Generate a directed acyclic graph whose vertices are the entries of the dynamic programming table computing the most similar alignment between the two sequences, and there is an edge from $d(i_1, j_1)$ to $d(i_2, j_2)$ if $d(i_2, j_2)$ sends an optimal value to $d(i_1, j_1)$ in the dynamic programming recursion (a similarity value or a gap penalty is added to $d(i_2, j_2)$ to get $d(i_1, j_1)$). The optimal alignments are the paths from $d(n, m)$ to $d(0, 0)$, where n and m are the length of the two sequences. Thus, the task is to sample uniformly a path between two vertices in an acyclic graph.

Exercise 6. Let $G = (V, E)$ be a planar graph. Fix an arbitrary total ordering on E, and let (v_i, v_j) be the smallest edge. Observe that the number of perfect matchings containing (v_i, v_j) is the number of perfect matchings in $G \setminus \{v_i, v_j\}$ (that is, we remove vertices v_i and v_j from G, together with all the edges incident to v_i or v_j), and the number of perfect matchings not containing (v_i, v_j) is the number of perfect matchings in $G \setminus \{(v_i, v_j)\}$ (that is, we remove the edge (v_i, v_j) from G).

Exercise 10. It is easy to see that if

$$g = ae^{-a(x-a)}$$

and

$$c = \frac{e^{\frac{-a^2}{2}}}{\sqrt{2\pi a}\Phi(-a)}$$

then

$$cg(x) \geq f(x)$$

for all $x \geq a$. Therefore, we can use g and f in a rejection sampling. When the generated random number is x, the acceptance probability is

$$\frac{f(x)}{cg(x)} = \frac{e^{\frac{-x^2}{2}}}{e^{\frac{-a^2}{2}}e^{-a(x-a)}} = e^{-\frac{(x-a)^2}{2}}.$$

The expected acceptance probability is

$$\int_{x=a}^{\infty} ae^{-a(x-a)}e^{-\frac{(x-a)^2}{2}} = \sqrt{\frac{\pi}{2}}ae^{\frac{a^2}{2}}erfc\left(\frac{a}{\sqrt{2}}\right),$$

where $erfc$ is the complementary error function of the normal distribution. It can be shown that the expected acceptance probability grows strictly monotonously with a, and it is ≈ 0.65567 when $a = 1$.

Exercise 12. Let B be the set of points for which $\pi(x) - p(x) \geq 0$. Observe the following.

1. It holds that

$$\sum_{x \in B}(\pi(x) - p(x)) = \max_{A \subset X} \sum_{x \in A}(\pi(x) - p(x)).$$

2. It also holds that

$$\sum_{x \in B}(\pi(x) - p(x)) = -\sum_{x \in \overline{B}}(\pi(x) - p(x)).$$

Since for any $x \in \overline{B}$, $-(\pi(x) - p(x)) = |\pi(x) - p(x)|$, and for any $x \in B$, $(\pi(x) - p(x)) = |\pi(x) - p(x)|$, it holds that

$$\sum_{x}|\pi(x) - p(x)| = 2\sum_{x \in B}(\pi(x) - p(x)) = 2\max_{A \subset X}\sum_{x \in A}(\pi(x) - p(x)).$$

Dividing both ends of this equality by 2, we get the equality to be proved.

Exercise 14. For example, the following approach works. Let $G = (V, E)$ be a graph. Set M to the empty set as a starting state of the Markov chain (observe that the empty matching is also a matching). Then a step in the Markov chain is the following.

1. Draw uniformly a random edge e from E.

2. If $e \in M$, then remove e from M.

3. If $e \notin M$ and e is not adjacent to any edge in M, then add e to M.

It is easy to see that for any matchings M_1 and M_2, the transition probability from M_1 to M_2 equals the transition probability from M_2 to M_1. Indeed, if M_1 and M_2 differ by more than one edge, then the transition probability is 0. Otherwise the presented/missing edge is chosen from E with the same probability, thus, the transition probabilities are the same.

We also have to show that the Markov chain is irreducible. This is clearly true since the empty matching can be obtained from any matching by deleting all the edges, and any matching can be obtained from the empty matching by adding the edges of the matching in question.

Exercise 16. Applying a random transposition will work, however, it is important to notice that a transposition changes the parity of the permutation. That is, if the transitions of a Markov chain are the random transpositions, then the Markov chain is periodic. Applying the lazy Markov chain technique solves this problem.

Chapter 7

Mixing of Markov chains and their applications in the theory of counting and sampling

Since it is very easy to design irreducible, aperiodic, and reversible Markov chains that converge to a prescribed distribution over a finite space, we focus on them. Such Markov chains have the property that they have a globally stable stationary distribution. Furthermore, their eigenvalues are all real.

Theorem 67. *If M is a reversible Markov chain on a finite state space, then all of its eigenvalues are real, and fall on the interval $[-1, 1]$.*

Proof. It comes from the general theory of Markov chains that any eigenvalue of a Markov chain falls onto the unit circle of the complex plane. Therefore, it is sufficient to show that each eigenvalue is real. Indeed, let T denote the transition matrix of M. Since M is reversible, there exists a distribution π such that it holds for any pair of states (x_i, x_j) that

$$\pi(x_i)T(x_j|x_i) = \pi(x_j)T(x_i|x_j) \tag{7.1}$$

and therefore it also holds that

$$\sqrt{\frac{\pi(x_i)}{\pi(x_j)}}T(x_j|x_i) = \sqrt{\frac{\pi(x_j)}{\pi(x_i)}}T(x_i|x_j). \tag{7.2}$$

That is, the matrix

$$\Pi^{-\frac{1}{2}}T\Pi^{\frac{1}{2}} \tag{7.3}$$

is symmetric, where Π is the diagonal matrix containing the $\pi(x_i)$ values in the diagonal. Any symmetric real matrix can be diagonalized and has only real eigenvalues. That is, it holds that

$$\Pi^{-\frac{1}{2}}T\Pi^{\frac{1}{2}} = W\Lambda W^{-1} \tag{7.4}$$

for some matrix W and diagonal, real matrix Λ. In that case, T can also be diagonalized and has all real eigenvalues, since

$$T = \Pi^{\frac{1}{2}}W\Lambda W^{-1}\Pi^{-\frac{1}{2}}. \tag{7.5}$$

$$\square$$

In this chapter, any Markov chain is considered to be irreducible, aperiodic, and reversible, and we will not mention this later on. We also fix the following notations that will be used throughout the chapter. The state space of the Markov chain is denoted by X. The transition matrix of a Markov chain is denoted by T, and the eigenvalues are denoted by $1 = \lambda_1 > \lambda_2 \geq \ldots \geq \lambda_r$, where $r = |X|$. We are going to show that λ_2, the second-largest eigenvalue, tells us if a Markov chains provides an FPAUS algorithm. There are techniques to give lower and upper bounds on the second-largest eigenvalue. We use this theory to prove a dichotomy theory for self-reducible counting problems. Any self-reducible counting problem is either in FPAUS or essentially cannot be approximated in polynomial time. To be able to prove it, we also have to prove another important theorem for self-reducible counting problems: any self-reducible counting problem is either in both FPAUS and FPRAS, or it is in neither of these classes.

7.1   Relaxation time and second-largest eigenvalue

Definition 54. *The second-largest eigenvalue modulus of a Markov chain is defined as*

$$\max\{\lambda_2, |\lambda_r|\} \tag{7.6}$$

and is denoted by ρ. It is also abbreviated as SLEM.

Definition 55. *The relaxation time of a Markov chain is defined as*

$$\tau_i(\varepsilon) := min\{n_0 | \forall n \geq n_0, d_{TV}(T^n \mathbb{1}_i, \pi) \leq \varepsilon\} \tag{7.7}$$

where the vector $\mathbb{1}_i$ contains 0 in each coordinate except in the ith coordinate, which is 1.

Theorem 68. *For a Markov chain, it holds that*

$$\tau_i(\varepsilon) \leq \frac{1}{1-\rho}\left(\log\left(\frac{1}{\pi(x_i)}\right) + \log\left(\frac{1}{\varepsilon}\right)\right) \tag{7.8}$$

and

$$\max_i\{\tau_i(\varepsilon)\} \geq \frac{\rho}{2(1-\rho)}\log\left(\frac{1}{2\varepsilon}\right). \tag{7.9}$$

The proof of the first inequality can be found in [54], while the proof of the second inequality can be found in [5]. Theorem 68 says that the relaxation time is proportional to the inverse of the difference between the largest eigenvalue (that is, 1) and the SLEM. The following theorem says that it is sufficient to consider the second-largest eigenvalue of a Markov chain.

Theorem 69. *Let M be a quadratic matrix with eigenvalues $\lambda_1, \ldots, \lambda_r$ and eigenvectors $v_1, \ldots, v_n$. Then the matrix*

$$\frac{M+I}{2} \tag{7.10}$$

has eigenvalues $\frac{\lambda_1+1}{2}, \ldots, \frac{\lambda_r+1}{2}$ and eigenvectors $v_1, \ldots, v_n$, where I is the identity matrix.

Proof. It trivially comes from the basic properties of linear algebraic operations. Indeed,

$$\frac{M+I}{2}v_i = \frac{1}{2}(M+I)v_i = \frac{1}{2}(Mv_i + Iv_i) = \frac{1}{2}(\lambda_i v_i + v_i) = \frac{\lambda_i+1}{2}v_i. \tag{7.11}$$

□

Corollary 13. *If M is a Markov chain, then the random process defined by the following algorithm (so-called lazy version of M) is also a Markov chain whose SLEM is $\frac{\lambda_2+1}{2}$ and converges to the same, globally stable stationary distribution.*

1. *Draw a random number u uniformly from the $[0,1]$ interval.*

2. *If $u \leq \frac{1}{2}$ then do nothing; the next state of the Markov chain is the current state. Otherwise, the next state is drawn following the Markov chain M.*

Indeed, this process is a Markov chain, since the series of random states satisfies the Markov property: where we are going depends on the current state and not where we came from. Its transition matrix is $\frac{T+I}{2}$, so we can apply Theorem 69. The largest eigenvalue and its corresponding eigenvector do not change, so the Markov chain still converges to the same distribution.

We are ready to state and prove the main theorem on the mixing time of Markov chains and FPAUS algorithms.

Theorem 70. *Let #A be a counting problem in #P, and let x denote a problem instance in #A with size n. Assume that the following holds:*

(a) *a solution of x can be constructed in $O(poly(n))$ time,*

(b) *there is a Markov chain with transition matrix T that converges to the uniform distribution of the solutions of x, and for its second-largest eigenvalue it holds that*

$$\frac{1}{1-\lambda_2} = O(poly(n)), \tag{7.12}$$

(c) *there is a random algorithm that for any solution y, draws an entry from the conditional distribution $T(\cdot|y)$ in polynomial time.*

Then #A is in FPAUS.

Proof. Let x be a problem instance of #A, and let $\varepsilon > 0$. Since #A is in #P, there is a constant $c > 1$ and a polynomial $poly_1$ such that the number of solutions of x is less than or equal to $c^{poly_1(n)}$. Indeed, any witness can be verified in polynomial time, and the witnesses are described using a fixed alphabet (the alphabet depends on only the problem and not the problem instance). To verify a solution, it must be read. What follows is that the number of solutions cannot be more than $|\Sigma|^{poly(n)}$, where Σ is the alphabet used to describe the solutions and $poly()$ is the natural or given polynomial upper bound on the running time to verify a solution.

Having said these, it is easy to show that the following algorithm is an FPAUS:

1. Construct a solution y of x.

2. Using y as the starting point of the Markov chain, do

$$\frac{2}{1-\lambda_2}\left(poly_1(n)\log(c) + \log\left(\frac{1}{\varepsilon}\right)\right) \tag{7.13}$$

   number of steps in the lazy version of the Markov chain.

3. Return with the last state of the Markov chain.

Indeed, the state that the algorithm returns follows a distribution that satisfies Equation (1.31), since

$$\tau_y(\varepsilon) \leq \frac{2}{1 - \lambda_2} \left(\log \left(\frac{1}{\pi(y)} \right) + \log \left(\frac{1}{\varepsilon} \right) \right) \leq$$
$$\frac{2}{1 - \lambda_2} \left(\text{poly}_1(n) \log(c) + \log \left(\frac{1}{\varepsilon} \right) \right). \tag{7.14}$$

The first inequality comes from Theorem 68 and from Corollary 13. The second inequality comes from the observation that $\frac{1}{\pi(y)}$ is the size of the solution space, since π is the uniform distribution, and we showed that $c^{poly_1(n)}$ is an upper bound of it. The running time of the algorithm is $O(poly(n, -\log(\varepsilon)))$, since the initial state y can be constructed in polynomial time, there are $poly(n, -\log(\varepsilon))$ number of steps in the Markov chain, and each of them can be performed in $O(poly(n))$ time. $\qquad \square$

This theorem justifies the following definition.

Definition 56. *Let $\#A$ be a counting problem in $\#P$. Let $\mathcal{M}$ be a class of Markov chains, such that for each problem instance x of $\#A$, it contains a Markov chain converging to the uniform distribution of witnesses of x. Let M_x denote this Markov chain, and let $\lambda_{2,x}$ denote its second-largest eigenvalue. We say that $\mathcal{M}$ is rapidly mixing if*

$$\frac{1}{1 - \lambda_{2,x}} = O(poly(|x|)). \tag{7.15}$$

Similarly, we can say that a Markov chain is slowly or torpidly mixing if

$$\frac{1}{1 - \lambda_{2,x}} = \Omega(c^{|x|}) \tag{7.16}$$

for some $c > 1$.

7.2   Techniques to prove rapid mixing of Markov chains

In this section, we are going to give bounds on the second-largest eigenvalue. There are three techniques to prove bounds on the second-largest eigenvalue and thus prove rapid mixing of Markov chains. The first one is a geometric technique. Cheeger's inequality says that a Markov chain is rapidly mixing if and only if it does not contain a bottleneck. If we can prove that a Markov chain walks in a convex body and some mild conditions hold, then we can prove rapid mixing, since convex bodies do not have a bottleneck.

The second technique is a combinatorial method. It sets up a system of paths, a distribution of paths between any pair of states. If we can show that in this path system none of the edges are used extensively, then we can prove rapid mixing. Indeed, if a Markov chain contains a bottleneck, then there are only a few edges in it, and in the above-mentioned path system, at least one of them would be used heavily.

The third technique is a probabilistic method. Consider two Markov chains that depend on each other, however, both of them are copies of the same Markov chain M. The dependence is such that once the two Markov chains reach the same state at a given step n_0, then for all $n \geq n_0$, they remain in the same state. If this coupling time n_0 happens quickly with a high probability, then the Markov chain is rapidly mixing.

7.2.1 Cheeger's inequalities and the isoperimetric inequality

Cheeger's inequalities connect the conductance of a Markov chain and its second-largest eigenvalue. First we define the conductance and then state the inequalities.

Definition 57. *The* capacity *of a subset $S \subseteq X$ of the state space is defined as*

$$\pi(S) := \sum_{x \in S} \pi(x). \tag{7.17}$$

The ergodic flow *of a subset $S \subseteq X$ of the state space is defined as*

$$F(S) := \sum_{\substack{x \in S \\ y \in \overline{S}}} \pi(x) T(y|x). \tag{7.18}$$

The conductance *of a Markov chain is*

$$\Phi := \min \left\{ \frac{F(S)}{\pi(S)} \middle| S \subset X, 0 < \pi(S) \leq \frac{1}{2} \right\}. \tag{7.19}$$

Theorem 71. Cheeger's inequality *The second-largest eigenvalue λ_2 satisfies the following inequalities:*

$$1 - 2\Phi \leq \lambda_2 \leq 1 - \frac{\Phi^2}{2}. \tag{7.20}$$

The proof can be found in [25], Chapter 6. The two inequalities in Equation (7.20) will be referred to as the left and right Cheeger's inequality. This theorem says that a Markov chain is rapidly mixing if and only if its conductance is large. We can prove the torpid mixing of a Markov chain by finding a subset whose ergodic flow is negligible compared to its capacity.

Example 20. *A fully balanced, rooted binary tree of depth n contains 2^n leaves and altogether $2^{n+1} - 1$ vertices. The vertices at depth k are labeled by $v_{k,1}, v_{k,2}, \ldots, v_{k,2^k}$.*

Let $\mathcal{M}$ be a set of Markov chains that contains a Markov chain for each fully balanced, rooted binary tree of depth n. The state space of the Markov chain M_n is the set of vertices of the fully balanced rooted binary tree of depth n. The transition probabilities are

$$P(v_{k+1,2j-1}|v_{k,j}) = P(v_{k+1,2j}|v_{k,j}) =$$

$$P(v_{k-1,\lceil \frac{j}{2} \rceil}|v_{k,j}) = \frac{1}{3} \quad k = 1, \ldots n-1 \quad (7.21)$$

$$P(v_{1,1}|v_{0,1}) = P(v_{1,2}|v_{0,1}) =$$

$$P(v_{0,1}|v_{0,1}) = \frac{1}{3} \tag{7.22}$$

$$P(v_{n-1,\lceil \frac{j}{2} \rceil}|v_{n,j}) = \frac{1}{3} \tag{7.23}$$

$$P(v_{n,j}|v_{n,j}) = \frac{2}{3}. \tag{7.24}$$

Show that the Markov chain is torpidly mixing, that is, $\lambda_{2,n}$ converges to 1 exponentially quick.

Solution. It is easy to see that the Markov chain is irreducible, aperiodic, and reversible with respect to the uniform distribution of the vertices. Consider the subtree rooted in $v_{1,1}$, and let S be the set which contains its vertices. This set contains $2^n - 1$ vertices, therefore, the capacity is less than half. However, the only way to flow out from this subset is to go from $v_{1,1}$ to the root, $v_{0,1}$. Therefore, the ergodic flow divided by the capacity is

$$\frac{\pi(v_{1,1})T(v_{0,1}|v_{1,1}))}{\pi(S)} = \frac{\frac{1}{2^{n+1}-1}\frac{1}{3}}{\frac{2^n-1}{2^{n+1}-1}} = \frac{1}{3(2^n - 1)}. \tag{7.25}$$

The conductance cannot be larger than this particular value, therefore, we get that

$$1 - \frac{2}{3(2^n - 1)} \leq 1 - \Phi \leq \lambda_{2,n}. \tag{7.26}$$

That is, $\lambda_{2,n}$ converges exponentially quickly to 1. In other words, if we rearrange Equation (7.26), then we get that

$$\frac{1}{1 - \lambda_{2,n}} \geq \frac{3(2^n - 1)}{2} = \Omega(2^n). \tag{7.27}$$

Therefore, the relaxation time grows exponentially quick. ∎

In the definition of conductance, there might be double exponentially many subsets to be considered. Indeed, the size of the state space might be an exponential of the input size, and the number of subsets with at most half probability might be also an exponential function of the base set, see also Exercise 8. In spite of this, surprisingly, we can prove rapid mixing of a Markov chain using the right Cheeger's inequality.

Example 21. *Let $\mathcal{M}$ be a set of Markov chains; the Markov chains in it are defined on the same state spaces as in Example 20. However, in this case, let the jumping probabilities be the following:*

$$P(v_{k+1,2j-1}|v_{k,j}) = P(v_{k+1,2j}|v_{k,j}) = \frac{1}{4} \quad k = 0, \ldots n-1 \quad (7.28)$$

$$P(v_{k-1,\lceil \frac{j}{2} \rceil}|v_{k,j}) = \frac{1}{2} \quad k = 1, \ldots n-1 \quad (7.29)$$

$$P(v_{0,1}|v_{0,1}) = \frac{1}{2} \quad (7.30)$$

$$P(v_{n-1,\lceil \frac{j}{2} \rceil}|v_{n,j}) = P(v_{n,j}|v_{n,j}) = \frac{1}{2}. \quad (7.31)$$

Show that the Markov chain is rapidly mixing, that is, $\lambda_{2,n}$ converges to 1 only polynomially quick.

Solution. Let $S_i := \{x_{i,1}, x_{i,2}, \ldots, x_{i,2^i}\}$. It is easy to see that $\pi(S_i) = \frac{1}{n+1}$, π is the uniform distribution restricted to any S_i, and the Markov chain is reversible with respect to π. It follows that for any $x_{i,j}$,

$$\pi(x_{i,j}) = \frac{1}{(n+1)2^i}. \quad (7.32)$$

We can also observe the following. Let S be an arbitrary subset of vertices. S can be decomposed into connected components, $S = \sqcup_k C_k$. Then

$$\frac{F(S)}{\pi(S)} \geq \min_k \left\{ \frac{F(C_k)}{\pi(C_k)} \right\}. \quad (7.33)$$

Indeed,

$$\frac{F(S)}{\pi(S)} = \frac{\sum_k F(C_k)}{\sum_k \pi(C_k)} \quad (7.34)$$

therefore, it is sufficient to show that for any $a_1, a_2, b_1, b_2 > 0$, the inequality

$$\frac{a_1 + a_2}{b_1 + b_2} \geq \min \left\{ \frac{a_1}{b_1}, \frac{a_2}{b_2} \right\} \quad (7.35)$$

holds. Without loss of generality, we can say that $\frac{a_1}{b_1} \leq \frac{a_2}{b_2}$. Then we have to show that

$$\frac{a_1 + a_2}{b_1 + b_2} \geq \frac{a_1}{b_1}. \quad (7.36)$$

Rearranging this, we get that

$$a_1 b_1 + a_2 b_1 \geq a_1 b_1 + a_1 b_2. \quad (7.37)$$

That is,

$$a_2 b_1 \geq a_1 b_2, \quad (7.38)$$

which holds, since $\frac{a_1}{b_1} \leq \frac{a_2}{b_2}$.

What follows is that the conductance is taken on a connected subgraph, which has to be a subtree. Let S contain these vertices. We distinguish two cases, based on whether or not S contains $v_{0,1}$. If S does not contain $v_{0,1}$, then let $v_{i,j}$ be the root of S. Observe that $\pi(v_{i,j}) \geq \frac{1}{n}\pi(S)$. Indeed, the equality

$$\pi(v_{i,j}) = \pi(v_{i+1,2j-1}) + \pi(v_{i+1,2j}) \tag{7.39}$$

holds for any vertex. Then

$$\Phi = \frac{F(S)}{\pi(S)} \geq \frac{\pi(v_{i,j})T\left(v_{i-1,\lceil\frac{j}{2}\rceil}\middle|v_{i,j}\right)}{\pi(S)} \geq \frac{1}{2n}. \tag{7.40}$$

If S contains $v_{0,1}$, then $\overline{S}$ is a disjoint union of subtrees. These subtrees have roots $v_{i_1,j_1}, v_{i_2,j_2}, \ldots$. For each of these $v_{i,j}$, $v_{i-1,\lceil\frac{j}{2}\rceil}$ is in S. Furthermore, it holds that

$$\pi\left(v_{i-1,\lceil\frac{j}{2}\rceil}\right) \geq \frac{2}{n}\pi(G_{v_{i,j}}) \tag{7.41}$$

where $G_{v_{i,j}}$ denotes the subtree rooted in $v_{i,j}$. Since $\pi(\overline{S}) \geq \frac{1}{2}$, it follows that

$$\sum_k \pi\left(v_{i_k-1,\lceil\frac{j_k}{2}\rceil}\right) \geq \frac{1}{n}. \tag{7.42}$$

Then

$$\Phi = \frac{F(S)}{\pi(S)} \geq \frac{\sum_k \pi\left(v_{i_k-1,\lceil\frac{j_k}{2}\rceil}\right)T\left(v_{i_k,j_k}\middle|v_{i_k-1,\lceil\frac{j_k}{2}\rceil}\right)}{\frac{1}{2}} \geq$$

$$\frac{\frac{1}{n}\frac{1}{4}}{\frac{1}{2}} = \frac{1}{2n}. \tag{7.43}$$

Therefore, the conductance is at least $\frac{1}{2n}$. Putting this into the right Cheeger inequality, we get that

$$\lambda_{2,n} \leq 1 - \frac{\Phi^2}{2} \leq 1 - \frac{1}{8n^2}. \tag{7.44}$$

That is, $\lambda_{2,n}$ tends to 1 only polynomially quick with n. ∎

We are going to use the same example Markov chain to demonstrate other proving methods. In many cases, the state space of a Markov chain has a more complex structure than in this example. However, if the states can be embedded into a convex body, then we might be able to prove rapid mixing, since a convex body does not have a bottleneck. This might be stated formally with the following theorem.

Theorem 72. *Let X be a convex body in $\mathbb{R}^n$, partitioned into two parts, U and W. Let C denote the $n-1$-dimensional surface cutting X into U and W. Then the inequality*

$$A(C) \geq \frac{\min\{V(U), V(W)\}}{diam(X)} \tag{7.45}$$

holds, where A denotes the area ($n-1$-dimensional measure), V denotes the volume, and $diam(X)$ denotes the diameter of X.

The proof can be found in [109]. Below we show an example of how to use this theorem to prove rapid mixing. In Chapter 8, we will prove that #LE is in FPAUS applying this theorem.

Example 22. *Let $\mathcal{M}$ be a set of Markov chains which contains a Markov chain M_n for each positive integer n. The state space of M_n includes the $0-1$ vectors of length n. There is a transition between two states if they differ in exactly one coordinate. Each transition probability is $\frac{1}{n}$. Prove that the Markov chain is rapidly mixing.*

Solution. It is easy to see that the Markov chain converges to the uniform distribution, and each state has probability $\left(\frac{1}{2}\right)^n$. We will denote this by π. Let $s = \{s_1, s_2, \ldots, s_n\}$ be a state in M_n. Assign a convex body to s defined by the following inequalities:

$$\frac{s_i}{2} \leq x_i \leq \frac{s_i}{2} + \frac{1}{2}. \tag{7.46}$$

It is easy to see that the unions of these convex bodies are hypercubes, and their union is the unit hypercube. A transition of the Markov chain can happen between two states whose hypercubes have a common surface which is an $n-1$-dimensional hypercube. The area of this surface is $\left(\frac{1}{2}\right)^{n-1}$. The volume of each small hypercube is $\left(\frac{1}{2}\right)^n$. The diameter of the unit hypercube is $\sqrt{n}$. Let S be the subset of the state space of the Markov chain that defines the conductance. Let U denote the body corresponding to S and let W denote the body corresponding to $\overline{S}$. Let C denote the surface separating U and W. Observe that C is a union of $n-1$-dimensional hypercubes that corresponds to the possible transitions between U and W. Then on one hand, we know that

$$\Phi = \frac{F(S)}{\pi(S)} = \frac{\sum_{x \in S, y \in \overline{S}} \pi(x)T(y|x)}{\sum_{x \in S} \pi(S)} = \frac{\frac{A(C)}{\left(\frac{1}{2}\right)^{n-1}}\pi\frac{1}{n}}{\frac{V(U)}{\left(\frac{1}{2}\right)^n}\pi} = \frac{A(C)}{2nV(U)}. \tag{7.47}$$

On the other hand, from Theorem 72, we know that

$$\frac{A(C)}{V(U)} \geq \frac{1}{\sqrt{n}}, \tag{7.48}$$

therefore we get that

$$\Phi \geq \frac{1}{2n\sqrt{n}}. \tag{7.49}$$

Combining this with the right Cheeger's inequality, we obtain that

$$\lambda_{2,n} \leq 1 - \frac{1}{8n^5}. \tag{7.50}$$

That is, the second-largest eigenvalue converges to 1 only polynomially quick, and the Markov chain is thus rapidly mixing. ∎

7.2.2   Mixing of Markov chains on factorized state spaces

Intuitively, when there is a partitioning of the state space of a Markov chain, the Markov chain is rapidly mixing on the partitions, and the Markov chain is rapidly mixing among the partitions, too, then the Markov chain is rapidly mixing on the whole space. In this section we provide a mathematiclly rigorous description of this intuitive observation and prove it. The theorem and its proof were given by Erdős, Miklós and Toroczkai [64].

First, we need a generalization of the Cheeger inequality (the lower-bound).

Lemma 14. *For any reversible Markov chain, and any subset S of its state space,*

$$\frac{1 - \lambda_2}{2} \min\{\pi(S), \pi(\overline{S})\} \leq \sum_{x \in S, y \in \overline{S}} \pi(x)T(y|x) . \tag{7.51}$$

Proof. The right-hand side of Equation (7.51) is symmetric due to the reversibility of the chain. Thus, if $\pi(S) > \frac{1}{2}$, then S and $\overline{S}$ can be switched. If $\pi(S) \leq \frac{1}{2}$, the inequality is simply a rearrangement of the Cheeger inequality (the left inequality in Theorem 71.). Indeed,

$$1 - 2\frac{\sum_{x \in S, y \in \overline{S}} \pi(x)T(y|x)}{\pi(S)} \leq 1 - 2\Phi \leq \lambda_2 . \tag{7.52}$$

Rearranging the two ends of the inequality in Equation (7.52), we get the inequality in Equation (7.51). □

Now we are ready to state and prove a general theorem on rapidly mixing Markov chains on factorized state spaces.

Theorem 73. *Let $\mathcal{M}$ be a class of reversible, irreducible and aperiodic Markov chains whose state space Y can be partitioned into disjoint classes $Y = \cup_{x \in X} Y_x$ by the elements of some set X. The problem size of a particular chain is denoted by n. For notational convenience we also denote the element $y \in Y_x$ via the pair (x, y) to indicate the partition it belongs to. Let T be the transition matrix of $M \in \mathcal{M}$, and let π denote the stationary distribution of*

M. Moreover, let π_X denote the marginal of π on the first coordinate that is, $\pi_X(x) = \pi(Y_x)$ for all x. Also, for arbitrary but fixed x let us denote by π_{Y_x} the stationary probability distribution restricted to Y_x, i.e., $\pi(y)/\pi(Y_x)$, $\forall y \in Y_x$. Assume that the following properties hold:

i *For all x, the transitions with x fixed form an aperiodic, irreducible and reversible Markov chain denoted by M_x with stationary distribution π_{Y_x}. This Markov chain M_x has transition probabilities as Markov chain M for all transitions fixing x, except loops, which have increased probabilities such that the transition probabilities sum up to 1. All transitions that would change x have 0 probabilities. Furthermore, this Markov chain is rapidly mixing, i.e., for its second-largest eigenvalue $\lambda_{M_x,2}$ it holds that*

$$\frac{1}{1 - \lambda_{M_x,2}} \leq \text{poly}_1(n).$$

ii *There exists a Markov chain M' with state space X and with transition matrix T' which is aperiodic, irreducible and reversible w.r.t. π_X, and for all x_1, y_1, x_2 it holds that*

$$\sum_{y_2 \in Y_{x_2}} T((x_2, y_2)|(x_1, y_1)) \geq T'(x_2|x_1). \tag{7.53}$$

Furthermore, this Markov chain is rapidly mixing, namely, for its second-largest eigenvalue $\lambda_{M',2}$ it holds that

$$\frac{1}{1 - \lambda_{M',2}} \leq \text{poly}_2(n).$$

Then M is also rapidly mixing as its second-largest eigenvalue obeys:

$$\frac{1}{1 - \lambda_{M,2}} \leq \frac{256\,\text{poly}_1^2(n)\text{poly}_2^2(n)}{\left(1 - \frac{1}{\sqrt{2}}\right)^4}$$

Proof. For any non-empty subset S of the state space $Y = \bigcup_x Y_x$ of M we define

$$X(S) := \{x \in X \mid \exists y, (x, y) \in S\}$$

and for any given $x \in X$ we have

$$Y_x(S) := \{(x, y) \in Y \mid (x, y) \in S\} = Y_x \cap S.$$

We are going to prove that the ergodic flow $F(S)$ (see Equation (7.18)) from any $S \subset Y$ with $0 < \pi(S) \leq 1/2$ cannot be too small and therefore, neither the conductance of the Markov chain will be small. We cut the state space into two parts $Y = Y^l \cup Y^u$, namely the lower and upper parts using the following definitions (see also Fig. 7.1): the partition $X = L \sqcup U$ is defined as

$$L \; := \; \left\{ x \in X \, \middle| \, \frac{\pi(Y_x(S))}{\pi(Y_x)} \leq 1/\sqrt{2} \right\},$$

$$U \; := \; \left\{ x \in X \, \middle| \, \frac{\pi(Y_x(S))}{\pi(Y_x)} > 1/\sqrt{2} \right\}.$$

Furthermore, we introduce:

$$Y^l := \bigcup_{x \in L} Y_x \quad \text{and} \quad Y^u := \bigcup_{x \in U} Y_x \,,$$

and finally let

$$S_l := S \cap Y^l \quad \text{and} \quad S_u := S \cap Y^u.$$

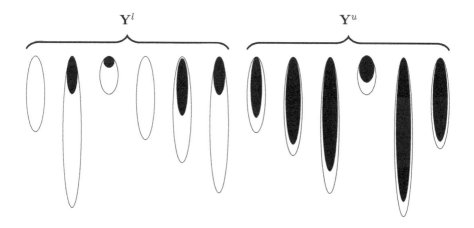

FIGURE 7.1: The structure of $Y = Y^l \sqcup Y^u$. A non-filled ellipse (with a simple line boundary) represents the space Y_x for a given x. The solid black ellipses represent the set S with some of them (the S_l) belonging to the lower part Y^l, and the rest (the S_u) belonging to the upper part (Y^u).

Since M' is rapidly mixing we can write (based on Theorem 71):

$$1 - 2\Phi_{M'} \leq \lambda_{M',2} \leq 1 - \frac{1}{\text{poly}_2(n)} \,,$$

or

$$\Phi_{M'} \geq \frac{1}{2\text{poly}_2(n)} \,.$$

Without loss of generality, we can assume that $\text{poly}_2(n) > 1$ for all positive n, a condition that we need later on for technical reasons. We use this lower bound of conductance to define two cases regarding the lower and upper part of S.

1. We say that the lower part S_l is not a negligible part of S when

$$\frac{\pi(S_l)}{\pi(S_u)} \geq \frac{1}{4\sqrt{2}\mathrm{poly}_2(n)}\left(1 - \frac{1}{\sqrt{2}}\right). \qquad (7.54)$$

2. We say that the lower part S_l is a negligible part of S when

$$\frac{\pi(S_l)}{\pi(S_u)} < \frac{1}{4\sqrt{2}\mathrm{poly}_2(n)}\left(1 - \frac{1}{\sqrt{2}}\right). \qquad (7.55)$$

Our plan is the following: the ergodic flow $F(S)$ is positive on any non-empty subset and it obeys:

$$F(S) = F'(S_l)\frac{\pi(S_l)}{\pi(S)} + F'(S_u)\frac{\pi(S_u)}{\pi(S)},$$

where

$$F'(S_l) := \frac{1}{\pi(S_l)}\sum_{x \in S_l, y \in \bar{S}} \pi(x)T(y|x)$$

and

$$F'(S_u) := \frac{1}{\pi(S_u)}\sum_{x \in S_u, y \in \bar{S}} \pi(x)T(y|x).$$

In other words, $F'(S_l)$ and $F'(S_u)$ are defined as the flow going from S_l and S_u and leaving S.

The value $F(S)$ cannot be too small, if at least one of $F'(S_l)$ or $F'(S_u)$ is big enough (and the associated fraction $\pi(S_l)/\pi(S)$ or $\pi(S_u)/\pi(S)$). In Case 1 we will show that $F'(S_l)$ itself is big enough. To that end it will be sufficient to consider the part which leaves S_l but not Y^l (this guarantees that it goes out of S, see also Fig. 7.2). For Case 2 we will consider $F'(S_u)$, particularly that part of it which goes from S_u to $Y^l \setminus S_l$ (and then going out of S, not only S_u, see also Fig. 7.3).

In **Case** 1, the flow going out from S_l within Y^l is sufficient to prove that the conditional flow going out from S is not negligible. We know that for any particular x, we have a rapidly mixing Markov chain M_x over the second coordinate y. Let their smallest conductance be denoted by Φ_X. Since all these Markov chains are rapidly mixing, we have that

$$\max_x \lambda_{M_x,2} \leq 1 - \frac{1}{\mathrm{poly}_1(n)}$$

or, equivalently:

$$\Phi_X \geq \frac{1}{2\mathrm{poly}_1(n)}.$$

However, in the lower part, for any particular x one has:

$$\pi_{Y_x}(Y_x(S)) = \frac{\pi(Y_x(S))}{\pi(Y_x)} \leq \frac{1}{\sqrt{2}},$$

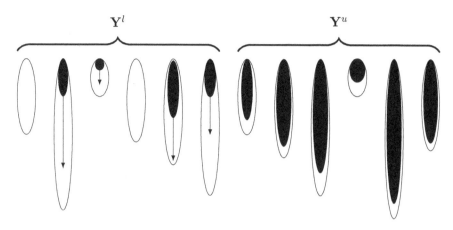

FIGURE 7.2: When S_l is not a negligible part of S, there is a considerable flow going out from S_l to within Y^l, implying that the conditional flow going out from S cannot be small. See text for details and rigorous calculations.

so for any fixed x belonging to L it holds that

$$\frac{1}{2\text{poly}_1(n)} \min\left\{\pi_{Y_x}(Y_x(S)), \left(1 - \frac{1}{\sqrt{2}}\right)\right\} \le$$
$$\le \sum_{(x,y)\in S, (x,y')\in \bar{S}} \pi_{Y_x}((x,y))T((x,y')|(x,y))$$

using the modified Cheeger inequality (Lemma 14). Observing that

$$\pi_{Y_x}((x,y)) = \frac{\pi((x,y))}{\pi(Y_x)} \,,$$

we obtain:

$$\frac{1}{2\text{poly}_1(n)}\pi(Y_x(S))\left(1 - \frac{1}{\sqrt{2}}\right) \le$$
$$\le \frac{1}{2\text{poly}_1(n)} \min\left\{\pi(Y_x(S)), \pi(Y_x)\left(1 - \frac{1}{\sqrt{2}}\right)\right\} \le$$
$$\le \sum_{(x,y)\in S, (x,y')\in \bar{S}} \pi((x,y))T((x,y')|(x,y)) \,.$$

Summing this for all the x's belonging to L, we deduce that

$$\pi(S_l)\frac{1}{2\text{poly}_1(n)}\left(1 - \frac{1}{\sqrt{2}}\right) \le$$
$$\sum_{x|Y_x(S)\subseteq S_l}\left(\sum_{(x,y)\in S, (x,y')\in \bar{S}} \pi((x,y))T((x,y')|(x,y))\right) \,.$$

Note that the flow on the right-hand side of Equation 7.56 is not only going out from S_l but also from the entire S. Therefore, we have that

$$F(S) \geq \frac{\pi(S_l)}{\pi(S)} \times \frac{1}{2\text{poly}_1(n)} \left(1 - \frac{1}{\sqrt{2}}\right).$$

Either $\pi(S_l) \leq \pi(S_u)$, which then yields

$$\frac{\pi(S_l)}{\pi(S)} = \frac{\pi(S_l)}{\pi(S_l) + \pi(S_u)} \geq \frac{\pi(S_l)}{2\pi(S_u)} \geq \frac{1}{8\sqrt{2}\text{poly}_2(n)} \left(1 - \frac{1}{\sqrt{2}}\right)$$

after using Equation 7.54, or $\pi(S_l) > \pi(S_u)$, in which case we have

$$\frac{\pi(S_l)}{\pi(S)} > \frac{1}{2} \geq \frac{1}{8\sqrt{2}\text{poly}_2(n)} \left(1 - \frac{1}{\sqrt{2}}\right).$$

(Note that $\text{poly}_2(n) > 1$.) Thus in both cases the following inequality holds:

$$F(S) \geq \frac{1}{8\sqrt{2}\text{poly}_2(n)} \left(1 - \frac{1}{\sqrt{2}}\right) \times \frac{1}{2\text{poly}_1(n)} \left(1 - \frac{1}{\sqrt{2}}\right).$$

In **Case** 2, the lower part of S is a negligible part of S. We have that

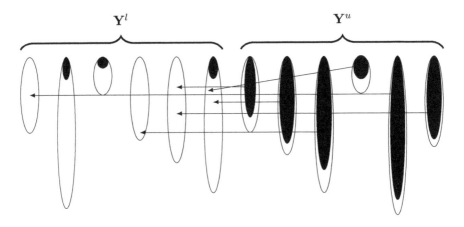

FIGURE 7.3: When S_l is a negligible part of S, there is a considerable flow going out from S_u into $Y^l \setminus S_l$. See text for details and rigorous calculations.

$$\pi_X(X(S_u)) \leq \frac{1}{\sqrt{2}}$$

otherwise $\pi(S_u) > 1/2$ would happen (due to the definition of the upper part), and then $\pi(S) > 1/2$, a contradiction.

Hence in the Markov chain M', based on the Lemma 14, we obtain for $X(S_u)$ that

$$\frac{1}{2\mathrm{poly}_2(n)} \min\left\{\pi_X(X(S_u)), \left(1 - \frac{1}{\sqrt{2}}\right)\right\} \leq \sum_{\substack{x' \in \overline{X(S_u)} \\ x \in X(S_u)}} \pi_X(x)T'(x'|x). \quad (7.56)$$

For all y for which $(x, y) \in S_u$, due to Equation (7.53), we can write:

$$T'(x'|x) \leq \sum_{y'} T((x', y')|(x, y)) .$$

Multiplying this by $\pi((x, y))$, then summing for all suitable y:

$$\pi(Y_x(S))T'(x'|x) \leq \sum_{y|(x,y)\in S_u} \sum_{y'} \pi((x, y))T((x', y')|(x, y))$$

(note that $x \in U$ and thus $Y_x(S) = Y_x(S_u)$) and thus

$$T'(x'|x) \leq \frac{\sum_{y|(x,y)\in S_u} \sum_{y'} \pi((x, y))T((x', y')|(x, y))}{\pi(Y_x(S))} .$$

Inserting this into Equation 7.56, we find that

$$\frac{1}{2\mathrm{poly}_2(n)} \min\left\{\pi_X(X(S_u)), \left(1 - \frac{1}{\sqrt{2}}\right)\right\} \leq$$

$$\leq \sum_{x \in X(S_u), x' \in \overline{X(S_u)}} \frac{\pi_X(x)}{\pi(Y_x(S))} \sum_{y|(x,y)\in S_u} \sum_{y'} \pi((x, y))T((x', y')|(x, y)).$$

Recall that $\pi_X(x) = \pi(Y_x)$, and thus $\frac{\pi_X(x)}{\pi(Y_x(S))} \leq \sqrt{2}$ for all $x \in X(S_u)$. Therefore we can write that

$$\frac{1}{2\mathrm{poly}_2(n)} \min\left\{\pi_X(X(S_u)), \left(1 - \frac{1}{\sqrt{2}}\right)\right\} \leq$$

$$\sqrt{2} \sum_{(x,y)\in S_u} \left(\sum_{(x',y')|x'\in\overline{X(S_u)}} \pi((x, y))T((x', y')|(x, y)) \right) .$$

Note that $\pi(S_u) \leq \pi_X(X(S_u)) < 1$, and since both items in the minimum taken in the LHS are smaller than 1, their product will be smaller than any of them. Therefore we have

$$\frac{1}{2\sqrt{2}\mathrm{poly}_2(n)} \pi(S_u) \left(1 - \frac{1}{\sqrt{2}}\right) \leq$$

$$\leq \sum_{(x,y)\in S_u} \left(\sum_{(x',y')|x'\in\overline{X(S_u)}} \pi((x, y))T((x', y')|(x, y)) \right) .$$

This flow is going out from S_u, and it is so large that at most half of it can be picked up by the lower part of S (due to reversibility and due to Equation 7.55), and thus the remaining part, i.e., at least half of the flow, must go out of S. Therefore:

$$\frac{\pi(S_u)}{\pi(S)} \times \frac{1}{4\sqrt{2}\text{poly}_2(n)} \left(1 - \frac{1}{\sqrt{2}}\right) \leq F(S) .$$

However, since S_u dominates S, namely, $\pi(S_u) > \frac{\pi(S)}{2}$, we have that

$$\frac{1}{8\sqrt{2}\text{poly}_2(n)} \left(1 - \frac{1}{\sqrt{2}}\right) \leq F(S).$$

Comparing the bounds from Case 1 and Case 2, for all S satisfying $0 < \pi(S) \leq \frac{1}{2}$, we can write:

$$\frac{1}{16\sqrt{2}\text{poly}_2(n)\text{poly}_1(n)} \left(1 - \frac{1}{\sqrt{2}}\right)^2 \leq F(S).$$

And thus, for the conductance of the Markov chain M (which is the minimum over all possible S)

$$\frac{1}{16\sqrt{2}\text{poly}_2(n)\text{poly}_1(n)} \left(1 - \frac{1}{\sqrt{2}}\right)^2 \leq \Phi_M.$$

Applying this to the Cheeger inequality, one obtains

$$\lambda_{M,2} \leq 1 - \frac{\left(\frac{1}{16\sqrt{2}\text{poly}_2(n)\text{poly}_1(n)} \left(1 - \frac{1}{\sqrt{2}}\right)^2\right)^2}{2}$$

and thus

$$\frac{1}{1 - \lambda_{M,2}} \leq \frac{256\text{poly}_1^2(n)\text{poly}_2^2(n)}{\left(1 - \frac{1}{\sqrt{2}}\right)^4}$$

which is what we wanted to prove. □

Martin and Randall [125] have developed a similar theorem. They assume a disjoint decomposition of the state space Ω of an irreducible and reversible Markov chain defined via the transition probabilities $P(y|x)$. They require that the Markov chain be rapidly mixing when restricted onto each partition Ω_i ($\Omega = \cup_i \Omega_i$) and furthermore, another Markov chain, the so-called projection Markov chain $\overline{P}(i|j)$ defined over the indices of the partitions be also rapidly mixing. If all these hold, then the original Markov chain is also rapidly mixing. For the projection Markov chain they use the normalized conditional flow

$$\overline{P}(j|i) = \frac{1}{\pi(\Omega_i)} \sum_{x \in \Omega_i, y \in \Omega_j} \pi(x)P(y|x) \qquad (7.57)$$

as transition probabilities. This can be interpreted as a weighted average transition probability between two partitions, while in our case, Equation (7.53) requires only that the transition probability of the lower bounding Markov chain is not more than the minimum of the sum of the transition probabilities going out from one member of the partition (subset Y_{x_1}) to the other member of the partition (subset Y_{x_2}) with the minimum taken over all the elements of Y_{x_1}. Obviously, it is a stronger condition that our Markov chain must be rapidly mixing, since a Markov chain is mixing slower when each transition probability between any two states is smaller. (The latter statement is based on a comparison theorem by Diaconis and Saloff-Coste [52].) Therefore, from that point of view, our theorem is weaker. On the other hand, the average transition probability (Equation (7.57)) is usually hard to calculate, and in this sense our theorem is more applicable. Note that Martin and Randall have also resorted in the end to using chain comparison techniques (Sections 2.2 and 3 in their paper) employing a Metropolis-Hastings chain as a lower bounding chain instead of the projection chain above. Our theorem, however, provides a direct proof of a similar statement.

7.2.3 Canonical paths and multicommodity flow

The key idea in this approach is that if we can set up (possibly random) paths between each pair of states in the Markov graph such that the (expected) usage of any edge is comparable with the size of the state space (and also some mild conditions hold), then there is no bottleneck in the Markov chain. That is, the conductance is high and the Markov chain is rapidly mixing. This idea was first introduced in the PhD thesis of Alastair Sinclair [157]. Diaconis and Stroock observed that path arguments can be used to directly bound the second-largest eigenvalue, and thus the mixing time, independently of the conductance of the Markov chain [54]. Later Sinclair improved these bounds extending the method to random paths (multicommodity flow) [158]. Since then several variants have appeared, from which we show a selection below.

First, we define the maximum load of a path system.

Definition 58. *Let G be the Markov graph of a Markov chain. We define the conductance of an edge $e = (v_i, v_j)$ as*

$$Q(e) := \pi(v_i)T(v_j|v_i). \tag{7.58}$$

For each ordered pair of states (v_x, v_y), we define a path $v_x = v_1, v_2, \ldots v_l = v_y$ that uses any edge at most once. The path is denoted by $\gamma_{x,y}$. Let Γ denote the so obtained path system. The maximum load of Γ is defined as

$$\vartheta_\Gamma := \max_e \frac{1}{Q(e)} \sum_{\gamma_{x,y} \ni e} \pi(v_x)\pi(v_y). \tag{7.59}$$

Sinclair proved the following theorem [157]:

Theorem 74. *Let* Γ *be a path system of a Markov chain. Then the inequality*

$$\Phi \geq \frac{1}{\vartheta_\Gamma} \tag{7.60}$$

holds.

We can combine this theorem with the right Cheeger's inequality to get the following corollary:

Corollary 15. *Let* Γ *be a path system of a Markov chain. Then the inequality*

$$\lambda_2 \leq 1 - \frac{1}{2\vartheta_\Gamma^2} \tag{7.61}$$

holds.

Diaconis and Stroock proved a theorem on mixing rates of Markov chains that does not use conductance. Their method gives better bound on the mixing time if the paths are relatively short. They use the Poincaré coefficient that we define below.

Definition 59. *Let the conductance of an edge be defined as above. The Q-measure of a path is defined as the sum of the resistance of the edges, that is, the inverse of the conductances:*

$$|\gamma_{x,y}|_Q := \sum_{e \in \gamma_{x,y}} \frac{1}{Q(e)}. \tag{7.62}$$

Let Γ *denote the so obtained path system. The* Poincaré *coefficient of* Γ *is defined as*

$$\kappa_\Gamma := \max_e \sum_{\gamma_{x,y} \ni e} |\gamma_{x,y}|_Q \pi(v_x)\pi(v_y). \tag{7.63}$$

The following theorem holds for the Poincaré coefficient.

Theorem 75. *Let* Γ *be a path system of a Markov chain. Then the inequality*

$$\lambda_2 \leq 1 - \frac{1}{\kappa} \tag{7.64}$$

holds.

The proof can be found in [54]. We show an example of how to use these theorems.

Example 23. *Let* $\mathcal{M}$ *be the set of Markov chains as in Example 21. Prove its rapid mixing using Theorem 74.*

Solution. Since the Markov graph is a tree, there is only one possible path system: the path connecting the pair of vertices in the graph. We have to show that for any edge e going from $v_{i,j}$ to $v_{i-1,\lceil\frac{j}{2}\rceil}$,

$$\frac{1}{Q(e)} \sum_{\gamma_{x,y} \ni e} \pi(v_x)\pi(v_y) \tag{7.65}$$

is polynomially bounded, and the same is true for the antiparallel edge e' going from $v_{i-1,\lceil\frac{j}{2}\rceil}$ to $v_{i,j}$.

Let S denote the set of vertices containing $v_{i,j}$ and the vertices below $v_{i,j}$. Then

$$\frac{1}{Q(e)} \sum_{\gamma_{x,y} \ni e} \pi(v_x)\pi(v_y) = \frac{1}{\pi(v_{i,j})\frac{1}{2}} \sum_{v_x \in S} \pi(v_x) \sum_{v_y \in \overline{S}} \pi(v_y). \tag{7.66}$$

We know that

$$\pi(v_{i,j}) \geq \frac{1}{n}\pi(S), \tag{7.67}$$

therefore

$$\frac{1}{\pi(v_{i,j})\frac{1}{2}} \sum_{v_x \in S} \pi(v_x) \sum_{v_y \in \overline{S}} \pi(v_y) \leq 2n \sum_{v_y \in \overline{S}} \pi(v_y) \leq 2n. \tag{7.68}$$

Similarly, we know that

$$\pi\left(v_{i-1,\lceil\frac{j}{2}\rceil}\right) \geq \frac{2}{n}\pi(S). \tag{7.69}$$

Therefore

$$\frac{1}{Q(e')} \sum_{\gamma_{x,y} \ni e'} \pi(v_x)\pi(v_y) = \frac{1}{\pi\left(v_{i-1,\lceil\frac{j}{2}\rceil}\right)\frac{1}{4}} \sum_{v_x \in S} \pi(v_x) \sum_{v_y \in \overline{S}} \pi(v_y) \leq$$

$$2n \sum_{v_y \in \overline{S}} \pi(v_y) \leq 2n. \tag{7.70}$$

Therefore, we get that

$$\vartheta_\Gamma \leq 2n, \tag{7.71}$$

and thus

$$\lambda_{2,n} \leq 1 - \frac{1}{8n^2}. \tag{7.72}$$

∎

Sometimes it is hard to handle $|\gamma_{x,y}|_Q$. Sinclair [158] modified the theorem of Diaconis and Stroock:

Theorem 76. *Let Γ be a path system. Define*

$$K_\Gamma := \max_e \frac{1}{Q(e)} \sum_{\gamma_{x,y} \ni e} |\gamma_{x,y}| \pi(v_x)\pi(v_y), \qquad (7.73)$$

where $|\gamma_{x,y}|$ is the length of the path, that is, the number of edges in it. Then the inequality

$$\lambda_2 \le 1 - \frac{1}{K_\Gamma} \qquad (7.74)$$

holds.

This theorem is related to the maximal load. Indeed, the following corollary holds.

Corollary 16. *For any path system Γ, the inequality*

$$\lambda_2 \le 1 - \frac{1}{l_\Gamma \vartheta_\Gamma} \qquad (7.75)$$

holds, where l_Γ is the length of the longest path in the path system.

When $l_\Gamma < \vartheta_\Gamma$, then this inequality provides better bounds. Obviously, if we would like to prove only polynomial mixing time, any of these theorems might be applicable. On the other hand, when we try to get sharp estimates, it is important to carefully select which theorem to use.

When the Markov graph has a complicated structure, or just the opposite, a very symmetric structure, it might be hard to design a path system such that none of the edges are overloaded. An overloaded edge might cause weak upper bound on λ_2 and thus fail to prove rapid mixing. In such cases, a random path system might help. First we define the flow functions.

Definition 60. *Let $\Pi_{i,j}$ denote the set of paths between v_i and v_j in a Markov graph, and let $\Pi = \cup_{i,j}\Pi_{i,j}$. A multicommodity flow in G is a function $f : \Pi \to \mathbb{R}^+ \cup \{0\}$ satisfying*

$$\sum_{\gamma \in \Pi_{i,j}} f(\gamma) = \pi(v_i)\pi(v_j). \qquad (7.76)$$

We can think of f as having a probability distribution of paths for each pair of vertices (v_i, v_j), and then these probabilities are scaled by $\pi(v_i)\pi(v_j)$.

We can also define for each edge e

$$f(e) := \sum_{\gamma \ni e} f(\gamma), \qquad (7.77)$$

and similar to ϑ_f, the maximal expected load as

$$\vartheta_f := \max_e \frac{f(e)}{Q(e)}. \qquad (7.78)$$

The probabilistic version of Theorem 74 and its corollary can also be stated.

Theorem 77. *Let f be a multicommodity flow. Then the inequality*

$$\Phi \geq \frac{1}{2\varphi_f} \tag{7.79}$$

holds.

Corollary 17. *Let f be a multicommodity flow. Then the inequality*

$$\lambda_2 \leq 1 - \frac{1}{8\vartheta_f^2} \tag{7.80}$$

holds.

We give an example of when the canonical path method does not give a good upper bound, however, the multicommodity flow approach will help find a good upper bound on the second-largest eigenvalue.

Example 24. *Consider the Markov chain whose state space contains the $2n$-long sequences containing n a characters and n b characters in an arbitrary order. The transitions are defined with the following algorithm. We choose an index i between 1 and $2n - 1$ uniformly. If the characters in position i and $i + 1$ are different, then we swap them, otherwise the Markov chain remains in the current state. Prove that the mixing time grows only polynomially with n.*

Solution. One might think that the following canonical path system gives a good upper bound on the second-largest eigenvalue.

Let $x = x_1 x_2 \ldots x_{2n}$ and $y = y_1 y_2 \ldots y_{2n}$. Construct a path γ_{xy} in the following way. If $x_1 \neq y_1$, find the smallest index i such that $x_i = y_1$. "Bubble down" x_i, that is, swap x_{i-1} and x_i, then x_{i-2} with x_{i-1} (which is the prior x_i), etc. Then compare x_2 and y_2; if they are different, then find the smallest i such that $x_i = y_2$, bubble down x_i, and so on.

However, this path does not provide a good upper bound. To see this, consider the edge in the Markov graph going from state

$$\underbrace{aa \ldots a}_{n/2-1} ba \underbrace{bb \ldots b}_{n-1} \underbrace{aa \ldots a}_{n/2}$$

to state

$$\underbrace{aa \ldots a}_{n/2} \underbrace{bb \ldots b}_{n} \underbrace{aa \ldots a}_{n/2}.$$

How many γ_{xy} paths are going through this edge? The first state x of the path γ_{xy} might be an arbitrary sequence that has a suffix of all a's of length $n/2$. There are n b characters and $n/2$ a characters in its corresponding prefix, that might be arranged in $\binom{1.5n}{n}$ different ways. Similarly, the last state y of the path γ_{xy} might be an arbitrary string with a prefix of all a's of length $n/2$. Again, there are $\binom{1.5n}{n}$ possibilities. Therefore, there are

$$\binom{1.5n}{n}^2 \approx \left(\sqrt{\frac{3\pi n}{\pi n 2\pi n}} \frac{\left(\frac{1.5n}{e}\right)^{1.5n}}{\left(\frac{0.5n}{e}\right)^{0.5n}\left(\frac{n}{e}\right)^n}\right)^2 =$$

$$\sqrt{\frac{3}{2\pi n}}\left(\frac{1.5^{1.5}}{0.5^{0.5}}\right)^{2n} = \sqrt{\frac{3}{2\pi n}}6.75^n \qquad (7.81)$$

paths that go through this edge. Here we used the Stirling formula to approximate factorials. On the other hand, there are only

$$\binom{2n}{n} \approx \sqrt{\frac{4\pi n}{2\pi n 2\pi n}} \frac{\left(\frac{2n}{e}\right)^{2n}}{\left(\frac{n}{e}\right)^n\left(\frac{n}{e}\right)^n} = \frac{1}{\sqrt{\pi n}}4^n \qquad (7.82)$$

states in the Markov chain. It is easy to see that the stationary distribution of the Markov chain is the uniform distribution. In the definition of max load in Equation (7.59), the stationary distribution probability in $Q(e)$ and one of the stationary probabilities in the summation cancel each other. There is another stationary distribution probability in the summation which is (neglecting the non-exponential terms) in order $\frac{1}{4^n}$. However, the summation is on an order of 6.75^n paths (again, neglecting the non-exponential terms). That is, the maximum load is clearly exponential for this canonical path system.

Consider now a multicommodity flow such that there are at most $(n!)^2$ paths between any pair of states with non-zero measure. First, we construct a multiset of paths of cardinality $(n!)^4$. Each path in this multiset has measure

$$f(\gamma) = \frac{\pi(v_i)\pi(v_j)}{(n!)^4}. \qquad (7.83)$$

Then the final flow between the two states is obtained by assigning a measure to each path that is proportional to its multiplicity in the multiset.

The paths with non-zero measures are defined in the following way. Let $x = x_1 x_2 \ldots x_{2n}$ and $y = y_1 y_2 \ldots y_{2n}$, and let $\sigma_{x,a}$, $\sigma_{x,b}$, $\sigma_{y,a}$ and $\sigma_{y,b}$ be four permutations. Index both the a characters and the b characters by the permutations in both x and y. Now transform the indexed x into the index y in the same way as we did in the canonical path reconstruction. The important difference is that there is exactly one a_i and b_j character in both x and y for each of the indexes i and j. Thus, if x_1 is not y_1, then find the unique character x_i in x which is y_1, and bubble it down into the first position. Do this procedure for each character in x to thus transform x into y. The path assigned to this transformation can be constructed by removing the indexes and also removing the void steps (when some neighbors a_i and a_j are swapped in the indexed version, but they are two neighbor a's when indexes are removed, there is nothing to swap there!). We are going to estimate the number of paths that can get through a given edge. For this, we are going to count the indexed versions of the paths keeping in mind that each of them has a measure given in Equation (7.83).

Surprisingly, this multicommodity flow provides an upper bound on the mixing time which is only polynomial with n. To see this, consider any edge e swapping characters z_i and z_{i+1} in a sequence z. It is easy to see that any indexing of the characters in z might appear on indexed paths getting through e. So fix an arbitrary indexing. We are bubbling down character z_{i+1}, so there is some $0 \leq k < i$ for which the indexes of the first k characters and the index of character z_{i+1} are in order as they will appear in y. The other indexes appear as they were in x. Then there are at most

$$\binom{2n}{k+1}(k+1)! \tag{7.84}$$

possible indexed x sequences that might be the starting sequence of an indexed path containing e. Indeed, the first k characters as well as character z_{i+1} must be put back to the other $2n - k - 1$ characters in arbitrary order with the exception that the position of z_{i+1} cannot be smaller than $i + 1$. The other $2n - k - 1$ characters can be put into an arbitrary order (z_{i+1} will be in position $k + 1$ in the destination sequence y), thus there are $(2n - k - 1)!$ possible y sequences. Thus, for a fixed k and a fixed indexing, the flow measure getting through e is at most

$$\frac{\frac{1}{\binom{2n}{n}^2}}{(n!)^4}\binom{2n}{k+1}(k+1)!(2n-k-1)! = \frac{1}{(2n)!}. \tag{7.85}$$

There are $(n!)^2$ possible indexes and at most $2n - 1$ possible ks. Therefore, the maximum flow getting through an edge is

$$f(e) \leq \frac{(2n-1)(n!)^2}{(2n)!}. \tag{7.86}$$

This should be multiplied by the inverse of $Q(e)$, so we get that the maximum load is

$$\vartheta_f \leq \frac{1}{\frac{1}{\binom{2n}{n}}\frac{1}{2n-1}}\frac{(2n-1)(n!)^2}{(2n)!} = (2n-1)^2. \tag{7.87}$$

Applying Theorem 77, we can set a lower bound on the conductance that proves the rapid mixing of the Markov chain. ∎

Finally, it is also possible to generalize Theorem 76 and its corollary.

Theorem 78. *Let f be a multicommodity flow. Define*

$$K_f := \max_e \frac{1}{Q(e)}\sum_{\gamma \ni e} f(\gamma)|\gamma|. \tag{7.88}$$

Then the inequality

$$\lambda_2 \leq 1 - \frac{1}{K_f} \tag{7.89}$$

holds.

Corollary 18. *Let f be a multicommodity flow. Define*

$$l_f := \max_{\gamma \in \Pi, f(\gamma) > 0} |\gamma|. \tag{7.90}$$

Then the inequality

$$\lambda_2 \leq 1 - \frac{1}{l_f \vartheta_f} \tag{7.91}$$

holds.

7.2.4   Coupling of Markov chains

Coupling is an old idea of Wolfgang Doeblin [55] that was revived in Markov chain theory thirty years later. The coupling of Markov chains is defined in the following way.

Definition 61. *Let M be a Markov chain on the state space X with transition probabilities $T(\cdot|\cdot)$. Let M' be a Markov chain on $X \times X$, and let T' denote its transition probabilities. We say that M' is a coupling of M if the following equalities hold:*

$$\sum_{y_2} T'((x_2, y_2)|(x_1, y_1)) = T(x_2|x_1) \tag{7.92}$$

$$\sum_{x_2} T'((x_2, y_2)|(x_1, y_1)) = T(y_2|y_1). \tag{7.93}$$

$$\tag{7.94}$$

Namely, both coordinates of M' behave like a copy of M. One possible way to achieve this is independency, that is

$$T'((x_2, y_2|(x_1, y_1))) = T(x_2|x_1)T(y_2|y_1), \tag{7.95}$$

although it is not the only way, and for the application described below is not desirable. The two coordinates might be forced to get closer without violating the equalities in Equation (7.92) and (7.92). Once $x_i = y_i$ is achieved, it is easy to force the two coordinates to stay together by defining

$$T'((x_2, x_2)|(x_1, x_1)) := T(x_2|x_1) \tag{7.96}$$

and for $y \neq x_2$

$$T'((x_2, y)|(x_1, x_1)) := T'((y, x_2)|(x_1, x_1)) = 0. \tag{7.97}$$

Such Markov chains can prove rapid mixing of M as stated by the following theorem proved by Aldous [6]:

Theorem 79. *Let M' be a coupling of M. Assume that $t : [0,1] \to \mathbb{N}$ is a function such that for any state (x_0, y_0), it satisfies the inequality:*

$$\left\langle T'^{t(\varepsilon)} 1_{(x_0,y_0)} \middle| 1_{x \neq y} \right\rangle \leq \varepsilon \tag{7.98}$$

where $1_{(x_0,y_0)}$ is the vector containing 0 everywhere except it is 1 on coordinate (x_0, y_0), and $1_{x \neq y}$ is the vector containing 1 everywhere except it is 0 on coordinates (x_i, x_i) for all $x_i \in X$. In other words, the probability that the two coordinates of the Markov chain M' are not the same after $t(\varepsilon)$ steps starting at an arbitrary state (x_0, y_0) is less than or equal to ε.

Then $t(\varepsilon)$ upper bounds the relaxation time $\tau(\varepsilon) := \max_i \{\tau_i(\varepsilon)\}$ of M.

We demonstrate how to prove rapid mixing with the coupling technique in the following example. As a warm up, the reader might consider solving Exercise 18 before reading the example.

Example 25. *Use the coupling technique to show that the Markov chain on the vertices of the d-dimensional unit hypercube $\{0,1\}^d$ is rapidly mixing. The transition probabilities are defined in the following way. The Markov chain remains in the same state with probability $\frac{1}{2}$, and with probability $\frac{1}{2}$ a random coordinate is selected and its value x is replaced with $1 - x$.*

Solution. Let M' be a coupling of M whose transition probabilities are defined in the following way. Let (x, y) denote the current state, where $x = (x_1, x_2, \ldots, x_d)$ and $y = (y_1, y_2, \ldots, y_d)$s. In each step, a random coordinate i is selected uniformly and a fair coin is tossed. If $x_i = y_i$, then both or none of them are flipped based on if the tossed coin is a head or tail. If $x_i \neq y_i$, then x_i is flipped and y_i is not changed if the tossed coin is a head, otherwise y_i is flipped and x_i is not changed. Let $n(t)$ denote the number of coordinates on which x and y are the same in the coupling after t steps of the Markov chain. Observe the following: if $n(t) < d$, then the inequality

$$P(n(t+1) = n(t) + 1) \geq \frac{1}{d}, \tag{7.99}$$

and

$$P\left(n(t+1) = n(t) + 1 \vee n(t+1) = n(t)\right) = 1. \tag{7.100}$$

Indeed, if $n(t) < d$, then with at least $\frac{1}{d}$ probability, a coordinate i is selected such that $x_i \neq y_i$, and after the step in M', the two coordinates become the same.

Clearly, $n(t) = d$ indicates that $x = y$ after step t. What follows is that

$$\left\langle T'^{t(\varepsilon)} 1_{(x_0,y_0)} \middle| 1_{x \neq y} \right\rangle \tag{7.101}$$

is upper bounded by the probability that in a binomial distribution with $n = t$ and $p = \frac{1}{d}$, the random variable is less than d. This latter probability is

$$\sum_{i=0}^{d-1} \binom{t}{i} \left(\frac{1}{d}\right)^i \left(1 - \frac{1}{d}\right)^{t-i}. \tag{7.102}$$

When $t \geq d^2$ (and thus $d \leq np$), we can use Chernoff's inequality to bound the sum in Equation (7.102):

$$\sum_{i=0}^{d-1} \binom{t}{i} \left(\frac{1}{d}\right)^i \left(1 - \frac{1}{d}\right)^{t-i} \leq \exp\left(-\frac{(t\frac{1}{d} - d)^2}{2t\frac{1}{d}}\right). \qquad (7.103)$$

Since this upper bounds the mixing time, we would like to find a t satisfying

$$\exp\left(-\frac{(t\frac{1}{d} - d)^2}{2t\frac{1}{d}}\right) \leq \varepsilon, \qquad (7.104)$$

that is,

$$\frac{(t - d^2)^2}{2td} \geq -\ln(\varepsilon). \qquad (7.105)$$

Solving this inequality, we get that the mixing time is upper bounded by

$$d^2 + d\ln\left(\frac{1}{\varepsilon}\right) + \sqrt{2d^3\ln\left(\frac{1}{\varepsilon}\right) + d^2\ln\left(\frac{1}{\varepsilon}\right)^2} \qquad (7.106)$$

which is polynomial in both d and $-\ln(\varepsilon)$.

∎

7.2.5   Mixing of Markov chains on direct product spaces

When a state space is a direct product of smaller spaces, and the Markov chain changes only one coordinate in one step, then the Markov chain is rapidly mixing if the number of coordinates is not large and the Markov chains restricted on each coordinate are rapidly mixing. Below we state this theorem precisely and prove it.

Theorem 80. *Let $\mathcal{M}$ be a class of Markov chains whose state space is a K-dimensional direct product of spaces, and the problem size of a particular chain is denoted by n (where we assume that $K = O(poly_1(n))$).*

Any transition of the Markov chain $M \in \mathcal{M}$ changes only one coordinate (each coordinate with equal probability), and the transition probabilities do not depend on the other coordinates. The transitions on each coordinate form irreducible, aperiodic Markov chains (denoted by $M_1, M_2, \ldots M_K$), which are reversible with respect to a distribution π_i. Furthermore, each of $M_1, \ldots M_K$ are rapidly mixing, i.e., with the relaxation time $\frac{1}{1-\lambda_{2,i}}$ is bounded by a $O(poly_2(n))$ for all i. Then the Markov chain M converges rapidly to the direct product of the π_i distributions, and the second-largest eigenvalue of M is

$$\lambda_{2,M} = \frac{K - 1 + \max_i \{\lambda_{2,i}\}}{K}$$

and thus the relaxation time of M is also polynomially bounded:

$$\frac{1}{1 - \lambda_{2,M}} = O(poly_1(n)poly_2(n)).$$

Proof. The transition matrix of M can be described as

$$\frac{\sum_{i=1}^{K} \left[\bigotimes_{j=1}^{i-1} \mathbf{I}_j\right] \otimes \mathbf{M}_i \otimes \bigotimes_{j=i+1}^{K} \mathbf{I}_j}{K}$$

where $\otimes$ denotes the usual tensor product from linear algebra, $\mathbf{M}_i$ denotes the transition matrix of the Markov chain on the ith coordinate, and $\mathbf{I}_j$ denotes the identical matrix with the same size as $\mathbf{M}_j$. Since all pairs of terms in the sum above commute, the eigenvalues of M are

$$\left\{ \frac{1}{K} \sum_{i=1}^{K} \lambda_{j_i,i} : 1 \leq j_i \leq |\Omega_i| \right\}$$

where Ω_i is the state space of the Markov chain M_i on the ith coordinate. The second-largest eigenvalue of M is then obtained by combining the maximal second-largest eigenvalue (maximal among all the second-largest eigenvalues of the component transition matrices) with the other largest eigenvalues, i.e., with all others being 1s:

$$\frac{K - 1 + \max_i \{\lambda_{2,i}\}}{K} .$$

If g denotes the smallest spectral gap, ie., $g = 1 - \max_i \{\lambda_{2,i}\}$, then from the above, the second-largest eigenvalue of M is

$$\frac{K - g}{K} = 1 - \frac{g}{K}$$

namely, the second-largest eigenvalue of M is only K times closer to 1 than the maximal second-largest eigenvalue of the individual Markov chains. $\square$

7.3 Self-reducible counting problems

Self-reducible counting problems have been introduced in Section 6.5. We already learned there that it is possible to sample uniformly in polynomial time those self-reducible objects whose counting problem is in FP. Here we show that self-reducible counting problems have two additional important properties. Any self-reducible counting problem has an FPRAS if and only if it has an FPAUS. They also have a dichotomy theorem: any self-reducible problem either has an FPRAS or cannot be approximated in polynomial time with a polynomial approximation factor. We are going to state these precisely and prove them in the following subsections. Both proofs use the tree structure of the solution space.

7.3.1   The Jerrum-Valiant-Vazirani theorem

Mark Jerrum, Leslie Valiant and Vijay Vazirani proved that any self-reducible problem can be approximated with an FPRAS if and only if it can be sampled with an FPAUS [103]. The proof in both directions uses the tree structure of the solution space. The FPAUS $\Rightarrow$ FPRAS direction is relatively easy, although technically tedious. The idea is that a small number of samples of the solutions of a problem instance labeling a vertex v following the (almost) uniform distribution are sufficient to estimate what fraction of solutions are below the children of v. The FPRAS $\Rightarrow$ FPAUS direction is trickier. FPRAS estimations do not lead directly to an FPAUS; a rejection sampling is needed to get an FPAUS. Below we state the theorem and prove it.

Theorem 81. *Let $\#A$ be a self-reducible counting problem. Then $\#A$ is in FPRAS if and only if $\#A$ is in FPAUS.*

Proof. The proof in both directions is constructive. To construct an FPAUS algorithm from an FPRAS algorithm, we use the rejection sampling technique. The global picture of the construction is the following. We construct a random sampler based on FPRAS estimations which generates a random solution from a non-uniform distribution. This distribution is far from being almost uniform, however, it has the property that in a rejection sampling algorithm, it has a fair probability to get accepted. If it is accepted, then the solution is drawn from exactly the uniform distribution. If it is not accepted, then we start over till the first acceptance or a given number of trials. If none of the proposals is accepted, we generate an arbitrary solution. This procedure generates a distribution which is almost uniform. More precisely, we could do this, if in the FPRAS estimations it was guaranteed to achieve a given approximation ratio. However, there is a small probability that the FPRAS estimations are out of the given boundaries, and in that case the prescribed procedure fails to work. However, we can handle this small probability, and we have an almost uniform sampler even if we recognize that with a small probability the procedure fails and generate a solution from an unknown distribution.

Let x be a problem instance of $\#A$, and let n denote the size of x. Since there is an ε in the definitons of FPRAS and FPAUS, and we have to use both of them, we will denote the ε in FPRAS by ε' in this proof. Let a problem instance x and $\varepsilon > 0$ be given. We are going to construct a rejection sampling algorithm with the following properties:

1. A solution of x is generated which is accepted with at least $\frac{1}{4e^2}$ probability.

2. Given that a solution is accepted, with less than $\frac{\varepsilon}{2}$ probability, the algorithm does not work properly; it generates a solution of x from an unknown distribution. On the other hand, with at least $1 - \frac{\varepsilon}{2}$ probability, the solution is from exactly the uniform distribution of solutions.

3. The running time of the algorithm is polynomial with both the size of x and $-\log(\varepsilon)$.

We claim that such rejection sampler is sufficient to get an FPAUS algorithm with parameter ε. Indeed, repeat the rejection sampler till the first acceptance but at most $4e^2 \log\left(\frac{2}{\varepsilon}\right)$ times. Then the probability that all samples are rejected is

$$\left(1 - \frac{1}{4e^2}\right)^{4e^2 \log\left(\frac{2}{\varepsilon}\right)} < \left(\frac{1}{e}\right)^{\log\left(\frac{2}{\varepsilon}\right)} = \frac{\varepsilon}{2}. \tag{7.107}$$

In that case, generate a solution from an arbitrary distribution, for example, accept the last solution, whatever is it. The running time of this procedure is polynomial with both the size of x and $-\log(\varepsilon)$. It generates a solution following a distribution which is a convex combination of three distributions:

1. With at most $\frac{\varepsilon}{2}$ probability, it is an unknown distribution due to rejecting all proposals in the rejection sampling.

2. With at most $\frac{\varepsilon}{2}$ probability, it is an unknown distribution because the algorithm does not work properly when accepting a proposal.

3. With at least $1 - \varepsilon$ probability, it is the uniform distribution.

It is easy to see that the total variation distance of the convex combination of these three distributions and the uniform distribution is at most ε.

So we only have to provide a rejection sampling method with the prescribed properties. We can do it in the following way. Let v be the root of the tree describing the solution space of x, and let $d = O(poly(n))$ denote the depth of the tree.

While v is not a leaf, iterate the following. Obtain the children of v and compute the problem instances labeling these vertices. This can be done in polynomial time due to the definition of self-reducibility. For each of these problem instances, estimate their number of solutions using the available FPRAS algorithm with $\varepsilon' = \frac{1}{2d}$ relative error and set δ to $\frac{\varepsilon}{2(d+1)}$. Due to the definition of FPRAS, this runs in polynomial time with both n and $-\log(\varepsilon)$. Then select a random child u from the distribution being proportional to the estimated number of solutions. Set v to u.

If the selected nodes are $v = v_0, v_1, \ldots v_m$, then the probability of generating the solution y is

$$p(y) = \prod_{i=0}^{m} P(v_i) \tag{7.108}$$

where $P(v_i)$ is the probability of selecting child v_i in the ith iteration.

The probability $p(y)$ can be calculated exactly, so in the rejection sampling, we can set f to p. The target distribution is the uniform one, so the $g \equiv 1$ function suffices as a function proportional to the target distribution. To set up a rejection sampling, we need an enveloping constant such that for any solution y, the equation

$$cg(y) = c \geq f(y) \tag{7.109}$$

holds. For this, we estimate $Y = |\{y|xRy\}|$ with the available FPRAS algorithm setting ε' to 1 and δ to $\frac{\varepsilon}{2(d+1)}$. Let this estimation be denoted by $\hat{Y}$. We claim that the inverse of it multiplied by $2e$ will be a good estimation for the enveloping constant with high probability. Indeed, if all FPRAS approximations are in the given boundary, then on one hand it holds that

$$c := \frac{2e}{\hat{Y}} \geq \frac{e}{Y}, \tag{7.110}$$

and on the other hand,

$$\frac{e}{Y} \geq \left(1 + \frac{1}{2d}\right)^{2d} \frac{1}{Y} \geq p(y). \tag{7.111}$$

The left-hand side inequality in Equation (7.111) is trivial. To see the right-hand side inequality, observe the following. If exact calculations were given instead of FPRAS in all steps during the construction of y, then p would be the uniform distribution, and thus $p(y)$ was $\frac{1}{Y}$. However, in case of FPRAS approximation, it could happen that the number of solutions below the node that was selected was overestimated by a $1 + \frac{1}{d}$ factor, while all other nodes were underestimated by the same factor. Still, the error introduced at that point cannot be larger than the square of this factor. There are at most n iterations, and although the errors here are multiplicative, it still cannot be larger than

$$\left(1 + \frac{1}{2d}\right)^{2d}. \tag{7.112}$$

What is the probability of acceptance? With similar considerations, it is easy to see that

$$c := \frac{2e}{\hat{Y}} \leq \frac{4e}{Y}, \tag{7.113}$$

and

$$\frac{1}{eY} \leq \frac{\frac{1}{Y}}{\left(1 + \frac{1}{2d}\right)^{2d}} \leq p(y). \tag{7.114}$$

Thus the ratio of $p(y)$ and c is smaller than $4e^2$, thus the acceptance ratio is at least $\frac{1}{4e^2}$. All these calculations hold only if all the FPRAS approximations are in the prescribed boundaries. The probability that any of the FPRAS approximations fall out of the boundaries is at most

$$1 - \left(1 - \frac{\varepsilon}{2(d+1)}\right)^{d+1} \leq \frac{\varepsilon}{2}, \tag{7.115}$$

due to the Bernulli 's inequality.

Therefore the prescribed properties hold for the rejection sampler, and thus, the overall procedure is indeed an FPAUS algorithm.

The global picture on how to construct an FPRAS algorithm from an FPAUS algorithm is the following. We sample solutions using the FPAUS algorithm and estimate which fraction of the solutions fall below the different children of the root. If there are m children and the number of samples falling below child v_i is n_i, then the estimated fraction is

$$\hat{f}_i = \frac{n_i}{\sum_{j=1}^{m} n_j}.$$

We select the child with the largest fraction, and iterate this procedure. We will arrive at a particular solution after k steps. Let $\hat{f}_1, \hat{f}_2, \ldots, \hat{f}_k$ denote the estimations of the largest fractions during this process. We give an estimation of the number of solutions as

$$\prod_{i=1}^{k} \frac{1}{\hat{f}_i}.$$

We claim that this is an FPRAS algorithm when the parameters in this procedure are chosen appropriately.

Let ε' and δ be the given parameters of the required FPRAS algorithm. For each internal node we visit during the procedure, we prescribe a total variation distance for the FPAUS algorithm and prescribe a number of samples. Generating the prescribed number of samples with the prescribed FPAUS, we give an estimation of the largest fraction of solutions that fall below a particular child u with the property that the relative error is at most $\frac{\varepsilon'}{d}$ with at least $1 - \frac{\delta}{d}$ probability, where d is still the depth of the tree. We construct the problem instance x_1 labeling u, and sample solutions of x_1 to estimate the largest fraction of the solutions that fall below a child. We iterate this procedure until we arrive at a leaf. Then the solution represented by the leaf is a fraction of the solution space, and the inverse of this fraction is the number of solutions. We have an estimation for this fraction. It is the product of the fractions estimated during the iterations. Since we have at most d estimations of fractions, this indeed leads to an FPRAS algorithm.

Due to the definition of self-reducibility, there are at most $|\Sigma|^{\sigma(x)}$ number of children of an internal node labeled by the problem instance x. For sake of simplicity, let the largest number of children be denoted by N. It is easy to see that $N = O(poly(n))$. Due to the pigeonhole rule, there is a child below which at least $\frac{1}{N}$ fraction of the solutions reside. Let u denote this child. If the samples came from the uniform distribution, the fraction of samples below u would be an unbiased estimator of the probability that a solution is below u. However, the samples are only almost uniform, and this causes a systematic bias. Fortunately, we can require that ε in the FPAUS algorithm is $\frac{\varepsilon'}{2d|\sigma|^{poly(n)}}$ and then the running time of the FPAUS algorithm is still polynomial in both n and $\frac{1}{\varepsilon'}$. Furthermore, the systematic error is smaller than $\frac{\varepsilon'}{2d}$. The number of samples must be set such that the probability that the measurement error is larger than $\frac{\varepsilon'}{2d}$ is smaller than $\frac{\delta}{d}$. Let the probability that a solution sampled by the FPAUS algorithm is below u be denoted by p. Then the number of solutions below u follows a binomial distribution with parameter p and expectation mp, where m is the number of samples. We can use Chernoff's inequality saying that for the actual number of samples Y below u, it holds that

$$P\left(Y < \frac{mp}{1 + \frac{\varepsilon'}{2d}}\right) \le P\left(Y < mp\left(1 - \frac{\varepsilon'}{3d}\right)\right) \le$$

$$exp\left(-\frac{1}{2p}\frac{mp - mp\left(1 - \frac{\varepsilon'}{3d}\right)^2}{m}\right). \tag{7.116}$$

The right-hand side should be bounded by $\frac{\delta}{2d}$ (the other half of the probability will go to the other tail):

$$exp\left(-\frac{1}{2p}\frac{mp - mp\left(1 - \frac{\varepsilon'}{3d}\right)^2}{m}\right) \le \frac{\delta}{2d}. \tag{7.117}$$

Solving Equation (7.117) we get that

$$m \ge \frac{-2\log\left(\frac{\delta}{2d}\right)}{p\frac{\varepsilon'}{3d}}. \tag{7.118}$$

It is easy to see that this bound for m is polynomial in n, $\frac{1}{\varepsilon'}$, and $-\log(\delta)$. Indeed, p is larger than $\frac{1}{N}$ and d is polynomial in n.

For the other tail, we can also use Chernoff's inequality:

$$P\left(Y > mp\left(1 + \frac{\varepsilon'}{2d}\right)\right) \le exp\left(-\frac{m(1 - p) - (m - mp(1 + \frac{\varepsilon'}{2d}))^2}{2(1 - p)m}\right). \tag{7.119}$$

Upper bounding this by $\frac{\delta}{2d}$ and solving the inequality, we get that

$$m \ge \frac{-2(1 - p)\log\left(\frac{\delta}{2d}\right)}{p^2 \frac{\varepsilon'^2}{(2d)^2}}. \tag{7.120}$$

It is again polynomial in all n, $\frac{1}{\varepsilon'}$, and $-\log(\delta)$. Therefore, a polynomial number of samples are sufficient to estimate the fraction of solutions below vertex u with prescribed relative error and prescribed probability. □

7.3.2 Dichotomy theory on the approximability of self-reducible counting problems

Alistair Sinclair and Mark Jerrum observed that very rough estimations of the number of solutions of self-reducible counting problems is sufficient to obtain an FPAUS algorithm [156]. What follows is that if we have a very rough estimation, then we also have an FPRAS, due to Theorem 81. That is, there is an approximable/non-approximable dichotomy for self-reducible problems: either they have a rough approximation, and then they also have an FPRAS, or they do not have even a rough approximation.

The key observation is that a Markov chain on a tree is always reversible, and it is easy to estimate its stationary distribution.

Lemma 19. *Let M be a Markov chain whose states are vertices of a tree, $G = (V, E)$, and for any two states $v, u \in V$, $P(u|v) \neq 0$ if and only if u and v are adjacent. Then there exists a function from $w : E \to \mathbb{R}^+$ with the following properties:*

1. For any $(u, v) \in E$ it holds that

$$T(u|v) = \frac{w((u, v))}{\sum_{u'|(u',v) \in E} w((u', v))} \tag{7.121}$$

and

$$T(v|u) = \frac{w((u, v))}{\sum_{v'|(u,v') \in E} w((v', u))}. \tag{7.122}$$

2. The Markov chain is reversible with respect to the distribution

$$\pi(v) := \frac{\sum_{u|(u,v) \in E} w((u, v))}{2 \sum_{e \in E} w(e)}. \tag{7.123}$$

Proof. We prove that the weight function w can be constructed iteratively. Let v be an arbitrary leaf, connected to u. Define $w((v, u)) := T(u|v)(= 1)$. Then consider the other edges of u. For each of u's neighbors $v' \neq v$, define

$$w((v', u)) := \frac{T(v'|u)w((v, u))}{T(v|u)}. \tag{7.124}$$

Then indeed for each neighbor v' of u, it holds that

$$T(v'|u) = \frac{w((u, v'))}{\sum_{v''|(u,v'') \in E} w(v''|u)}, \tag{7.125}$$

since

$$\sum_{v''|(u,v'') \in E} w(v''|u) = \frac{\sum_{v''|(u,v'') \in E} T(v''|u)w((v, u))}{T(v|u)} = \frac{w((v, u))}{T(v|u)}, \tag{7.126}$$

and thus

$$T(v'|u) = \frac{w((u,v'))}{\sum_{v''|(u,v'')\in E} w(v''|u)} = \frac{\frac{T(v'|u)w((v,u))}{T(v|u)}}{\frac{w((v,u))}{T(v|u)}}. \tag{7.127}$$

Now for any neighbor v' of u, it holds that either v' is a leaf or for one of the edges of v', the weight of the edge is defined and for all other edges, the weight is not defined. If v' is a leaf, then for its edge weight and transition probability, Equation (7.121) naturally holds. For any internal nodes, the weights of the other edges can be defined similarly to u.

We can iterate this procedure until all vertices are reached. Due to the tree structure, it is impossible that a vertex is visited twice.

It is easy to see that the measure π defined in Equation (7.123) is indeed a distribution. The detailed balance also holds, since

$$\pi(v)T(u|v) = \frac{\sum_{u'|(u',v)\in E} w((u',v))}{2\sum_{e\in E} w(e)} \frac{w((u,v))}{\sum_{u'|(u',v)\in E} w((u',v))} =$$

$$\frac{w((u,v))}{2\sum_{e\in E} w(e)} = \frac{\sum_{v'|(v',u)\in E} w((v',u))}{2\sum_{e\in E} w(e)} \frac{w((u,v))}{\sum_{v'|(v',u)\in E} w((v',u))} =$$

$$\pi(u)T(v|u). \tag{7.128}$$

$\square$

The idea of Sinclair and Jerrum was to reverse the construction: we can define weights for the edges that will define a corresponding Markov chain. If all weights of the edges connecting leaves to the remaining tree are the same, then the stationary distribution is the uniform one on the leaves. We need two further properties: i) the probability of the leaves in the stationary distribution must be non-negligible, ii) the Markov chain must be rapidly mixing. Both properties hold if the weights come from a very rough estimation of the number of solutions, as stated and proved in the following theorem.

Theorem 82. *Let $\#A$ be a self-reducible counting problem. Let $\mathcal{C}$ be a polynomial time computable function such that for any problem instance x of $\#A$, $\mathcal{C}(x)$ gives an approximation for the number of solutions of x with an approximation factor $\mathcal{F}(x) = poly(|x|)$. Let $\mathcal{M}$ be a set of Markov chains such that for each problem instance x in $\#A$, it contains a Markov chain M. The state space of M is the vertices of the tree representing the solution space of x, and the transition probabilities are defined in the following way. For each edge (u,v), where u is a child of v, define $w((u,v)) := \mathcal{C}(x_u)$, where x_u is the problem instance labeling u, if u is an internal edge, and let $w((u,v))$ be 1 if u is a leaf. The transition probability $T(u|v) \neq 0$ if and only if u and v are neighbors. In that case, it is defined as*

$$T(u|v) := \frac{w((u,v))}{\sum_{u'|(u',v)\in E} w((u',v))}. \tag{7.129}$$

Then $\mathcal{M}$ is a rapidly mixing Markov chain, its stationary distribution on the leaves is the uniform distribution, and the inverse of the probabilities of the leaves in the stationary distribution is polynomially bounded.

Proof. It is easy to see that the stationary distribution is the uniform one on the leaves, since weights of edges incident to leaves are the same. It is also easy to show that

$$\sum_{v \in L} \pi(v) \geq \frac{1}{1 + 2g(x)(1 + \mathcal{F}(x))}, \tag{7.130}$$

where $g(x)$ is the function from the definition of self-reducibility measuring the length of the solutions, and L is the set of the leaves in the tree. If $\mathcal{C}$ measured exactly the number of solutions, then the following inequality would be true:

$$\sum_{v \in V_d} \sum_{u|(u,v) \in E} \mathcal{C}(x_u) \leq 2|\{x|xRy\}| \tag{7.131}$$

where V_d is the set of internal vertices at depth d. Indeed, if d is such that there is no leaf at depth d, then equality would hold. For d such that there are leaves at depth d, the solutions represented by those leaves are out of the summation. Since $\mathcal{C}$ has an approximation factor $\mathcal{F}(x)$, it still holds that

$$\sum_{v \in V_d} \sum_{u|(u,v) \in E} \mathcal{C}(x_u) \leq (1 + \mathcal{F}(x))2|\{x|xRy\}|. \tag{7.132}$$

Thus we have that

$$\sum_{v \in V \setminus L} \sum_{u|(u,v) \in E} \mathcal{C}(x_u) \leq \sum_{d=0}^{g(x)} \sum_{v \in V_d} \sum_{u|(u,v) \in E} \mathcal{C}(x_u) \leq$$
$$2g(x)(1 + \mathcal{F}(x))|\{y|xRy\}|. \tag{7.133}$$

We get that

$$\sum_{v \in L} \pi(v) = \frac{|\{y|xRy\}|}{2\sum_{e \in E} w(e)} =$$

$$\frac{|\{y|xRy\}|}{|\{y|xRy\}| + \sum_{v \in V \setminus L} \sum_{u|(u,v) \in E} \mathcal{C}(x_u)} \geq$$

$$\frac{|\{y|xRy\}|}{|\{y|xRy\}| + 2g(x)(x)|\{y|xRy\}|} = \frac{1}{1 + 2g(x)(1 + \mathcal{F}(x))}. \tag{7.134}$$

To prove the rapid mixing, we can use an idea similar to that in Example 21. The conductance of the Markov chain is taken on a connected subtree S. There are two cases: the root of S is the root of the whole tree or not. If the root of S is not the root of the whole tree, then we show that the ergodic flow on the root is already comparable with $\pi(S)$. Let v denote the root of S, let x_v denote the problem instance labeling v, let u denote the parent of v, and let V_d denote the set of vertices in S in depth d. Observe that

$$T(u|v) \geq \frac{1}{1 + \mathcal{F}(x)^2}. \tag{7.135}$$

Indeed, if $\mathcal{C}$ measured exactly the number of solutions, then the transition probability from a vertex to its root was exactly $\frac{1}{2}$. Since there is an $\mathcal{F}(x)$ approximation factor, in the worst case it could happen that the number of solutions for the problem instance x_v is maximally underestimated, and the number of solutions for each child of v is maximally overestimated. In that case,

$$T(u|v) = \frac{w((u,v))}{\sum_{u'|(u',v)\in E} w((u',v))} =$$

$$\frac{\frac{|\{y|x_v Ry\}|}{1+\mathcal{F}(x)}}{\frac{|\{y|x_v Ry\}|}{1+\mathcal{F}(x)} + |\{y|x_v Ry\}|(1+\mathcal{F}(x))} = \frac{1}{1+\mathcal{F}(x)^2}. \tag{7.136}$$

Then

$$\Phi = \frac{F(S)}{\pi(S)} \geq \frac{\pi(v)T(u|v)}{\sum_{v\in S}\pi(v)} \geq \frac{\frac{|\{y|x_v Ry\}|}{1+\mathcal{F}(x)}\frac{1}{1+\mathcal{F}(x)^2}}{\sum_{v'\in S}\sum_{u|(u,v')\in E} w((u,v'))} \geq$$

$$\frac{\frac{|\{y|x_v Ry\}|}{1+\mathcal{F}(x)}\frac{1}{1+\mathcal{F}(x)^2}}{|\{y|x_v Ry\}| + |\{y|x_v Ry\}|2g(x)(1+\mathcal{F}(x))} =$$

$$\frac{1}{(1+\mathcal{F}(x))(1+\mathcal{F}(x)^2)(1+2g(x)(1+\mathcal{F}(x)))}. \tag{7.137}$$

If the root of S is the root of the whole tree, then the complement of S is the union of disjoint subtrees. Similar to the previous calculations, the ergodic flow going out of their root is at most a polynomial factor smaller than the probability of the trees. Since the Markov chain is reversible, the ergodic flow of S equals the ergodic flow of the complement. Thus, the ergodic flow of S is at most a polynomial factor smaller than the probability of the complement of S, which cannot be smaller than the probability of S. Thus, the inverse of the conductance in this case is also polynomially bounded.

Since the conductance is polynomially bounded, the Markov chain is rapidly mixing. $\qquad\square$

Corollary 20. *Let $\#A$ be a self-reducible counting problem. Let $\mathcal{C}$ be a polynomial time computable function such that for any problem instance x of $\#A$, $\mathcal{C}(x)$ gives an approximation for the number of solutions of x with a polynomial approximation factor. Then $\#A$ is in FPAUS and thus is in FPRAS.*

Proof. We know that there is a rapidly mixing Markov chain on the vertices of the tree representing the solution space of the problem instance x, such that its stationary distribution restricted to the leaves of the tree is the uniform distribution; furthermore, the inverse of the probabilities of the leaves in the stationary distribution is polynomially bounded. Then we can sample a vertex of the tree from a distribution being very close to its stationary distribution in polynomial time. The probability the sample is a not a leaf is less than

$$1 - \frac{1}{poly(|x|)} \tag{7.138}$$

for some polynomial. Then the probability that none of the samples is a leaf from $O(poly(|x|, -\log(\varepsilon)))$ number of samples is less than $\frac{\varepsilon}{2}$. In that case we choose an arbitrary solution of x. The number of steps of the Markov chain for one sample should be set such that the total variation distance of the distribution after the given number of steps restricted to the leaves and the uniform distribution of the leaves should be smaller than $\frac{\varepsilon}{2}$. Then the following procedure will be an FPAUS. The procedure makes the prescribed number of steps in the Markov chain, returns the current state, and iterates it till the first sampled leaf but at most $O(poly(|x|, -\log(\varepsilon)))$ times; and in case of no sampled leaves, it returns an arbitrary solution. Since #A is self-reducible, if it is in FPAUS, it is also in FPRAS. □

7.4 Further reading and open questions

7.4.1 Further reading

- There are several variants of the distinguished path technique. Jason Schweinsberg introduced a variant where he proved that for any subset B of the state space V, it holds that

$$\frac{1}{1 - \lambda_2} \leq \frac{4L}{\pi(B)} \max_{e \in E} \frac{1}{Q(e)} \sum_{x \in V, y \in B | e \ni \gamma_{x,y}} \pi(x)\pi(y). \tag{7.139}$$

  The statement also has a probabilistic (multicommodity flow) version. Schweinsberg used this theorem to give an asymptotically sharp upper bound on the relaxation time of a Markov chain walking on the leaf-labeled, unrooted binary tree [152].

- The coupling from the past technique was developed by James Propp and David Wilson [143]. The technique provides samples following exactly the stationary distribution of a Markov chain. The price we have to pay for the perfect samples is that the running time of the method is a random variable. We consider a particular algorithm efficient if the expected running time grows polynomially with the size of the problem instance. An example of such a fast perfect sampler was published by Mark Huber. His method perfectly samples linear extensions of posets [95].

- Russ Bubley and Martin Dyer introduced the path coupling technique, where two Markov chains are coupled via a path of intermediate states. Using this technique, they proved fast mixing of a Markov chain on linear extensions of a poset. Their method gave a better upper bound on the mixing time that could be achieved using a geometric approach [29].

- Cheeger's inequalities say that a Markov chain is rapidly mixing if and only if there is no bottleneck in them. However, bottlenecks might not only be geometric when there is only a few edges between two subsets of vertices in the Markov graph. They might also be probabilistic bottlenecks where two subsets of vertices are connected with many edges, however, the average transition probabilities are very small. Examples of such bottlenecks are presented in [79] and [131]. Goldberg and Jerrum showed that the so-called Burnside process converges slowly. It is a random walk on a bipartite graph whose vertices are the members of a permutation group and combinatorial objects on which the group acts. Two vertices are connected if the group element fixes the combinatorial object. Restricting the walk to the combinatorial objects (that is, considering only every second step on the bipartite graph) yields a Markov chain whose Markov graph is fully connected. Indeed, the identity of the group fixes all combinatorial objects. Still, the Markov chain might be torpidly mixing for some permutation groups, as proved by Goldberg and Jerrum. Miklós, Mélykúti and Swenson considered a Markov chain whose state space is the set of shortest reversal sorting paths of signed permutations [131]. The corresponding Markov graph is fully connected, however, the Markov chain is torpidly mixing since the majority of the transitions have negligible probability.

- The Lazy Markov chain technique is somewhat paradoxical in the sense that we have to slow down a Markov chain to prove its rapid mixing. To avoid it, we have to prove that the smallest eigenvalue is sufficiently separated from -1. Diaconis and Strock proved an inequality on the smallest eigenvalue [54]. Greenhill used a similar inequality to prove rapid mixing of the non-lazy version of a Markov chain sampling regular directed graphs [83].

- The logarithmic Sobolev inequality was given by Gross [86]. Diaconis and Saloff-Coste showed its application in mixing time of Markov chains [53]. They gave an inequality for the mixing time where $\log\left(\frac{1}{\pi(x_i)}\right)$ in Equation (7.8) is replaced with $\log\left(\log\left(\frac{1}{\pi(x_i)}\right)\right)$ while $1 - \lambda_2$ is replaced with the so-called logarithmic Sobolev constant α. Rothaus proved that $\alpha \geq 2(1 - \lambda_2)$ [147]. What follows is that upper bounds on the mixing time using logarithmic Sobolev inequalities might be several orders smaller than the upper bound given in Equation (7.8). These upper bounds might be asymptotically optimal in several cases.

7.4.2   Open problems

- It is easy to see that if a self-reducible counting problem is in FP, then there exists a polynomial running time algorithm that generates samples following exactly the uniform distribution. However, we do not know if the opposite is true. Namely, if a self-reducible counting problem has an exact uniform sampler in polynomial time, does that imply that it is in FP?

- We do not know if there exists a counting problem which is in FPRAS but not in FPAUS, or in FPAUS but not in FPRAS.

- We do not know if there is a problem in #P-complete for which a random solution can be generated from exactly the uniform distribution in deterministic polynomial time.

7.5   Exercises

1. Let T be the transition matrix of a Markov chain constructed by the Metropolis-Hastings algorithm. Prove that there are matrices W and Λ for which
$$T = W\Lambda W^{-1}$$
and Λ is diagonal.

2. Let T be the transition matrix of a reversible Markov chain. Prove that
$$\frac{T + I}{2}$$
can be diagonalized.

3. * Let p_1, p_2 and p_3 be three distributions over the same domain, and let α be a real number in $[0, 1]$. Show that
$$d_{TV}(p_1, \alpha p_2 + (1 - \alpha)p_3) \leq \alpha d_{TV}(p_1, p_2) + (1 - \alpha)d_{TV}(p_1, p_3).$$

4. * Let M be a Markov chain reversible with respect to the distribution π. Show that if M is periodic with time period 2, then -1 is an eigenvalue of M. Construct its corresponding eigenvector, too.

5. Show that a reversible Markov chain cannot be periodic with a time period larger than 2.

6. Let $G = (V, E)$ be a simple graph, and let $d(v)$ denote the degree of v. Let M be a Markov graph whose state space is V, and for each $(u, v) \in E$ it has a transition from u to v with probability $\frac{1}{d(u)}$ and a transition from v to u with probability $\frac{1}{d(v)}$. Show that G is reversible, and find its stationary distribution.

7. Let $\tau(\varepsilon)$ denote $\max_i\{\tau_i(\varepsilon)\}$. Show that

$$\tau(\varepsilon) \leq \tau\left(\frac{1}{2e}\right)\ln\left(\frac{1}{\varepsilon}\right).$$

8. ○ Let X be a finite space and let π be any non-vanishing distribution on it. Show that the set

$$\left\{S \subset X \,\middle|\, 0 < \pi(S) \leq \frac{1}{2}\right\}$$

has size at least $2^{|X|-1} - 1$.

9. ○ Prove that the number of connected subgraphs of a graph might be an exponential function of the number of vertices of the graph.

10. What is the diameter of the d-dimensional hypercube?

11. What is the diameter of the convex body whose vertices are those vertices of the d-dimensional unit hypercube $\{0, 1\}^d$ which contain an even number of 1s on their coordinates.

12. Consider the Markov chain on a d-dimensional lattice that contains k vertices in each dimension. The state space is the k^d vertices, and there is a transition between neighbor vertices, each of them with probability $\frac{1}{2d}$. Furthermore, vertices at the border have non-zero transition loop probabilities, that is, the Markov chain has non-zero probability to stay in those states. Use the Cheeger inequality and Theorem 72 to show that the mixing time of this Markov chain grows polynomial with both d and k.

13. ○ Consider a set of Markov chains $\mathcal{M}$ that contains a Markov chain M_n for each positive integer n. The state space of M_n contains the permutations S_n, and there is a transition between any two permutations differing in two neighbor positions. Each transition has probability $\frac{1}{n-1}$. Use the canonical path method to prove that $\mathcal{M}$ is rapidly mixing.

14. Consider a set of Markov chains $\mathcal{M}$ that contains a Markov chain M_n for each positive integer n. The state space of M_n contains the permutations S_n, and there is a transition between any two permutations differing in two (not necessarily neighbor) positions. Each transition has probability $\frac{2}{n(n-1)}$. Use the canonical path method to prove that $\mathcal{M}$ is rapidly mixing.

15. * Let G be a simple graph which is a union of paths and cycles. Let M be a Markov chain whose state space contains the (not necessarily perfect) matchings of G. There is a transition from a matching x_1 to another matching x_2 if they differ in exactly one edge. The transition probability is

$$\min\left\{\frac{1}{2k}, \frac{1}{2n}\right\}$$

where k is the number of edges in the matching that contains more edges and n is the number of possible extensions of the other matching with one more edge. In each state, the Markov chain remains in the state with the remaining probability. Show that this is indeed a Markov chain, that is, the sum of the prescribed probabilities never exceeds 1, furthermore, show that the chain is reversible with respect to the uniform distribution.

16. ◦ Using the canonical path method, show that the Markov chain in Exercise 15 is rapidly mixing, that is, the mixing time only grows polynomial with the size of the ground graph G.

17. ◦ Let M be a Markov chain whose states are the monotone paths on the 3D lattice from $(0,0,0)$ to (a,b,c) $(a,b,c \in \mathbb{Z}^+)$. Each path can be described as the series of steps in the directions of the 3 axes of the 3D coordinate system. For example, $xxzy$ means 2 steps in the first dimension, one step in the third dimension, and one step in the second dimension. There is a transition between two states if they differ in two consecutive steps. The transition probability between any two such states is

$$\frac{1}{a+b+c-1}.$$

The chain remains in the same state with the remaining probabilities. Show that this Markov chain converges to the uniform distribution. Using multicommodity flow, show that the mixing time grows only polynomially with $a+b+c$.

18. * Alice and Bob are playing against the devil. The devil put a hat on Bob's head and a hat on Alice's head. The hat might be white or black and the two hats might be the same or different colors. Alice and Bob might discuss a strategy before the game, but after that they cannot communicate and can see only the hat of the opposite player. From this information, they have to guess the color of *their own* hat.

(a) What should their strategy be to guarantee that with probability 1, one of them guesses correctly?

(b) Work out a strategy such that with probability 1, one of them guess correctly. Furthermore, for both Alice and Bob, it holds that their guess is "white" with probability 0.5 regardless of the color of the other person's hat.

19. Let $\mathcal{M}$ be a set of Markov chains that contains a Markov chain M_n for each positive integer n. The state space of M_n is the set of Dyck words of length $2n$, and the transition probabilities are defined by the following algorithm.

   (a) Draw uniformly a random number between 1 and $2n - 1$.

   (b) If swapping the characters in positions i and $i + 1$ is also a Dyck word, then swap them. Otherwise, do nothing.

   Using the coupling argument, prove that $\mathcal{M}$ is rapidly mixing.

20. ○ Use the coupling technique to show that the Markov chain in Example 21 is rapidly mixing.

21. Consider the Markov chain on the d-dimensional toroidal lattice that contains k vertices on its circles in each dimension. The state space of the Markov chain contains the k^d vertices of the toroidal lattice, and the transition probabilities are between neighbor vertices; each transition probability is $\frac{1}{2d}$. Using the coupling technique, prove that the mixing time grows polynomially with both d and k.

22. * Show that counting the spanning trees of a graph is a self-reducible counting problem.

23. Show that counting the shortest paths between two vertices of a graph is a self-reducible counting problem.

24. * Let D be a degree sequence of length n, and let F be a subset of edges of the complete graph K_n. We say that $G = (V, E)$ is a realization of D avoiding F if $E \cap F = \emptyset$ (and obviously, G is a realization of D). Show that counting the realizations of a degree sequence avoiding a star is a self-reducible counting problem.

25. Show that counting the alignments of two sequences is a self-reducible counting problem.

7.6   Solutions

Exercise 3. The proof comes directly from the definition of total variation distance and basic properties of the absolute value function. Indeed,

$$d_{TV}(p_1, \alpha p_2 + (1-\alpha)p_3) = \frac{1}{2}\sum_x |p_1(x) - \alpha p_2(x) - (1-\alpha)p_3(x)| \le$$

$$\frac{1}{2}\sum_x (|\alpha p_1(x) - \alpha p_2(x)| + |(1-\alpha)p_1(x) - (1-\alpha)p_3(x)|) =$$

$$\alpha d_{TV}(p_1, p_2) + (1-\alpha)d_{TV}(p_1, p_3).$$

Exercise 4. Let U and V be the two classes of the bipartite Markov graph. Consider the vector $\tilde{\pi}$ defined as

$$\tilde{\pi}(u) := \pi(u)$$

for all $u \in U$ and

$$\tilde{\pi}(v) = -\pi(v)$$

for all $v \in V$. It is easy to see that $\tilde{\pi}$ is an eigenvector with eigenvalue -1. Indeed, observe that for any $u \in U$

$$\sum_{v \in V} \pi(v)T(u|v) = \pi(u)$$

since π is an eigenvector with eigenvalue 1, and thus

$$\sum_{v \in V} \tilde{\pi}(v)T(u|v) = \sum_{v \in V} -\pi(v)T(u|v) = -\pi(u).$$

Observe that similar equalities hold for all $v \in V$. That is,

$$T\tilde{\pi} = -\tilde{\pi},$$

which means that $\tilde{\pi}$ is indeed an eigenvector with eigenvalue -1.

Exercise 8. Observe that for any $S \subseteq X$, the equation

$$\pi(S) + \pi(\overline{S}) = 1$$

holds.

Exercise 9. Consider, for example, the caterpillar binary tree, that is, the binary tree in which the internal vertices form a path.

Exercise 13. A good canonical path can be defined by bubbling down the appropriate elements of the permutation to the appropriate positions. We can set an upper bound on the maximum load of an edge similar to the calculations in the solution of Example 24. The difference is that we do not have to index the member of the permutations; observe that the indexed sequences in Example 24 behave like the permutations.

Exercise 15. Consider the Markov chain M' that with $\frac{1}{2}$ probability, deletes a uniformly selected edge from the presented edges and with $\frac{1}{2}$ probability, adds an edge uniformly selected from those edges with which the matching can be extended. Observe that the given transition probabilities are the transition probabilities of the Markov chain that we get by applying the Metropolis-Hastings algorithm on M' and setting the target distribution to the uniform one.

Exercise 16. The canonical path can be constructed similar to the Markov chain presented in Section 8.2.5.

Exercise 17. The construction of the multicommodity flow is similar to the one presented in the solution of Example 24.

Exercise 18.

(a) Alice will say the color of Bob's hat, Bob will say the opposite color of Alice's hat. If they have hats of the same color, then Alice will guess properly, if they have hats of different colors, then Bob will guess correctly.

(b) They toss a fair coin. If it is a head, they play the previous strategy, if it is a tail, they flip the roles.

Exercise 20. Couple the two Markov chains when they are both at the root of the tree. Show that the waiting time for this event only grows polynomially with the size of the problem.

Exercise 22. Let $G = (V, E)$ be a simple graph. Fix an arbitrary total order on E. Let e_i be the smallest edge in E. Then in any spanning tree T of G, either $e \in T$ or $e \notin T$. If $e_i \in T$, then let T^{e_i} be the tree obtained from T by contracting e, and let G^{e_i} be the graph obtained from G by contracting e_i. Then T^{e_i} is a spanning tree of G^{e_i}, furthermore any spanning tree $\tilde{T}^{e_i}$ of G^{e_i} can be obtained by contracting e in a spanning tree $\tilde{T}$ of G. Since for any $T \neq \tilde{T}$, it also holds that $T^{e_i} \neq \tilde{T}^{e_i}$, contracting e_i in those spanning trees of G that contain e is a bijection between those trees and the spanning trees of G^{e_i}.

Similarly, there is a bijection between the spanning trees of G not containing e_i and the spanning trees of $G \setminus \{e_i\}$. This provides a way to encode the spanning trees of G. If the encoding starts

$$e_i \in T,$$

then the extensions are the spanning trees of G^{e_i}; if the encoding starts

$$e_i \notin T,$$

then the extensions are the spanning trees of $G \setminus \{e_i\}$. It is easy to see that the granulation function σ has $O(\log(n))$ values: encoding index i takes $\log(n)$

bits, where $n = |V|$. It is also easy to see that all g, σ and ϕ are polynomial time computable functions, the encoding of both G^{e_i} and $G \setminus \{e_i\}$ are shorter than the encoding of G, and property 3d in the definition of self-reducibility also holds due to the above-mentioned bijections.

Exercise 24. Fix an arbitrary total order on the vertices. Let $D = d(v_1), d(v_2), \ldots d(v_n)$ denote the degree sequence. Let v_1 be the smallest vertex and assume that there is already given a star S with center v_0 as a forbidden set of edges. S might be empty. Let v_i be the smallest vertex such that $(v_1, v_i) \notin S$. Then the realizations of D avoiding S and not containing edge (v_1, v_i) are the realizations of D avoiding $S \cup \{(v_1, v_i)\}$ and the realizations of D avoiding S and containing edge (v_1, v_i) are the realizations of the degree sequence $D' = d(v_1) - 1, d(v_2), \ldots, d(v_i) - 1, \ldots, d(v_n)$ avoiding $S \cup \{(v_1, v_i)\}$. Finally, the realizations of D avoiding the star with $n - 1$ leaves and center v_1 are the realizations of $D" = d(v_2), \ldots, d(v_n)$.

Chapter 8

Approximable counting and sampling problems

There are #P-complete counting problems that are very easy to approximate. We already discussed in Chapter 6 that an FPAUS can be given for sampling satisfying assignments of a disjunctive normal form. Since counting the satisfying assignments of a DNF is a self-reducible counting problem, we can conclude that #DNF is also in FPRAS.

The core of the mentioned FPAUS is a rejection sampler for which the inverses of the acceptance probabilities are polynomial bounded. In this chapter, we show two further examples of FPAUS algorithms based on rejection samplers.

We know only Markov chain approaches for generating almost uniform samples of solutions to problem instances in #P-complete counting problems. The second part of this chapter introduces examples for such Markov chains.

8.1 Sampling with the rejection method

8.1.1 #Knapsack

There are several variants of the Knapsack problem. Here we introduce a simple variant whose counting version is in FPAUS and FPRAS.

Problem 17.
Name: #KNAPSACK.
Input: a series of integer numbers $w_1 \leq w_2 \leq \ldots \leq w_n$ and an integer W.
Output: the number of subsets of indexes $I \subseteq [n]$ satisfying that

$$\sum_{i \in I} w_i \leq W. \tag{8.1}$$

The counting problem #Knapsack is known to be #P-complete [136]. Following the work of Dyer [56], we show that it is in FPAUS and FPRAS. The problem is very similar to the money change problem, and indeed, it is easy to see that an (algebraic) dynamic programming algorithm can be constructed with $O(nW)$ running time. However, this is exponential in the *number of digits* of W. If W is a small number, say, less than or equal to n^2, then the algebraic dynamic programming approach still runs in polynomial time with n. Therefore, we might assume that $W > n^2$.

To handle this case, we scale down the weights to get a feasible solution. We define new weights as

$$w_i' := \left\lfloor \frac{n^2}{W} w_i \right\rfloor \tag{8.2}$$

and we are interested in the subset of weights whose sum does not exceed n^2. Let $\mathcal{S}$ denote the set of solutions of the original problem, and let $\mathcal{S}'$ denote the set of solutions of the modified problem. It is easy to see that $\mathcal{S} \subseteq \mathcal{S}'$. Indeed, for any $I \in \mathcal{S}$, it holds that

$$\sum_{i \in I} w_i' \leq \frac{n^2}{W} \sum_{i \in I} w_i \leq \frac{n^2}{W} W = n^2. \tag{8.3}$$

On the other hand, $\mathcal{S}' \setminus \mathcal{S}$ might be non-empty. However, we are going to show that

$$\frac{|\mathcal{S}'|}{n+1} \leq |\mathcal{S}|. \tag{8.4}$$

Let $I' \in \mathcal{S}' \setminus \mathcal{S}$ and let I be obtained from I' by removing the largest index k. We show that I is in $\mathcal{S}$. The first observation is that $w_k > \frac{W}{n}$. Indeed, if w_k was less than or equal to $\frac{W}{n}$, then it would hold that

$$\sum_{i \in I'} w_i \leq \sum_{i \in I'} \frac{W}{n} \leq n \frac{W}{n} = W,$$

contradicting that $I' \notin \mathcal{S}$. In particular, $w'_k \geq n$.

To simplify the notation, we define

$$\delta_i := w_i \frac{n^2}{W} - w'_i,$$

that is,

$$w_i = \frac{W}{n^2} \left(w'_i + \delta_i \right).$$

We are ready to show that $I' \setminus \{k\}$ is in $\mathcal{S}$. Indeed,

$$
\begin{aligned}
\sum_{i \in I' \setminus \{k\}} w_i &= \frac{W}{n^2} \sum_{i \in I' \setminus \{k\}} (w'_i + \delta_i) \\
&= \frac{W}{n^2} \left(\sum_{i \in I'} w'_i - w'_k + \sum_{i \subset I' \setminus \{k\}} \delta_i \right) \\
&\leq \frac{W}{n^2} \left(\sum_{i \in I'} w'_i - w'_k + n \right) \quad \text{since } \forall \delta_i \leq 1 \\
&\leq \frac{W}{n^2} \sum_{i \in I'} w'_i \quad \text{since } w'_k \geq n \\
&\leq \frac{W}{n^2} n^2 = W.
\end{aligned}
$$

Since any $I \in \mathcal{S}$ can be obtained in at most n different ways from an $I' \in \mathcal{S}' \setminus \mathcal{S}$, it follows that

$$|\mathcal{S}' \setminus \mathcal{S}| \leq n|\mathcal{S}|.$$

Therefore

$$|\mathcal{S}'| \leq (n+1)|\mathcal{S}|.$$

Uniformly sampling from $\mathcal{S}'$ can be done in polynomial time with n. The probability of sampling a solution from $\mathcal{S}$ is larger than $\frac{1}{n+1}$. Since checking if a solution is indeed in $\mathcal{S}$ can be done in $O(n \log(W))$, we can use rejection sampling to obtain an FPAUS algorithm whose running time is polynomial with n and $\log(W)$ (that is, polynomial with the size of the input) and also polynomial with $-\log(\varepsilon)$, where ε is the allowed deviation from the uniform distribution measured in total variation distance.

8.1.2 Edge-disjoint tree realizations without common internal vertices

We define the following counting problem.

Problem 18.
Name: #MinDegree1-2Trees-DegreePacking.

Input: two degree sequences $D = d_1, d_2, \ldots, d_n$, $F = f_1, f_2, \ldots, f_n$ such that $\sum_i d_i = \sum_i f_i = 2n - 2$ and for all i, $\min\{d_i, f_i\} = 1$.
Output: the number of edge-disjoint tree realizations of D and F.

We will refer to D and F as tree degree sequences without common internal nodes. To prove that #MinDegree1-2Trees-DegreePacking is in FPAUS and FPRAS, we need the following observations. The first is a well-known fact, and its proof can be found in many textbooks on enumerative combinatorics.

Observation 2. *The number of trees with degree sequence $d_1, d_2, \ldots, d_n$ is*

$$\frac{(n-2)!}{\prod_{k=1}^{n}(d_k - 1)!}. \tag{8.5}$$

We can use this observation to prove the following lemma.

Lemma 21. *Let D and F be tree degree sequences without common internal nodes. Let T_D and T_F be two, independent random tree realizations following the uniform distribution. Then the expected number of common edges is 1.*

Proof. First, we make an observation. Let T be a random tree following the uniform distribution of the realizations of the degree sequence $d_1, d_2, \ldots, d_n$. Then the probability that v_i and v_j are adjacent in T is

$$\frac{d_i + d_j - 2}{n - 2}. \tag{8.6}$$

To see this, map the trees in which v_i and v_j are adjacent to the trees of $n-1$ vertices with degree sequence

$$D' = d_1, d_2, \ldots, d_{i-1}, d_{i+1}, \ldots, d_{j-1}, d_{j+1}, \ldots, d_n, d_i + d_j - 2$$

by contracting v_i and v_j. It is easy to see that each tree realization of D' is an image $\binom{d_i + d_j - 2}{d_i - 1}$ times. Therefore, the number of tree realizations of D in which v_i is adjacent to v_j is

$$\frac{(n-3)!}{(d_i + d_j - 3)! \prod_{k \neq i,j}(d_k - 1)!} \frac{(d_i + d_j - 2)!}{(d_i - 1)!(d_j - 1)!}. \tag{8.7}$$

The probability that v_i and v_j are adjacent is the ratio of the values in Equations (8.7) and (8.5), which is indeed

$$\frac{d_i + d_j - 2}{n - 2}. \tag{8.8}$$

We define the following sets.

$$\mathcal{A} := \{v_i | d_i = 1 \wedge f_i > 1\} \tag{8.9}$$

$$\mathcal{B} := \{v_j | d_j > 1 \wedge f_j = 1\} \tag{8.10}$$

$$\tag{8.11}$$

Observe that there might be parallel edges of the two tree realizations only between these two sets. The expected number of edges is

$$\sum_{v_i \in \mathcal{A}} \sum_{v_j \in \mathcal{B}} \frac{(d_i - 1)(f_j - 1)}{(n-2)^2} = \sum_{v_i \in \mathcal{A}} \frac{d_i - 1}{n-2} \sum_{v_j \in \mathcal{B}} \frac{f_j - 1}{n-2}$$

$$\sum_{i=1}^{n} \frac{d_i - 1}{n-2} \sum_{j=1}^{n} \frac{f_j - 1}{n-2} = 1 \tag{8.12}$$

since $d_i = 1$ for all $v_i \in \overline{\mathcal{A}}$ and $f_j = 1$ for all $v_j \in \overline{\mathcal{B}}$, and the sum of the degrees decreased by 1 is $n - 2$ in any tree degree sequence. $\qquad\square$

We are ready to prove the following theorem.

Theorem 83. *The counting problem #MinDegree1-2Trees-DegreePacking is in FPAUS and FPRAS*

Proof. Let D and F be two degree sequences without common internal nodes. If there is a vertex v_i such that $d_i = n-1$ or $f_i = n-1$, then $d_i + f_i = n$, and it implies that D and F does not have edge-disjoint tree realizations. So we can assume that there are vertices $v_{i_1}, v_{i_2}, v_{j_1}$ and v_{j_2} such that $d_{i_1}, d_{i_2} > 1$ and $f_{j_1}, f(j_2) > 1$. The number of tree pairs (T_D, T_F) in which (v_{i_1}, v_{j_1}) and (v_{i_1}, v_{j_1}) are edges in both T_D and T_F is

$$\frac{(n-4)!}{(d_{i_1} - 2)!(d_{i_2} - 2)! \prod_{k \neq i_1, i_2}(d_k - 1)!} \times$$
$$\frac{(n-4)!}{(f_{j_1} - 2)!(f_{j_2} - 2)! \prod_{k \neq j_1, j_2}(f_k - 1)!} \tag{8.13}$$

Since the expected number of common edges is 1, the number of edge-disjoint realizations is also at least this number. What follows is that the probability that two random trees are edge disjoint is at least

$$p := \frac{(d_{i_1} - 1)(d_{i_2} - 1)(f_{j_1} - 1)(f_{j_2} - 1)}{(n-2)^2(n-3)^2}. \tag{8.14}$$

Sampling a random tree from the uniform distribution of realizations of a degree sequence can be done in polynomial time. The key observation is that the probability that a fixed leaf is adjacent to a vertex v_i is $\frac{d_i - 1}{n-2}$. Then we can select a random neighbor of a given leaf, and then generate a random tree of the remaining degree sequence.

We claim that the following algorithm is an FPAUS. Given a $\varepsilon > 0$, generate a random pair of tree realizations of D and F till the first edge-disjoint realization, but at most $\frac{-\log(\varepsilon)}{p}$ times. If none of the pairs are edge-disjoint, generate an arbitrary solution. Such a solution can be done in polynomial time [117]. The probability that none of the trials are edge-disjoint realizations is

$$(1 - p)^{\frac{-\log(\varepsilon)}{p}}. \tag{8.15}$$

It is easy to show that it is indeed smaller than ε. The generated edge-disjoint realization follows a distribution π which is a convex combination of two distributions: the uniform distribution and an unknown. Since the coefficient of the unknown distribution is smaller than ε, the total variation distance between π and the uniform distribution is less than ε.

Let ξ denote the indicator variable that two random trees are edge disjoint. Then the number of edge-disjoint tree realizations is

$$E[\xi]\frac{(n-2)!}{\prod_{k=1}^{n}(d_k-1)!}\frac{(n-2)!}{\prod_{k=1}^{n}(f_k-1)!}. \tag{8.16}$$

Since the inverse of the expectation is polynomial bounded, the necessary number of samples for an FPRAS estimation is polynomial in all n, $\frac{1}{\varepsilon}$, and $-\log(\delta)$. $\qquad\square$

8.2 Sampling with Markov chains

8.2.1 Linear extensions of posets

Let $(A, \le)$ be a finite poset, $|A| = n$, and let M be a Markov chain whose states are the total orderings of $(A, \le)$. The transitions are defined by the following algorithm. Let $a_{i_1}, a_{i_2}, \ldots, a_{i_n}$ be the current total ordering. With probability $\frac{1}{2}$ do nothing, namely, we are operating a Lazy Markov chain. With probability $\frac{1}{2}$, select an index k uniformly between 1 and $n-1$. If a_{i_k} and $a_{i_{k+1}}$ are uncomparable, swap them, otherwise do nothing. It is easy to see that the Markov chain is reversible with respect to the uniform distribution.

We are going to prove that the Markov chain is rapidly mixing, that is, the inverse of the spectral gap is polynomial bounded. We use the right Cheeger inequality and geometric considerations. Recall that the poset polytope is given by the inequalities

$$0 \quad \le \quad x_i \le 1 \tag{8.17}$$
$$x_i \quad \le \quad x_j \quad \forall \, a_i \le a_j. \tag{8.18}$$

Since any poset polytope contains the points $(0, 0, \ldots, 0)$ and $(1, 1, \ldots, 1)$, the diameter of a poset polytope of n elements is $\sqrt{n}$. We can define the poset polytope of any total ordering. Since there are $n!$ total orderings of an antichain of size n, and the polytopes of its total orderings has the same volume, the volume of a total ordering polytope of n elements is $\frac{1}{n!}$.

The transitions of the Markov chain corresponds to the common facet of the two total ordering polytopes. It is defined by the inequality system

$$0 \le x_{i_1} \le x_{i_2} \le \ldots \le x_{i_k} = x_{i_{k+1}} \le \ldots \le x_{i_n}. \tag{8.19}$$

It is almost a poset polytope of a total ordering of $n - 1$ elements except it is stretched by a factor of $\sqrt{2}$ in the direction $x_{i_k} = x_{i_{k+1}}$. Hence the surface ($n - 1$-dimensional volume) of the facet is $\frac{\sqrt{2}}{(n-1)!}$.

Now consider the subset S of the total orderings that defines the conductance of M. Let U be the union of the poset polytopes of total orderings in S, and let W be the union of the poset polytopes of total orderings in $\overline{S}$. Finally, let C denote the surface between U and W. Then we know that

$$\frac{\Phi S}{\pi S} = \frac{\sum_{x \in S} \sum_{y \in \overline{S}} \pi(x) T(y|x)}{\sum_{x \in S} \pi(x)} = \frac{\frac{A(C)}{\sqrt{2}} \pi(\cdot) \frac{1}{2(n-1)}}{\frac{V(U)}{\frac{1}{n!}} \pi(\cdot)} = \frac{A(C)}{V(U)} \frac{1}{\sqrt{2} 2 n (n-1)} \tag{8.20}$$

where $\pi(\cdot)$ denotes the probability of any element of the state space, V denotes the volume, and A denotes the area. Indeed, S contains $\frac{V(U)}{\frac{1}{n!}}$ total orderings, and the number of transitions between S and $\overline{S}$ is $\frac{A(C)}{\frac{\sqrt{2}}{(n-1)!}}$. The poset polytope is a convex body, since it is defined by a set of linear inequalities. Therefore, we can apply Theorem 72 saying that

$$\frac{A(C)}{V(U)} \geq \frac{1}{\sqrt{n}}. \tag{8.21}$$

Indeed, $\min\{V(U), V(W)\} = V(U)$, and the diameter is $\sqrt{n}$. Thus we get that

$$\Phi \geq \frac{1}{\sqrt{2} 2 n (n-1) \sqrt{n}}. \tag{8.22}$$

Combining it with the right Cheeger inequality, we get that

$$\lambda_2 \leq 1 - \frac{\Phi^2}{2} \leq 1 - \frac{1}{16 n^3 (n-1)^2} \tag{8.23}$$

that is,

$$\frac{1}{1 - \lambda_2} \leq 16 n^3 (n-1)^2. \tag{8.24}$$

The number of linear extensions is clearly at most $n!$. Therefore, by applying Theorem 68, we get for the relaxation time that

$$\tau_i(\varepsilon) \leq 16 n^3 (n-1)^2 \left(\log(n!) + \log\left(\frac{1}{\varepsilon}\right) \right). \tag{8.25}$$

This mixing time is clearly polynomial in both n and $-\log(\varepsilon)$. Finding one total ordering of a poset can be done in polynomial time as well as performing one step in a Markov chain can be done in polynomial time. What follows is that there is an FPAUS for almost uniformly sampling linear extensions of a poset. Since the number of linear extensions is a self-reducible counting problem, we get the following theorem.

Theorem 84. *The counting problem #LE is in FPRAS.*

8.2.2 Counting the (not necessarily perfect) matchings of a graph

We showed in Subsection 4.2.4 that counting the matchings in a graph is #P-complete and it remains #P-complete even in planar graphs. Jerrum and Sinclair [99] showed that the problem is in FPAUS and since the problem is self-reducible, it is also in FPRAS.

The FPAUS is provided by a rapidly mixing Markov chain. The authors proved the rapid convergence of a more general Markov chain. Let $G = (V, E)$ be an arbitrary graph, and let $w : E \to \mathbb{R}^+$ be edge weights. If $M \subseteq E$ is a matching, the weight of the matching is

$$W(M) := \prod_{e \in M} w(e). \tag{8.26}$$

Consider a Markov chain whose state space is the set of the possible matchings of G. We will denote this set with $\mathcal{M}(G)$. If M is the current state, choose an edge $e = (u, v)$ uniformly at random, and do the following:

(a) If $e \in M$, then move to $M \setminus \{e\}$ with probability $\frac{1}{1+w(e)}$.

(b) If $e \notin M$ and there is no edge in M which is incident to u or v, then move to $M \cup \{e\}$ with probability $\frac{w(e)}{1+w(e)}$.

(c) If $e' = (u, w) \in M$ and there is no edge in M which is incident to v, then move to $M \cup \{e\} \setminus \{e'\}$ with probability $\frac{w(e)}{w(e)+w(e')}$.

(d) Otherwise, do nothing.

The lazy version of this Markov chain is clearly irreducible and aperiodic. It is easy to show that it is reversible with respect to the distribution

$$\pi(M) \propto W(M). \tag{8.27}$$

Indeed, there is a transition from M to M' if M and M' differ by an edge or they differ by two adjacent edges. In the first case, w.l.o.g. we can assume that $M' = M \cup \{e\}$. Then

$$\pi(M')T(M|M') = \pi(M)w(e)T(M|M') =$$
$$\pi(M)w(e)\frac{1}{1 + w(e)} = \pi(M)T(M'|M). \tag{8.28}$$

In the second case, w.l.o.g. we can assume that $M' = M \cup \{e\} \setminus \{e'\}$. Then

$$\pi(M')T(M|M') = \pi(M)\frac{w(e)}{w(e')}T(M|M') =$$
$$\pi(M)\frac{w(e)}{w(e')}\frac{w(e')}{w(e) + w(e')} = \pi(M)T(M'|M). \tag{8.29}$$

We will apply the canonical path method to prove rapid mixing (Theorem 76). We fix an arbitrary total ordering of the edges, and construct a path system Γ, which contains a path for any pair of matchings.

Let X and Y be two matchings. We create a canonical path from X to Y in the following way, on which Z will denote the current realization. The symmetric difference $X \Delta Y$ is a union of disjoint alternating paths and alternating cycles. The total ordering of the edges defines a total ordering of the components. We transform X to Y by working on the components in increasing order. Let C_i be the ith component. There are two cases: C_i is either an alternating path or an alternating cycle. If it is a path, then consider its edges started from the end of the path containing the smaller edge, and denote them by $e_1, e_2, \ldots, e_{l_i}$. If e_1 is presented in Z, remove it. If it is not presented in Z, then add it, and remove e_2. Then continue the transformation by removing and adding edges till all the edges in Y are added, and all the edges in X are removed (the last step might be adding the last edge of the path).

If C_i is a cycle, then remove the smallest edge $e \in C_i \cap Z$. Then $C_i \setminus \{e\}$ is a path, and work with it as described above.

To give an upper bound on K_Γ, we introduce

$$M := X \Delta Y \Delta Z \tag{8.30}$$

for each Z. Let $e = (Z, Z')$ be an edge in the Markov graph. How many canonical paths use this edge? We claim that X and Y can be unequivocally reconstructed from the edge (Z, Z') and M. Indeed, observe that

$$M \Delta Z = X \Delta Y, \tag{8.31}$$

therefore $X \Delta Y$ can be reconstructed from e and M. From this, we can identify which is the component C_i that is being changed (recall that there is a total ordering of the components that does not depend on Z). Observe that all edges in $C_j \cap Z$, $j < i$ come from Y and all edges in $C_j \setminus Z$ come from X. Similarly, for all $k > i$, all edges in $C_k \cap Z$ come from X, and all edges in $C_k \setminus Z$ come from Y. Furthermore, from the edge e, we can also tell which edges in C_i come from X and which edges come from Y. Since X, Y and Z are identical on $Z \setminus (X \Delta Y)$, we can determine X and Y from e and M.

What does M look like? We claim that M is almost a matching except it might contain a couple of adjacent edges. Indeed, M is a matching on $M \setminus C_i$, and might contain two adjacent edges on C_i if one of them is already removed from Z and the other is going to be added to Z'. Let M^* be M if M is a matching (which is the case if the edge (Z, Z') represents the first alteration on the current component), and otherwise let M^* be $M \setminus \{f\}$, where f is the edge of the adjacent pair of edges in M which is going to be added by the operation represented by e.

Also observe that

$$\pi(M^*) = \frac{1}{w(f)} \frac{\pi(X)\pi(Y)}{\pi(Z)}. \tag{8.32}$$

It is also clear that any path from any X to Y has length at most $\frac{3n}{5}$, where n is the number of vertices. The smallest transition probability is $\frac{1}{1+w_{\max}}$, where $w_{\max}$ is the largest weight. In this way, we can get an estimation of the load of an edge $e = (Z, Z')$. Let F denote the set containing the edge in G that is added in the operation represented by e if such edge exists. If the operation represented by e deletes only an edge, then F is the empty set. Let $w(F) := \prod_{f \in F} w(f)$, where the empty product is defined as 1. Then we have the following upper bound on the load of e.

$$\frac{1}{Q(e)} \sum_{\gamma_{X,Y} \ni e} |\gamma_{X,Y}| \pi(X)\pi(Y) \leq \frac{3n}{5(1 + w_{\max})} \sum_{\gamma_{X,Y}} \frac{\pi(X)\pi(Y)}{\pi(Z)} \leq$$

$$\frac{3n(1 + w_{\max})}{5} \sum_{M \in \mathcal{M}(G)} \pi(M)w(F) \leq \frac{3n(1 + w_{\max})}{5} w_{\max}. \qquad (8.33)$$

Therefore, K_Γ in Theorem 76 is $O(nw_{\max}^2)$. According to the theorem, the Markov chain is rapidly mixing if the maximum weight is upper bounded by a polynomial function of n.

8.2.3 Sampling realizations of bipartite degree sequences

The number of bipartite degree sequence realizations is the following problem.

Problem 19.
Name: #BDSR.
Input: a degree sequence $D = \{d_{1,1}, d_{1,2}, \ldots, d_{1,n}\}, \{d_{2,1}, d_{2,2}, \ldots, d_{2,m}\}$.
Output: the number of vertex-labelled simple bipartite graphs $G = (U, V, E)$, such that for all i, $d(u_i) = d_{1,i}$, and for all j, $d_{v_j} = d_{2,j}$.

The decision version problem is the bipartite degree sequence problem. If there is at least one simple graph whose degrees are D, then D is called *graphical*, and such a simple graph G is called a *realization* of D.

The bipartite degree sequence problem is in P. It is the straight corollary of the Gale-Ryser theorem (see Exercise 42 in Chapter 2).

A *swap operation* removes two edges $e_1 = (u_1, v_1)$ and $e_2 = (u_2, v_2)$, $u_1 \neq u_2$, $v_1 \neq v_2$ from a (bipartite) graph G, and adds two edges (u_1, v_2) and (u_2, v_1) (given that neither of these edges are presented in G). Such a swap operation is denoted by

$$(u_1, v_1) \times (u_2, v_2).$$

Clearly, a swap operation does not change the degrees of the vertices in G. These small perturbations are sufficient to transform any realizations of a bipartite degree sequence to any another one, as the following theorem states.

Theorem 85. *Let G and G' be two realizations of the same bipartite degree sequence, then there exists a finite series of graphs*

$$G = G_1, G_2, \ldots, G_k = G'$$

such that for all $i = 1, 2, \ldots, k-1$, G_i can be transformed into G_{i+1} with a single swap operation. Furthermore, for the number of swap operations it holds that

$$k - 1 < \frac{|E(G)\Delta E(G')|}{2}.$$

Proof. Take the symmetric difference $H = G\Delta G'$. Since G and G' are realizations of the same bipartite degree sequence, each component of H is a Eulerian graph, and can be decomposed into alternating cycles (the edges in a cycle come from G and G' alternating) since the graphs are bipartite. Fix a cycle decomposition of $H = \bigcup_{j=1}^{m} C_j$. Clearly,

$$G' = G\Delta C_1 \Delta C_2 \Delta \ldots \Delta C_m.$$

Therefore, it is sufficient to show how to transform G into $G\Delta C$, where C is an alternating cycle (the edges in C are alternatingly presented and absent along the cycle). Take a walk along C starting in the smallest vertex u_1 and start walking on the edge in G'. Let $u_1, v_1, u_2, v_2, \ldots u_l, v_l$ be the vertices of C along this walk. Observe that the edge (u_1, v_1) is not in G, by the definition of the walk. Find the smallest i such that (u_1, v_i) is an edge in G. Such i exists, since (u_1, v_l) is an edge in G. Then the swap operations

$$(u_1, v_i) \times (u_i, v_{i-1}), (u_1, v_{i-1}) \times (u_{i-1}, v_{i-2}), \ldots, (u_1, v_2) \times (u_2, v_1)$$

can be applied on G (in the given order). If $i = l$, then G is transformed into $G\Delta C$. Otherwise the vertices $u_1, v_i, u_{i+1}, v_{i+1}, \ldots v_l$ represent an alternating cycle in the modified graph. This cycle is smaller than C, and (u_1, v_i) is not an edge in the modified graph. We can apply the same procedure on this shorter cycle, and after a finite number of iterations, i will be l, and thus, G will be transformed into $G\Delta C$.

We showed how to transform G into $G\Delta C_1$ with a finite series of swap operations. Then since $G\Delta C_1\Delta C_2 = (G\Delta C_1)\Delta C_2$, G can be transformed into $G\Delta C_1\Delta C_2$ with a finite series of swap operations, and generally into $G' = G\Delta C_1\Delta C_2\Delta \ldots \Delta C_m$ with a finite series of swap operations.

The number of swap operations necessary to transform G into $G\Delta C$ is $\frac{|C|-2}{2}$. Therefore, the number of necessary swap operations to transform G into G' is indeed less than $\frac{|E(G)\Delta E(G')|}{2}$. $\qquad\square$

We will call the series of swap operations transforming G into $G\Delta C$ the *sweeping process*, and u_1 is the *cornerstone* of the sweeping. Theorem 85 provides a way to design a Markov chain by exploring the realizations of a bipartite degree sequence. Indeed, it is easy to see that the lazy version of the following Markov chain is reversible, aperiodic and irreducible. The state space contains the realizations of a bipartite degree sequence. The transition probability from G_1 to G_2 is

$$\frac{1}{\binom{n}{2}\binom{m}{2}}$$

if G_1 can be transformed into G_2 with a single swap operation, and 0 otherwise. The transition probability from G to itself (the probability that the Markov chain remains in the same state G) is

$$1 - \frac{c(G)}{\binom{n}{2}\binom{m}{2}}$$

where $c(G)$ denotes the number of possible swap operations on G.

It is conjectured that this Markov chain is rapidly mixing for any degree sequence. However, it is proved only for some special classes of degree sequences. Here we introduce the sketch of the proof for *half regular* bipartite degree sequences. First, we define these degree sequences.

Definition 62. *A bipartite degree sequence* $D = \{d_{1,1}, d_{1,2}, \ldots, d_{1,n}\}$, $\{d_{2,1}, d_{2,2}, \ldots, d_{2,m}\}$ *is half regular if for all* i_1 *and* i_2, $d_{1,i_1} = d_{1,i_2}$.

The proof is based on the multicommodity flow method (Theorem 78). The multicommodity flow is defined in the following way. Let X and Y be the realizations of the same half regular bipartite degree sequence D. The cardinalities of the two vertex classes are denoted by n and m. Let $H = X \Delta Y$. For each vertex w in H, there are $\left(\frac{d(w)}{2}\right)!$ ways to pair the edges of X to the edges of Y incident to v, where $d(w)$ denotes the degree of w in H. For each possible pairing on each vertex in H, we define a cycle decomposition on H. Let Φ denote a fixed ensemble of pairings, and let $\varphi_v(e)$ denote the pair of e in Φ on vertex v. Let u_i be the smallest index vertex in H that does not have degree 0. Let v_j be the smallest index vertex for which (u_i, v_j) is in $H \cap Y$. Let (u_i, v_j) be denoted by e. Then define a circuit that starts with e, ends with $\varphi_{u_i}(e)$, and contains edges

$$e = e_1, e_2, \ldots, e_l$$

where for each $e_k = (u_k, v_k)$, e_{k+1} is defined as $\varphi_{v_k}(e_k)$. Denote the so-defined circuit by $\mathcal{C}_1$. If $H \setminus \mathcal{C}_1$ is not the empty graph, repeat the same on $H \setminus \mathcal{C}_1$ to get a circuit $\mathcal{C}_2$. The process is iterated till $H \setminus (\mathcal{C}_1 \cup \mathcal{C}_2 \cup \ldots \cup \mathcal{C}_s)$ is the empty graph. Then each $\mathcal{C}_i$ is decomposed into cycles $C_{i,1}, C_{i,2}, C_{i,j_i}$. The cycle $C_{i,1}$ is the cycle between the first and second visit of w, where w is the first revisited vertex in $\mathcal{C}_i$ (note that w might be both in U and V). Then $C_{i,2}$ is defined in the same way in $\mathcal{C}_i \setminus C_{i,1}$, etc. The path from X to Y is defined by processing the cycles

$$C_{1,1}, C_{1,2}, \ldots C_{1,j_1}, C_{2,1}, \ldots C_{s,s_j}$$

applying the sequence of swap operations as described in the proof of Theorem 85. For the so-obtained path γ, we define

$$f(\gamma) := \frac{\pi(X)\pi(Y)}{\prod_{w \in V(H)} \left(\frac{d(w)}{2}\right)!}.$$

Let the number of realizations of D be N. Since π is the uniform distribution, and the length of any path is less than nm according to Theorem 85, it holds that

$$\frac{1}{Q(e)} \sum_{\gamma \ni e} f(\gamma)|\gamma| < \frac{\binom{n}{2}\binom{m}{2}nm}{N \prod_{w \in V(H)} \left(\frac{d(w)}{2}\right)!} \sum_{\gamma \ni e} 1.$$

Therefore, if the number of paths in the path system going through any edge is less than

$$poly(n,m)N \prod_{w \in V(H)} \left(\frac{d(w)}{2}\right)!$$

for some $poly(n,m)$, then the swap Markov chain is rapidly mixing. Let Z be a realization on a path γ going from X to Y obtained from the ensemble of pairings Φ, and let e be the transition from Z to Z'. Let M_G denote the adjacency matrix of a bipartite graph G, and let

$$\hat{M} := M_X + M_Y - M_Z. \tag{8.34}$$

Miklós, Erdős and Soukup [130] proved that the path γ and thus X and Y can be unequivocally obtained from $\hat{M}$, Φ, e and $O(\log(nm))$ bits of information. The proof is quite involved and thus omitted here. The corollary is that the number of paths going through a particular $e = (Z, Z')$ is upper bounded by

$$poly(n,m)|\mathcal{M}_Z| \prod_{v \in V(H)} \left(\frac{d(v)}{2}\right)! \tag{8.35}$$

where $\mathcal{M}_Z$ are the set of possible $\hat{M}$ matrices defined in Equation (8.34). Therefore, it is sufficient to show that the number of $\hat{M}$ matrices is upper bounded by

$$poly(n,m)N.$$

To prove this, we show that any possible $\hat{M}$ is in Hamming distance 12 from an adjacency matrix of a realization of the degree sequence D. This is sufficient to prove that

$$|\mathcal{M}_Z| \leq poly(n,m)N$$

since $\hat{M} \in \{-1, 0, 1, 2\}^{n \times m}$, and the number of matrices in $\{-1, 0, 1, 2\}^{n \times m}$ which are at most Hamming distance 12 from a given adjacency matrix of a realization of D is

$$\sum_{i=1}^{12} \binom{nm}{i} 3^i.$$

First observe that $\hat{M}$ has the same row and column sums as the adjacency matrix of any realization of D, and it might contain at most 3 values which are not 0 or 1. Indeed, if

$$Z = X \Delta C_1 \Delta C_2 \Delta \ldots \Delta C_k$$

then $\hat{M}$ is a 0-1 matrix, and thus is an adjacency matrix of a realization of D. If Z is a realization during processing a cycle C_i, then there might be 3 chords of C_i whose corresponding values in $\hat{M}$ are not 0 or 1. Particularly, there might be two 2 values and one -1. A value 2 might appear when Z does not contain an edge which is presented both in X and Y, and a -1 might appear when Z contains an edge which is neither in X nor in Y. Furthermore these "bad" values are in the same line corresponding to the cornerstone of the sweeping process.

Assume that $\hat{m}_{i,j} = 2$. There must be an i' such that $\hat{m}_{i',j} = 0$. Since D is half-regular, each row in the adjacency matrix of a realization has the same sum. Due to the pigeonhole rule, there is a j', such that $\hat{m}_{i,j'} < \hat{m}_{i',j'}$. Since all bad values are in the same row, it follows that $\hat{m}_{i,j'} \leq 0$ and $\hat{m}_{i',j'} \geq 0$. If $\hat{m}_{i,j'} = -1$ and $\hat{m}_{i',j'} = 0$, then there must be another j'' such that $\hat{m}_{i,j''} < \hat{m}_{i',j''}$, and since there are at most one -1 in $\hat{M}$, it follows that $\hat{m}_{i,j''} = 0$ and $\hat{m}_{i',j''} = 1$. Recall j'' to j' in this case. Changing $\hat{m}_{i,j}$ to 1, $\hat{m}_{i',j'}$ to 1, increasing $\hat{m}_{i,j'}$ by 1, and decreasing $\hat{m}_{i',j'}$ by 1 does not change the row and column sums, however, it eliminates at least one "bad" value from $\hat{M}$. Let's call this modified matrix $\hat{M}'$. It is easy to see that $\hat{M}'$ has a Hamming distance 4 from $\hat{M}$.

If there is another 2 value in $\hat{M}'$, we can use the same procedure. The so-obtained $\hat{M}''$ has Hamming distance 4 from $\hat{M}'$, and thus has at most Hamming distance 8 from $\hat{M}$. It is easy to see that a -1 can also be eliminated by changing 4 appropriately selected entries. The so-obtained matrix is an adjacency matrix of a realization of D. Since the at most 3 "bad" values can be eliminated by changing at most $3 \times 4 = 12$ entries in $\hat{M}$, $\hat{M}$ is at most Hamming distance 12 from an adjacency matrix of a realization of D. Therefore, the number of paths going through on a particular edge is upper bounded by the value in Equation (8.35), and thus

$$\frac{1}{Q(e)} \sum_{\gamma \ni e} f(\gamma)|\gamma| \; < \; \frac{\binom{n}{2}\binom{m}{2}nm}{N \prod_{w \in V(H)} \left(\frac{d(w)}{2}\right)!} \sum_{\gamma \ni e} 1 \leq poly(nm). \quad (8.36)$$

By Theorem 78, this proves the rapid mixing of the Markov chain.

8.2.4   Balanced realizations of a JDM

A symmetric matrix J with non-negative integer elements is the *joint degree matrix* (JDM) of an undirected simple graph G if the element $J_{i,j}$ gives the number of edges between the class V_i of vertices all having degree i and the class V_j of vertices all with degree j in the graph. In this case we also say that J is *graphical* and that G is a *graphical realization* of J. Note that there can be many different graphical realizations of the same JDM.

Given a JDM, the number of vertices $n_i = |V_i|$ in class i is obtained from:

$$n_i = \frac{J_{i,i} + \sum_{j=1}^{k} J_{i,j}}{i}, \quad (8.37)$$

where k denotes the maximum number of degrees. This implies that a JDM also uniquely determines the degree sequence, since we have obtained the number of nodes of given degrees for all possible degrees. For sake of uniformity we consider all vertex classes V_i for $i = 1, \ldots, k$; therefore we consider empty classes with $n_i = 0$ vertices as well. A necessary condition for J to be graphical is that all the n_i-s are integers. Let n denote the total number of vertices. Naturally, $n = \sum_i n_i$ and it is uniquely determined via Equation (8.37) for a given graphical JDM. The necessary and sufficient conditions for a given JDM to be graphical are provided in the following theorem

Theorem 86. [50] *A $k \times k$ matrix J is a graphical JDM if and only if the following conditions hold:*

1. *For all $i = 1, \ldots, k$*

$$n_i := \frac{J_{i,i} + \sum_{j=1}^{k} J_{i,j}}{i}$$

   is integer.

2. *For all $i = 1, \ldots, k$*

$$J_{i,i} \leq \binom{n_i}{2}.$$

3. *For all $i = 1, \ldots, k$ and $j = 1, \ldots, k$, $i \neq j$,*

$$J_{i_j} \leq n_i n_j.$$

Let $d_j(v)$ denote the number of edges such that one end-vertex is v and the other end-vertex belongs to V_j, i.e., $d_j(v)$ is the degree of v in V_j. The vector consisting of the $d_j(v)$-s for all j is called the *degree spectrum* of vertex v. We introduce the notation

$$\Theta_{i,j} = \begin{cases} 0, & \text{if } n_i = 0, \\ \frac{J_{i,j}}{n_i}, & \text{otherwise,} \end{cases}$$

which gives the average number of neighbors of a degree-i vertex in vertex class V_j. Then a realization of the JDM is *balanced* iff for every i and all $v \in V_i$ and all j, we have

$$|d_j(v) - \Theta_{i,j}| < 1.$$

The following theorem is proven in paper [50] as Corollary 5:

Theorem 87. *Every graphical JDM admits a balanced realization.*

A restricted swap operation (RSO) takes two edges (x, y) and (u, v) with x and u from the same vertex class and swaps them with two non-edges (x, v) and (u, y). The RSO preserves the JDM, and in fact forms an irreducible Markov chain over all its realizations [50]. An RSO Markov chain restricted to *balanced realizations* can be defined as follows:

Definition 63. *Let J be a JDM. The state space of the RSO Markov chain consists of all the balanced realizations of J. It was proved by Czabarka et al. [50] that this state space is connected under restricted swap operations. The transitions of the Markov chain are defined in the following way. With probability $1/2$, the chain does nothing, so it remains in the current state (we consider a lazy Markov chain). With probability $1/2$ the chain will choose four, pairwise disjoint vertices, v_1, v_2, v_3, v_4 from the current realization (the possible choices are order dependent) and check whether v_1 and v_2 are chosen from the same vertex class, and furthermore whether the*

$$E \setminus \{(v_1, v_3), (v_2, v_4)\} \cup \{(v_1, v_4), (v_2, v_3)\}$$

swap operation is feasible. If this is the case then our Markov chain performs the swap operation if it leads to another balanced JDM realization. Otherwise the Markov chain remains in the same state. (Note that exactly two different orders of the selected vertices will provide the same swap operation, since the roles of v_1 and v_2 are symmetric.) Then there is a transition with probability

$$\frac{1}{n(n-1)(n-2)(n-3)}$$

between two realizations iff there is a RSO transforming one into the other.

Here we prove that such a Markov chain is rapidly mixing. The convergence of a Markov chain is measured as a function of the input data size. Here we note that the size of the data is the number of vertices (or number of edges, they are polynomially bounded functions of each other) and not the number of digits to describe the JDM. This distinction is important as, for example, one can create a 2×2 JDM with values $J_{2,2} = J_{3,3} = 0$ and $J_{2,3} = J_{3,2} = 6n$, which has $\Omega(n)$ number of vertices (or edges) but it needs only $O(\log(n))$ number of digits to describe (except in the unary number system). Alternatively, one might consider the input is given in unary.

Formally, we state the rapid mixing property via the following theorem:

Theorem 88. *The RSO Markov chain on balanced JDM realizations is a rapidly mixing Markov chain, namely, for the second-largest eigenvalue λ_2 of this chain, it holds that*

$$\frac{1}{1 - \lambda_2} = O(\text{poly}(n))$$

where n is the number of vertices in the realizations of the JDM.

Note that the expression on the LHS is called, with some abuse of notation, the *relaxation time*: it is the time is needed for the Markov chain to reach its stationary distribution. The proof is based on the special structure of the state space of the balanced JDM realizations. This special structure allows the following proof strategy: if we can prove that some auxiliary Markov chains are rapidly mixing on some sub-spaces obtained from decomposing the above-mentioned specially structured state space, then the Markov chain on the whole space is also rapidly mixing. We are going to prove the rapid mixing of these auxiliary Markov chains, as well as give the proof of the general theorem, that a Markov chain on this special structure is rapidly mixing, hence proving our main Theorem 88.

In order to describe the structure of the space of balanced JDM realizations, we first define the almost semi-regular bipartite and almost regular graphs.

Definition 64. *A bipartite graph $G(U, V; E)$ is almost semi-regular if for any $u_1, u_2 \in U$ and $v_1, v_2 \in V$*

$$|d(u_1) - d(u_2)| \leq 1$$

and

$$|d(v_1) - d(v_2)| \leq 1.$$

Definition 65. *A graph $G(V, E)$ is almost regular, if for any $v_1, v_2 \in V$*

$$|d(v_1) - d(v_2)| \leq 1.$$

It is easy to see that the restriction of any graphical realization of the JDM to vertex classes $V_i, V_j, i \neq j$ can be considered as the coexistence of two almost regular graphs (one on V_i and the other on V_j), and one almost semi-regular bipartite graph on the vertex class pair V_i, V_j. More generally, the collection of these almost semi-regular bipartite graphs and almost regular graphs completely determines the balanced JDM realization. Formally:

Definition 66 (Labeled union). *Any balanced JDM realization can be represented as a set of almost semi-regular bipartite graphs and almost regular graphs. The realization can then be constructed from these factor graphs as their* labeled union: *the vertices with the same labels are collapsed, and the edge set of the union is the union of the edge sets of the factor graphs.*

It is useful to construct the following auxiliary graphs. For each vertex class V_i, we create an auxiliary bipartite graph, $\mathcal{G}_i(V_i, U; E)$, where U is a set of "super-nodes" representing all vertex classes V_j, including V_i. There is an edge between $v \in V_i$ and super-node u_j representing vertex class V_j iff

$$d_j(v) = \lceil \Theta_{i,j} \rceil ,$$

i.e., iff node v carries the ceiling of the average degree of its class i toward the other class j. (For sake of uniformity, we construct these auxiliary graphs for all $i = 1, \ldots, k$, even if some of them have no edge at all. Similarly, all super-nodes are given, even if some of them have no incident edge.) We claim that these k auxiliary graphs are *half-regular*, i.e., each vertex in V_i has the same degree (the degrees in the vertex class U might be arbitrary). Indeed, the vertices in V_i all have the same degree in the JDM realization, therefore, the number of times they have the ceiling of the average degree toward a vertex class is constant in a balanced realization.

Let Y denote the space of all balanced realizations of a JDM and just as before, let k denote the number of vertex classes (some of them can be empty). We will represent the elements of Y via a vector y whose $k(k+1)/2$ components are the k almost regular graphs and the $k(k-1)/2$ almost regular bipartite graphs from their labeled union decomposition, as described in Definition 66 above. Given an element $y \in Y$ (i.e., a balanced graphical realization of the JDM) it has k associated auxiliary graphs $\mathcal{G}_i(V_i, U; E)$, one for every vertex class V_i (some of them can be empty graphs). We will consider this collection of auxiliary graphs for a given y as a k-dimensional vector x, where $x = (\mathcal{G}_1, \ldots, \mathcal{G}_k)$.

For any given y we can determine the corresponding x (so no particular y can correspond to two different xs), however, for a given x there can be several y's with that same x. We will denote by Y_x the subset of Y containing all the y's with the same (given) x and by X the set of all possible induced x vectors. Clearly, the x vectors can be used to define a disjoint partition on Y: $Y = \bigcup_{x \in X} Y_x$. For notational convenience we will consider the space Y as pairs (x, y), indicating the x-partition to which y belongs. This should not be confused with the notation for an edge, however, this should be evident from the context. A restricted swap operation might fix x, in which case it will make a move only within Y_x, but if it does not fix x, then it will change both x and y. For any x, the RSOs moving only within Y_x form a Markov chain. On the other hand, tracing only the x's from the pairs (x, y) is not a Markov chain: the probability that an RSO changes x (and thus also y) depends also on the current y not only on x. However, the following theorem holds:

Theorem 89. *Let (x_1, y_1) be a balanced realization of a JDM in the above mentioned representation.*

  i *Assume that (x_2, y_2) balanced realization is derived from the first one with one restricted swap operation. Then, either $x_1 = x_2$ or they differ in exactly one coordinate, and the two corresponding auxiliary graphs differ only in one swap operation.*

  ii *Let x_2 be a vector differing only in one coordinate from x_1, and furthermore, only in one swap within this coordinate, namely, one swap within one coordinate is sufficient to transform x_1 into x_2. Then there exists at least one y_2 such that (x_2, y_2) is a balanced JDM realization and (x_1, y_1) can be transformed into (x_2, y_2) with a single RSO.*

Proof. (i) This is just the reformulation of the definitions for the (x, y) pairs. (ii) (See also Fig. 8.1) By definition there is a degree $i, 1 \le i \le k$ such that auxiliary graphs $x_1(\mathcal{G}_i)$ and $x_2(\mathcal{G}_i)$ are different and one swap operation transforms the first one into the second one. More precisely there are vertices $v_1, v_2 \in V_i$ such that the swap transforming $x_1(\mathcal{G}_i)$ into $x_2(\mathcal{G}_i)$ removes edges

(v_1, U_j) and (v_2, U_k) (with $j \neq k$) and adds edges (v_1, U_k) and (v_2, U_j). (The capital letters show that the second vertices are super-vertices.) Since the edge (v_1, U_j) exists in the graph $x_1(G_1)$ and (v_2, U_j) does not belong to graph $x_1(G_i)$, therefore $d_j(v_1) > d_j(v_2)$ in the realization (x_1, y_1). This means that there is at least one vertex $w \in V_j$ such that w is connected to v_1 but not to v_2 in the realization (x_1, y_1). Similarly, there is at least one vertex $r \in V_k$ such that r is connected to v_2 but not to v_1 (again, in realization (x_1, y_1)). Therefore, we have a required RSO on nodes v_1, v_2, w, r. $\qquad\square$

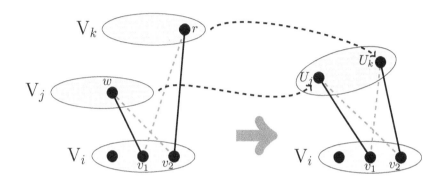

FIGURE 8.1: Construction of the auxiliary bipartite graph $\mathcal{G}_i$ and a RSO $\{(v_1, w), (v_2, r)\} \mapsto \{(v_1, r), (v_2, w)\}$ taking (x_1, y_1) into (x_2, y_2).

Thus any RSO on a balanced realization yielding another balanced realization either does not change x or changes x exactly on one coordinate (one auxiliary graph), and this change can be described with a swap, taking one auxiliary graph into the other.

We are going to apply Theorem 73 to prove that the RSO Markov chain is rapidly mixing on the balanced JDM realizations. We partition its state space according to the vectors x of the auxiliary graph collections (see Definition 66 and its explanations). The following result will be used to prove that all derived (marginal) Markov chains M_x are rapidly mixing. Next, we announce two theorems that are direct extensions of statements for fast mixing swap Markov chains for regular degree sequences (Cooper, Dyer and Greenhill [45]) and for half-regular bipartite degree sequences (Erdős, Kiss, Miklós and Soukup [62]).

Theorem 90. *The swap Markov chain on the realizations of almost regular degree sequences is rapidly mixing.*

Theorem 91. *The swap Markov chain on the realizations of almost half-regular bipartite degree sequences is rapidly mixing.*

We are now ready to prove the main theorem.

Proof. (Theorem 88) We show that the RSO Markov chain on balanced realizations fulfills the conditions in Theorem 73. First we show that condition (i) of Theorem 73 holds. When restricted to the partition Y_x (that is with x fixed), the RSO Markov chain over the balanced realizations walks on the union of almost semi-regular and almost regular graphs. By restriction here we mean that all probabilities which would (in the original chain) leave Y_x are put onto the shelf-loop probabilities. Since an RSO changes only one coordinate at a time, independently of other coordinates, all the conditions in Theorem 80 are fulfilled. Thus the relaxation time of the RSO Markov chain restricted onto Y_x is bounded from above by the relaxation time of the chain restricted onto that coordinate (either an almost semi-regular bipartite or an almost regular graph) on which this restricted chain is the slowest (the smallest gap). However, based on Theorems 90 and 91, all these restrictions are fast mixing, and thus by Theorem 80 the polynomial bound in (i) holds. (Here $K = \frac{k(k+1)}{2}$, see Definition 66 and note that an almost semi-regular bipartite graph is also an almost half-regular bipartite graph.)

Next we show that condition (ii) of Theorem 73 also holds. The first coordinate is the union of auxiliary bipartite graphs, all of which are half-regular. The M' Markov chain corresponding to Theorem 73 is the swap Markov chain on these auxiliary graphs. Here each possible swap has a probability

$$\frac{1}{n(n-1)(n-2)(n-3)}$$

and by Theorem 89 it is guaranteed that condition 7.53 is fulfilled. Since, again all conditions of Theorem 80 are fulfilled (mixing is fast within any coordinate due to Theorems 90 and 91), the M' Markov chain is also fast mixing. The condition in Equation (7.53) holds due to Theorem 89. Since all conditions in Theorem 73 hold, the RSO swap Markov chain on balanced realizations is also rapidly mixing. □

8.2.5   Counting the most parsimonious DCJ scenarios

Here we introduce a counting and sampling problem coming from bioinformatics introduced by Miklós and Tannier [134]. First we give a mathematical model of genomes and the operations changing the genomes.

Definition 67. *A* genome *is a directed, edge-labeled graph, in which each vertex has total degree (in-degree plus out-degree) 1 or 2, and each label is unique. Each edge is called a* marker. *The beginning of an edge is called its* tail *and the end of an edge is called its* head; *the joint name of heads and tails is* extremities. *Vertices with total degree 2 are called* adjacencies, *and vertices with total degree 1 are called* telomeres.

It is easy to see that any genome is a set of disjoint paths and cycles, where neither the paths nor the cycles are necessarily directed. The components of

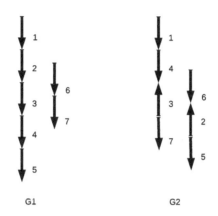

FIGURE 8.2: An example of two genomes with 7 markers.

the genome are called *chromosomes*. An example of a genome is drawn in Figure 8.2. All adjacencies correspond to an unordered set of two marker extremities. All telomeres correspond to one marker extremity. For example, the $(h1, t4)$ describes the vertex of genome 2 in Figure 8.2 in which the head of marker 1 and the tail of marker 4 meet, and similarly, $(h7)$ is the telomere where marker 7 ends. A genome is fully described by a list of such descriptions of adjacencies and telomeres. Two genomes with the same edge label set are *co-tailed* if they have the same telomeres. This is the case for the two genomes of Figure 8.2.

Definition 68. *A DCJ or double cut and join operation transforms one genome into another by modifying the adjacencies and telomeres in one of the following 4 ways:*

- *Take two adjacencies (a, b) and (c, d) and create two new adjacencies (a, c) and (b, d). The adjacency descriptors are not ordered: namely, the two new adjacencies might instead be (a, d) and (b, c).*

- *Take an adjacency (a, b) and a telomere (c), and create a new adjacency and a new telomere from the 3 extremities: either (a, c) and (b) or (b, c) and (a).*

- *Take two telomeres (a) and (b), and create a new adjacency (a, b).*

- *Take an adjacency (a, b) and create two new telomeres (a) and (b).*

Given two genomes G_1 and G_2 with the same label set, it is always possible to transform one into the other by a sequence of DCJ operations [185]. Such a sequence is called a DCJ *scenario* for G_1 and G_2. The minimum length of a scenario is called the DCJ *distance* and is denoted by $d_{\text{DCJ}}(G_1, G_2)$.

Definition 69. *The* Most Parsimonious DCJ (MPDCJ) *scenario problem for two genomes G_1 and G_2 is to compute $d_{\text{DCJ}}(G_1, G_2)$. The #MPDCJ problem asks for the number of scenarios of length $d_{\text{DCJ}}(G_1, G_2)$, denoted by* #MPDCJ(G_1, G_2).

For example, the DCJ distance between the two genomes of Figure 8.2 is three and there are nine different most parsimonious scenarios.

MPDCJ is an optimization problem, which has a natural corresponding decision problem asking if there is a scenario with a given number of DCJ operations. So we may write that #MPDCJ $\in$ #P, which means that #MPDCJ asks for the number of witnesses of the decision problem: "Is there a scenario for G_1 and G_2 of size $d_{\text{DCJ}}(G_1, G_2)$?"

Before turning to approximating the number of solutions, first we give an overview of how to find one solution. Here, the following combinatorial object plays a central role.

Definition 70. *The* adjacency graph $G(V_1 \cup V_2, E)$ *of two genomes G_1 and G_2 with the same edge label set is a bipartite multigraph with V_1 being the set of adjacencies and telomeres of G_1, V_2 being the set of adjacencies and telomeres of G_2. The number of edges between $u \in V_1$ and $v \in V_2$ is the number of extremities they share.*

Observe that the adjacency graph is a bipartite multigraph which falls into disjoint cycles and paths. The paths might belong to one of three types:

1. an odd path, containing an odd number of edges and an even number of vertices;

2. an even path with two endpoints in V_1; we will call them W-shaped paths; or

3. an even path with two endpoints in V_2; we will call them M-shaped paths.

In addition, cycles with two edges and paths with one edge are called *trivial* components. We can use the adjacency graph to obtain the DCJ distance between two genomes.

Theorem 92. *[185, 17]*

$$d_{\text{DCJ}}(G_1, G_2) = N - \left(C + \frac{I}{2}\right) \qquad (8.38)$$

where N is the number of markers, C is the number of cycles in the adjacency graph of G_1 and G_2, and I is the number of odd paths in the adjacency graph of G_1 and G_2.

Since calculating C and I is easy, MPDCJ is clearly in P and has a linear running time algorithm.

A DCJ operation on a genome G_1 which decreases the DCJ distance to a genome G_2 is called a *sorting* DCJ for G_1 and G_2. It is possible to characterize the effect of a sorting DCJ on the adjacency graph of genomes G_1 and G_2. It acts on the vertex set V_1 and has one of the following effects [24]:

- splitting a cycle into two cycles,

- splitting an odd path into a cycle and an odd path,

- splitting an M-shaped path into a cycle and an M-shaped path,

- splitting an M-shaped path into two odd paths,

- splitting a W-shaped path into a cycle and a W-shaped path,

- merging the two ends of a W-shaped path, thus transforming it into a cycle, or

- combining an M-shaped and a W-shaped path into two odd paths.

Note that trivial components are never affected by these operations, and all but the last type of DCJ operations act on a single component of the adjacency graph. The last type of DCJ acts on two components, which are M- and W-shaped paths.

In this context, *sorting* a component K means applying a sequence of sorting DCJ operations to vertices of K (in V_1) so that the resulting adjacency graph has only trivial components involving the extremities of K. In a minimum length DCJ scenario, every component is sorted independently, except M- and W-shaped paths, which can be sorted together. If in a DCJ scenario one operation acts on both an M- and a W-shaped path, we say that they are sorted *jointly*; otherwise we say that they are sorted *independently*.

It is conjectured that #MPDCJ is in #P-complete, although the problem is solvable in polynomial time for special cases when the genomes are co-tailed [24, 139], or more generally in the absence of M- and W-shaped paths. So the hard part is dealing with M- and W-shaped paths. We show here that for the general case, we may restrict ourselves to this hard part, and suppose that there are only M- and W-shaped paths in the adjacency graph.

Given two genomes G_1 and G_2 with the same label set, let AG be the adjacency graph of G_1 and G_2. Denote by G_1^* the genome that we get by sorting all cycles and odd paths in the adjacency graph of G_1 and G_2. By definition, the adjacency graph between G_1 and G_1^* has no M- and W-shaped paths, while the adjacency graph between G_1^* and G_2 has only trivial components and M- and W-shaped paths. Furthermore, it is easy to see that

$$d_{\mathrm{DCJ}}(G_1, G_2) = d_{\mathrm{DCJ}}(G_1, G_1^*) + d_{\mathrm{DCJ}}(G_1^*, G_2). \tag{8.39}$$

We would like to consider the following subproblem.

Definition 71. *The* #MPDCJ$_{MW}$ *problem* asks for the number of *DCJ scenarios between two genomes when their adjacency graph contains only trivial components and M- and W-shaped paths.*

The correspondence between solutions for #MPDCJ$_{MW}$ and #MPDCJ is stated by the following lemma.

Lemma 22. *It holds that*

$$\text{\#MPDCJ}(G_1, G_2) = \frac{d_{\text{DCJ}}(G_1, G_2)!}{d_{\text{DCJ}}(G_1^*, G_2)! \prod_i (c_i - 1)! \prod_j (l_j - 1)!}$$
$$\times \quad \prod_i c_i^{c_i - 2} \prod_j l_j^{l_j - 2} \times \text{\#MPDCJ}_{MW}(G_1^*, G_2) \quad (8.40)$$

where i indexes the cycles of the adjacency graph of G_1 and G_2, c_i denotes the number of vertices in vertex set V_1 belonging to the ith cycle, j indexes the odd paths of the adjacency graph, and l_j is the number of vertices in vertex set V_1 belonging to the jth odd path.

Proof. As *M-* and *W-*shaped paths and other components are always treated independently, we have

$$\text{\#MPDCJ}(G_1, G_2) = \binom{d_{\text{DCJ}}(G_1, G_2)}{d_{\text{DCJ}}(G_1^*, G_2)}$$
$$\times \quad \text{\#MPDCJ}(G_1, G_1^*) \times \text{\#MPDCJ}_{MW}(G_1^*, G_2).$$

For the genomes G_1 and G_1^*, whose adjacency graphs do not contain *M-* and *W-*shaped paths, we have from [24] and [139] that

$$\text{\#MPDCJ}(G_1, G_1^*) = \prod_i c_i^{c_i - 2} \prod_j l_j^{l_j - 2} \times \frac{d_{DCJ}(G_1, G_1^*)!}{\prod_i (c_i - 1)! \prod_j (l_j - 1)!}.$$

These two equations together with Equation (8.39) give the result.   □

The following theorem says that the hardness of the #MPDCJ problem is the same as the #MPDCJ$_{MW}$ problem.

Theorem 93.

$$\text{\#MPDCJ}_{MW} \in \text{FP} \quad \Longleftrightarrow \quad \text{\#MPDCJ} \in \text{FP} \quad (8.41)$$
$$\text{\#MPDCJ}_{MW} \in \text{\#P-complete} \quad \Longleftrightarrow \quad \text{\#MPDCJ} \in \text{\#P-complete} \quad (8.42)$$
$$\text{\#MPDCJ}_{MW} \in \text{FPRAS} \quad \Longleftrightarrow \quad \text{\#MPDCJ} \in \text{FPRAS} \quad (8.43)$$
$$\text{\#MPDCJ}_{MW} \in \text{FPAUS} \quad \Longleftrightarrow \quad \text{\#MPDCJ} \in \text{FPAUS} \quad (8.44)$$

Proof. Both the multinomial factor and the two products in Equation (8.40) can be calculated in polynomial time. Thus the transformation between the solutions to the two different counting problems is a single multiplication or division by an exactly calculated number. This proves that #MPDCJ$_{MW}$

is in FP if and only if #MPDCJ is in FP, as well as #MPDCJ$_{MW}$ is in #P-complete if and only if #MPDCJ is in #P-complete.

Such a multiplication and division keeps the relative error when the solution of one of the problems is approximated. This proves that #MPDCJ$_{MW}$ is in FPRAS if and only if #MPDCJ is in FPRAS.

Concerning the last equivalence, the $\Leftarrow$ part is trivial because #MPDCJ$_{MW}$ is a particular case of #MPDCJ. Now we prove that #MPDCJ$_{MW}$ $\in$ FPAUS $\Rightarrow$ #MPDCJ $\in$ FPAUS. Suppose an FPAUS exists for #MPDCJ$_{MW}$, and let G_1 and G_2 be two arbitrary genomes. The following algorithm gives a FPAUS for #MPDCJ.

- Draw a DCJ scenario between G_1^* and G_2 following a distribution p satisfying

$$d_{TV}(p, U) \leq \epsilon$$

  where U is the uniform distribution over all possible most parsimonious DCJ scenarios between G_1^* and G_2.

- Generate a DCJ scenario between G_1 and G_1^*, following the uniform distribution. This scenario can be sampled exactly uniformly in polynomial time: (1) there are only cycles and odd paths in the adjacency graph of G_1 and G_1^*, so the number of scenarios can be calculated in polynomial time; (2) there is a polynomial number of sorting DCJ steps on each component, and a sorting DCJ operation results in an adjacency graph that also only has cycles and odd paths.

- Draw a sequence of 0s and 1s, containing $d_{DCJ}(G_1^*, G_2)$ 1s and $d_{DCJ}(G_1, G_1^*)$ 0s, uniformly from all $\binom{d_{DCJ}(G_1, G_2)}{d_{DCJ}(G_1^*, G_2)}$ such sequences.

- Merge the two paths constructed at the two first steps, according to the drawn sequence of 0s and 1s.

Note that the DCJ scenario obtained transforms G_1 into G_2. Let us denote the distribution of paths generated by this algorithm by p', and the uniform distribution over all possible DCJ scenarios between G_1 and G_2 by U'. Let X_s denote the set of all possible scenarios drawn by the above algorithm using a specific scenario s between G_1^* and G_2. Then

$$\sum_{s' \in X_s} |p'(s') - U'(s')| = |p(s) - U(s)|. \tag{8.45}$$

Using Equation (8.45), we get that

$$d_{TV}(p', U') = \frac{1}{2} \sum_{s} \sum_{s' \in X_s} |p'(s') - U'(s')| = \frac{1}{2} \sum_{s} |p(s) - U(s)| = d_{TV}(p, U).$$
$$\tag{8.46}$$

This proves that the above algorithm is an FPAUS for #MPDCJ, proving the left-to-right direction in Equation (8.44). $\qquad\square$

We will show that $\#\text{MPDCJ}_{MW}$ is in FPAUS, and thus $\#\text{MPDCJ}$ is in FPAUS. As MPDCJ is a self-reducible problem, the FPAUS implies the existence of an FPRAS. The FPAUS algorithm for $\#\text{MPDCJ}_{MW}$ will be defined via a rapidly mixing Markov chain. First, we have to recall and prove some properties on the number of independent and joint sortings of M- and W-shaped paths.

Our goal is to show that the number of DCJ scenarios in which an M- and a W-shaped path are sorted independently is a significant fraction of the total number of scenarios sorting these M- and W-shaped paths (independently or jointly). We build on the following results by [24].

Theorem 94. *[24]*

- *The number of minimum-length* DCJ *scenarios sorting a cycle with $k > 1$ vertices in G_1 is k^{k-2}.*

- *The number of minimum-length* DCJ *scenarios sorting an odd path with $k > 1$ vertices in G_1 is k^{k-2}.*

- *The number of minimum-length* DCJ *scenarios sorting a W-shaped path with $k > 1$ vertices in G_1 is k^{k-2}.*

- *The number of minimum-length* DCJ *scenarios sorting an M-shaped path with $k > 0$ vertices in G_1 is $(k+1)^{k-1}$.*

Theorem 95. *The number of* DCJ *scenarios that independently sort a W- and an M-shaped path is $\binom{k_1+k_2-1}{k_1-1}k_1^{k_1-2}(k_2+1)^{k_2-1}$ where k_1 and k_2 are the number of vertices of G_1 in the W- and M-shaped paths, respectively.*

Proof. It is a consequence of the previous theorem. The W-shaped path is sorted in $k_1 - 1$ operations, and the M-shaped path is sorted in k_2 operations. Thus there are $\binom{k_1+k_2-1}{k_1-1}$ ways to merge two scenarios. $\qquad\square$

Theorem 96. *The number of* DCJ *scenarios that jointly sort a W- and an M-shaped path is less than $2(k_1+k_2)^{k_1+k_2-2}$, where k_1 and k_2 are the number of vertices of G_1 in the W- and M-shaped paths, respectively.*

Proof. Let t_1^W and t_2^W be the two telomeres of the W-shaped path, and t_1^M and t_2^M be the two telomeres of the M-shaped path. Let G_1' and G_2' be constructed from genomes G_1 and G_2 by adding a gene g, with extremities g^h and g^t, and replacing the telomeres of G_1 by adjacencies $(t_1^W, g^t), (t_2^W, g^h)$ and the telomeres of G_2 by adjacencies $(t_1^M, g^t), (t_2^M, g^h)$. In addition let G_1'' and G_2'' be constructed from G_1 and G_2 also by adding a gene g, and replacing the telomeres of G_1 by adjacencies $(t_1^W, g^t), (t_2^W, g^h)$ and the telomeres of G_2 by adjacencies $(t_1^M, g^h), (t_2^M, g^t)$. In both cases the M- and W-shaped paths in G_1 and G_2 are transformed into a cycle with $k_1 + k_2$ adjacencies in both genomes. Call the cycles C' for the first case, and C'' for the second.

We prove that any scenario that jointly sorts the W- and M-shaped paths has a corresponding distinct scenario either sorting the cycle C' or sorting the cycle C''. This proves the theorem, because there are $(k_1 + k_2)^{k_1+k_2-2}$ scenarios sorting each cycle.

A scenario jointly sorting the M- and W-shaped paths can be cut into two parts: the first contains DCJ operations which act only on the M- or only on the W-shaped path; the second part starts with a DCJ operation transforming an M- and a W-shaped path into two odd paths, and continues with operations independently sorting the two odd paths.

In the first part, either a DCJ operation acts on two adjacencies of the M- or W-shaped path, and the corresponding operation acts on the same two adjacencies on C' or C'', or it acts on an adjacency and a telomere of the W-shaped path, and the corresponding operation acts on two adjacencies of C' or C'', one of them containing an extremity of g. So there is a correspondence between being a telomere in the W-shaped path, and being adjacent to an extremity of g in C' or C''.

Now the corresponding operation of the DCJ transforming the two paths into two odd paths has to create two cycles from C' or C''. Choose C' or C'' so that it is the case. Now sorting an odd path exactly corresponds to sorting a cycle, by replacing a telomere in the path with an adjacency containing the extremity of the telomer and an extremity of g in the cycle.

So two different scenarios jointly sorting the M- and W-shaped paths correspond to two different scenarios sorting either C' or C''. Then the number of scenarios jointly sorting the M- and W-shaped paths is less than $2(k_1 + k_2)^{k_1+k_2-2}$.

$\square$

Theorem 97. *Let $T(k_1, k_2)$ denote the number of DCJ scenarios jointly sorting a W- and an M-shaped path with, respectively, k_1 and k_2 vertices G_1. Let $I(k_1, k_2)$ denote the number scenarios independently sorting the same paths. We have that*

$$\frac{T(k_1, k_2)}{I(k_1, k_2)} = O\left(\frac{k_1^{1.5} k_2^{1.5}}{(k_1 + k_2)^{1.5}}\right) \tag{8.47}$$

$$\frac{I(k_1, k_2)}{T(k_1, k_2)} = O(k_1 + k_2). \tag{8.48}$$

Proof. To prove Equation (8.47), it is sufficient to show that

$$\frac{2(k_1 + k_2)^{k_1+k_2-2}}{\binom{k_1+k_2-1}{k_1-1} k_1^{k_1-2}(k_2 + 1)^{k_2-1}} = O\left(\frac{k_1^{1.5} k_2^{1.5}}{(k_1 + k_2)^{0.5}}\right). \tag{8.49}$$

Using Stirling's formula, we get on the left-hand side of Equation (8.49)

$$\frac{2\sqrt{2\pi(k_1 - 1)}\left(\frac{k_1-1}{e}\right)^{k_1-1}\sqrt{2\pi(k_2)}\left(\frac{k_2}{e}\right)^{k_2}(k_1 + k_2)^{k_1+k_2-2}}{\sqrt{2\pi(k_1 + k_2 - 1)}\left(\frac{k_1+k_2-1}{e}\right)^{k_1+k_2-1} k_1^{k_1-2}(k_2 + 1)^{k_2-1}}. \tag{8.50}$$

After simplifications and algebraic rearrangement, we get

$$2\sqrt{\frac{2\pi(k_1-1)k_2}{k_1+k_2-1}}\left(\frac{k_1+k_2}{k_1+k_2-1}\right)^{k_1+k_2-1} \times$$

$$\times \left(\frac{k_1-1}{k_1}\right)^{k_1-1}\left(\frac{k_2}{k_2+1}\right)^{k_2}\left(\frac{k_1(k_2+1)}{k_1+k_2}\right). \tag{8.51}$$

from which Equation (8.49) follows by applying $(1+1/n)^n$ tends to e, and $(1-1/n)^n$ tends to $1/e$.

To prove Equation (8.48) consider the subset of DCJ scenarios jointly sorting the W- and M-shaped paths, and starting with a DCJ operation which acts on a telomere of the W-shaped path, and on an adjacency which is linked with a telomere of the M-shaped path. The result is two odd paths with, respectively, k_1 and k_2 adjacencies and telomeres in G_1. They can be sorted in, respectively, k_1-1 and k_2-1 steps, in $k_1^{k_1-2}$ and $k_2^{k_2-2}$ different ways. Since we can combine any two particular solutions in $\binom{k_1+k_2-2}{k_1-1}$ ways, $\frac{I(k_1,k_2)}{T(k_1,k_2)}$ is bounded by

$$\frac{\binom{k_1+k_2-1}{k_1-1}k_1^{k_1-2}(k_2+1)^{k_2-1}}{\binom{k_1+k_2-2}{k_1-1}k_1^{k_1-2}k_2^{k_2-2}}. \tag{8.52}$$

After minor algebraic simplification, this expression is equal to

$$\frac{k_1+k_2+1}{k_2}\left(1+\frac{1}{k_2}\right)^{k_2-1}k_2, \tag{8.53}$$

which is clearly $O(k_1+k_2)$. $\qquad\square$

We are going to sample DCJ scenarios in an indirect way. To do this, we define a Markov chain on (not necessarily perfect) matchings of a complete graph converging to a prescribed distribution. The construction and the proof that the obtained Markov chain is rapidly mixing is very similar to the case introduced in Subsection 8.2.2. We cannot directly apply the results in Subsection 8.2.2, since the weights of a matching are defined differently, furthermore, the individual weights of edges might be exponential functions of the problem size.

Assume that there are n W-shaped paths and m M-shaped paths, and consider the complete bipartite graph $K_{n,m}$. Let $\mathcal{M}$ be a matching of $K_{n,m}$, which might range from the empty graph up to any maximum matching. A DCJ scenario is said to be $\mathcal{M}$-compatible when an M-shaped and a W-shaped path are sorted jointly if and only if they are connected by an edge of $\mathcal{M}$.

We denote by $\{P_i\}_i$ the set of degree 0 vertices in $\mathcal{M}$, and by $\{M_iW_i\}_i$ the set of edges in $\mathcal{M}$. Let $l(P_i)$ be the minimum length of a DCJ scenario independently sorting P_i, and $l(M_iW_i)$ be the minimum length of a DCJ scenario jointly sorting M_i and W_i. We can calculate $N(M_i,W_i)$, the number of joint sortings of M_i and W_i, in polynomial time [24]. Denote by $N(P_i)$ the number of independent sortings of a path P_i. The number of $\mathcal{M}$-compatible scenarios is

$$f(\mathcal{M}) = \left(\frac{(\sum_i l(M_i W_i) + \sum_i l(P_i))!}{l(M_i, W_i)!, \ldots, l(P_i)!} \right) \Pi_i N(M_i, W_i) \Pi_i N(P_i),$$

and we can compute it in polynomial time. Define a distribution θ over the set of all matchings of the complete bipartite graph $K_{n,m}$ as

$$\theta(\mathcal{M}) \propto f(\mathcal{M}). \tag{8.54}$$

We first show that sampling DCJ scenarios from the uniform distribution is equivalent to sampling matchings of $K_{n,m}$ from the distribution θ.

Theorem 98. *Let a distribution q over the scenarios of n W-shaped paths and m M-shaped paths be defined by the following algorithm.*

- *Draw a random matching $\mathcal{M}$ of $K_{n,m}$ following a distribution p.*

- *Draw a random $\mathcal{M}$-compatible DCJ scenario from the uniform distribution of all $\mathcal{M}$-compatible ones.*

Then

$$d_{TV}(p, \theta) = d_{TV}(q, U) \tag{8.55}$$

where θ is the distribution defined in Equation (8.54), and U denotes the uniform distribution over all DCJ scenarios.

Proof. It holds that

$$d_{TV}(q, U) = \frac{1}{2} \sum_{x \ scenario} |q(x) - U(x)|.$$

We may decompose this sum into

$$\frac{1}{2} \sum_{(\mathcal{M} \ matching \ of \ K_{n,m})} \sum_{(x \ \mathcal{M}-compatible \ scenario)} |q(x) - U(x)|.$$

$\sum_{(x \ \mathcal{M}-compatible \ scenario)} q(x)$ is $p(\mathcal{M})$ since x is drawn uniformly among the scenarios compatible with $\mathcal{M}$, and $\sum_{(x \ \mathcal{M}-compatible \ scenario)} U(x)$ is $\theta(\mathcal{M})$. Furthermore, both $q(x)$ and $U(x)$ are constant for a particular matching $\mathcal{M}$, thus

$$\frac{1}{2} \sum_{(\mathcal{M} \ matching \ of \ K_{n,m})} \sum_{(x \ \mathcal{M}-compatible \ scenario)} |q(x) - U(x)| =$$

$$= \frac{1}{2} \sum_{(\mathcal{M} \ matching \ of \ K_{n,m})} |p(\mathcal{M}) - \theta(\mathcal{M})| = d_{TV}(p, \theta), \tag{8.56}$$

yielding the result. $\qquad\qquad\qquad\qquad\qquad\qquad\qquad\qquad\qquad\qquad\square$

So we are going to define an MCMC on matchings of $K_{n,m}$ converging to θ. The rapid convergence of this MCMC will imply that #MPDCJ$_{WM}$ admits an FPAUS, and hence #MPDCJ $\in$ FPAUS, and then #MPDCJ $\in$ FPRAS. The primer Markov chain walks on the matchings of $K_{n,m}$ and is defined by the following steps: suppose the current state is a matching $\mathcal{M}$, and

- with probability $1/2$, the next state of the Markov chain is the current state $\mathcal{M}$;

- with probability $1/2$, draw a random $i \sim U[1,n]$ and $j \sim U[1,m]$; if $ij \in \mathcal{M}$, then remove ij from $\mathcal{M}$; else if $\deg_{\mathcal{M}}(i) = 0$ and $\deg_{\mathcal{M}}(j) = 0$, then add ij to $\mathcal{M}$.

It is easy to see that this Markov chain is irreducible and aperiodic. We apply the standard Metropolis-Hastings algorithm on this chain, namely, when we are in state $\mathcal{M}$, we propose the next state $\mathcal{M}_{new}$ according to the primer Markov chain, and accept the proposal with probability

$$\min\left\{1, \frac{f(\mathcal{M}_{new})}{f(\mathcal{M})}\right\}. \tag{8.57}$$

The obtained Markov chain is reversible and converges to the distribution θ defined in Equation (8.54). Also observe that the defined chain is a lazy Markov chain, thus all its eigenvalues are positive real numbers.

An important property of this Markov chain is that

Observation 3. *The inverses of the non-zero transition probabilities are polynomially bounded.*

Indeed, the transition probability from $\mathcal{M}$ to $\mathcal{M}_{new}$, if non zero, is at least

$$\frac{1}{2 \times n \times m} \frac{f(\mathcal{M}_{new})}{f(\mathcal{M})}.$$

$\mathcal{M}$ and $\mathcal{M}_{new}$ vary by at most one edge M_iW_i, and on this edge, according to Theorem 97, the ratio of number of scenarios jointly and independently sorting M_i and W_i is polynomial. Furthermore, the combinatorial factors appearing in $f(\mathcal{M})$ and $f(\mathcal{M}_{new})$ due to merging the sorting steps on different components are the same. So $\frac{f(\mathcal{M}_{new})}{f(\mathcal{M})}$ as well as its inverse are polynomially bounded.

We now prove the rapid convergence of this Markov chain using a multicommodity flow technique. To prove that the Markov chain we defined on bipartite matchings has a polynomial relaxation time, we need to construct a path system Γ on the set of matchings of $K_{n,m}$, such that κ_Γ is bounded by a polynomial in N, the number of markers in G_1 and G_2.

In our case the path system between two matchings $\mathcal{X}$ and $\mathcal{Y}$ is a unique path with probability 1. Here is how we construct it.

Fix a total order on the vertex set of $K_{n,m}$. Take the symmetric difference of $\mathcal{X}$ and $\mathcal{Y}$, denoted by $\mathcal{X}\Delta\mathcal{Y}$. It is a set of disjoint paths and cycles. Define an order on the components of $\mathcal{X} \Delta \mathcal{Y}$, such that a component C is smaller than a component D if the smallest vertex in C is smaller than the smallest vertex in D. Now we orient each component in the following way: the beginning of each path is its extremity with the smaller vertex. The starting vertex of a cycle is its smallest vertex, and the direction is going toward its smaller neighbor.

We transform $\mathcal{X}$ to $\mathcal{Y}$ by visiting the components of $\mathcal{X} \Delta \mathcal{Y}$ in increasing order. Let the current component be C, and the current matching is $\mathcal{Z}$ (at first $\mathcal{Z} = \mathcal{X}$). If C is a path or cycle starting with an edge in $\mathcal{X}$, then the transformation steps are the following: delete the first edge of C from $\mathcal{Z}$, delete the third edge of C from $\mathcal{Z}$, add the second edge of C to $\mathcal{Z}$, delete the 5th edge of C from $\mathcal{Z}$, add the 4th edge of C to $\mathcal{Z}$, etc.

If C is a path or cycle starting with an edge in $\mathcal{Y}$, then the transformation steps are the following: delete the second edge of C from $\mathcal{Z}$, add the first edge of C to $\mathcal{Z}$, delete the 4th edge of C from $\mathcal{Z}$, add the third edge of C to $\mathcal{Z}$, etc.

This path has length at most nm, and κ_Γ can be written:

$$\kappa_\Gamma \le nm \max_{e=(u,v)\in E} \sum_{(x,y)\in V\times V : e \in \Gamma_{x,y}} \frac{\theta(x)\theta(y)}{Q(e)}.$$

By Property 3, the inverse of the transition probabilities is bounded by a polynomial in N, so we get

$$\kappa_\Gamma \le O(\text{poly}(N)) \max_{e=(u,v)\in E} \sum_{(x,y)\in V\times V : e \in \Gamma_{x,y}} \frac{\theta(x)\theta(y)}{\theta(u)}. \tag{8.58}$$

We then have to show that $\sum \frac{\theta(x)\theta(y)}{\theta(u)}$ can be bounded by a polynomial in N. Let $\mathcal{Z} \to \mathcal{Z}'$ be an edge on the path from $\mathcal{X}$ to $\mathcal{Y}$. We define

$$\widehat{\mathcal{M}} := \mathcal{X} \Delta \mathcal{Y} \Delta \mathcal{Z}. \tag{8.59}$$

Lemma 23. *The couple $\widehat{\mathcal{M}}$ and $\mathcal{Z} \to \mathcal{Z}'$ determines $\mathcal{X}$ and $\mathcal{Y}$.*

Proof. It is obvious that

$$\widehat{\mathcal{M}} \Delta \mathcal{Z} = \mathcal{X} \Delta \mathcal{Y}, \tag{8.60}$$

hence, $\mathcal{Z}$ and $\widehat{\mathcal{M}}$ determine the symmetric difference of $\mathcal{X}$ and $\mathcal{Y}$. From the transition $\mathcal{Z} \to \mathcal{Z}'$, we can trace back which transition steps have been already made in the following way. The order of the components of $\mathcal{X} \Delta \mathcal{Y}$ is determined, and from the transition $\mathcal{Z} \to \mathcal{Z}'$ we know the current component. We also know the beginning and the direction of the component, be it either a path or a cycle, hence, we know which edges have been changed in the component so far, and which ones not yet. From these, we can reconstruct $\mathcal{X}$ and $\mathcal{Y}$. $\square$

Lemma 24. *A matching can be obtained from $\widehat{\mathcal{M}}$ by deleting at most two edges.*

Proof. On each component in $\mathcal{X}\Delta\mathcal{Y}$, we delete at most two edges before putting back one. Hence $\widehat{\mathcal{M}}$ contains at most either 4 consecutive edges along a path or 2 pairs of edges, and all remaining edges are independent. Therefore it is sufficient to delete at most two edges from $\widehat{\mathcal{M}}$ to get a matching. □

Denote this matching by $\widetilde{\mathcal{M}}$.

Lemma 25. *It holds that*

$$\frac{\theta(\mathcal{X})\theta(\mathcal{Y})}{\theta(\mathcal{Z})} = O(\text{poly}(N))\theta(\widetilde{\mathcal{M}}). \tag{8.61}$$

Proof. We prove that

$$\frac{f(\mathcal{X})f(\mathcal{Y})}{f(\mathcal{Z})f(\widetilde{\mathcal{M}})} = O(\text{poly}(N)). \tag{8.62}$$

It proves the lemma, as $\theta(\cdot)$ and $f(\cdot)$ differ only by a normalizing constant. $\widetilde{\mathcal{M}}\Delta\mathcal{Z}$ differs at most in two edges from $\mathcal{X}\Delta\mathcal{Y}$. These edges appear in $\mathcal{X}\Delta\mathcal{Y}$, but not in $\widetilde{\mathcal{M}}\Delta\mathcal{Z}$. The two vertices of any missing edges correspond to components which are independently sorted either in $\mathcal{Z}$ or $\widetilde{\mathcal{M}}$, but jointly in either $\mathcal{X}$ or $\mathcal{Y}$. Amongst these two vertices, one of them corresponds to a W-shaped component $\mathcal{A}$, the other to an M-shaped component $\mathcal{B}$. Let k_1 be the number of adjacencies and telomeres of G_1 in $\mathcal{A}$, and k_2 the number of adjacencies and telomeres of G_1 in $\mathcal{B}$. The ratio on the left-hand side of Equation (8.62) due to such difference is

$$\frac{T(k_1, k_2)}{(k_1 + k_2 + 1)!} \Big/ \frac{I(k_1)I'(k_2)}{k_1!(k_2 + 1)!} \tag{8.63}$$

where $I(x)$ denotes the independent sorting of a W-shaped component of size x, and $I'(x)$ denotes the independent sorting of an M-shaped component of size x. However, it is polynomially bounded, since

$$\frac{I(k_1)I'(k_2)}{\binom{k_1+k_2+1}{k_1}} = I(k_1, k_2) \tag{8.64}$$

and we can apply Theorem 97. □

These results together lead to the following theorem:

Theorem 99. *The Metropolis-Hastings Markov chain on the matchings defined above converges rapidly to θ.*

Proof. From Lemma 25, Equation (8.58) may be written

$$\kappa_\Gamma \leq O(\text{poly}(N)) \max_{e=(u,v)\in E} \sum_{(x,y)\in V\times V: e\in\Gamma_{x,y}} \theta(\widetilde{\mathcal{M}}).$$

By Lemmas 23 and 24, a matching $\widetilde{\mathcal{M}}$ may appear only a polynomial number of times in this sum. So

$$\kappa_\Gamma \leq O(\mathrm{poly}(N)) \sum_{\widetilde{\mathcal{M}}} \theta(\widetilde{\mathcal{M}}),$$

and as $\sum_{\widetilde{\mathcal{M}}} \theta(\widetilde{\mathcal{M}}) = 1$, κ_Γ is bounded by a polynomial in N. This proves the theorem. $\square$

Using this result, we can prove the following theorem:

Theorem 100. *It holds that* $\#\mathrm{MPDCJ}_{MW} \in \mathrm{FPAUS}$.

Proof. The above-defined Markov chain on partial matchings is an aperiodic, irreducible and reversible Markov chain, with only positive eigenvalues. Furthermore, a step can be performed in running time, that is, polynomial with the size of the graph. We claim that for any start state i, $\log(1/\theta(i))$ is polynomially bounded with the size of the corresponding genomes G_1^* and G_2. Indeed, there are $O(N^2)$ DCJ operations, the length of the DCJ paths is less than N, and thus the number of sorting DCJ paths are $O(N^{2N})$, and the inverse of the probability of any partial matching is less than this. Thus, the relaxation time is polynomial in both N and $\log(1/\epsilon)$. This means that in fully polynomial running time (polynomial both in N and $-\log(\epsilon)$) a random partial matching can be generated from a distribution p satisfying

$$d_{TV}(p, \theta) \leq \epsilon. \tag{8.65}$$

But then a random DCJ path can be generated in fully polynomial running time following a distribution q satisfying

$$d_{TV}(q, U) \leq \epsilon \tag{8.66}$$

according to Theorem 98. This is what we wanted to prove. $\square$

Now we are ready to conclude by our main theorem:

Theorem 101. $\#\mathrm{MPDCJ} \in \mathrm{FPRAS}$

Proof. $\#\mathrm{MPDCJ}_{MW} \in \mathrm{FPAUS}$ according to Theorem 100. Then $\#\mathrm{MPDCJ} \in \mathrm{FPAUS}$ according to Theorem 93. Since $\#\mathrm{MPDCJ}$ is a self-reducible counting problem, it is in FPRAS. $\square$

8.2.6 Sampling and counting the k-colorings of a graph

A k-coloring of a graph $G = (V, U)$ is a mapping $c : V \to \{1, 2, \ldots, k\}$ such that for all $(u, v) \in E$, $c(u) \neq c(v)$. The exact counting of k-colorings of a d-regular graph is #P-complete when $k \geq d+1$ and $d \geq 3$ [33]. However, it is possible to design an FPAUS algorithm for sampling k-colorings of a graph

G when $k \geq 2\Delta + 1$, where Δ is the maximal degree in G [102]. It is easy to see that the following Markov chain (that we will call the Glauber dynamics Markov chain) converges to the uniform distribution of the k-colorings of a graph, G. The Markov chain uniformly samples a random vertex $v \in V$ and a random color $c' \in \{1, 2, \ldots, k\}$. If there is a vertex $u \in \Gamma(v)$ such that $c(u) = c'$, then the chain remains in the same state, otherwise, the color of v is changed to c'.

Jerrum proved the rapid mixing of this chain using coupling (Theorem 79). The Markov chain walks on the direct product of the k-colorings of G; a pair of such colorings is denoted by (X_t, Y_t). The coloring in X_t is denoted by $c_{X_t}(\cdot)$, similarly, the coloring in Y_t is denoted by $c_{Y_t}(\cdot)$. The coupling of the Markov chain is defined by the following algorithm.

1. Select a random vertex v uniformly.

2. Compute a permutation σ depending on v, G, X_t and Y_t described in detail below.

3. Choose a random color c' uniformly.

4. If there is a $u \in \Gamma(v)$ such that $c_{X_t}(u) = c'$, then X_{t+1} will be X_t. Otherwise the coloring of v will be c' in X_{t+1}. Similarly, if there is a $u \in \Gamma(v)$ such that $c_{Y_t}(u) = \sigma(c')$, then Y_{t+1} will be Y_t. Otherwise the coloring of v will be $\sigma(c')$ in Y_{t+1}.

It is clear that whatever permutation is used in step 2, $\sigma(c')$ follows the uniform distribution, therefore the given Markov chain is indeed a coupling. Let $A = A_t \subseteq V$ be such that for all $u \in A$, $c_{X_t}(u) = c_{Y_t}(u)$. Similarly, let $D = D_t \subseteq V$ be such that for all $u \in D$, $c_{X_t}(u) \neq c_{Y_t}(u)$. Let $d'(v)$ be the number of edges incident to v that have one endpoint in A and one endpoint in D. Observe that

$$\sum_{v \in A} d'(v) = \sum_{v \in D} d'(v) = m' \qquad (8.67)$$

where m' is the number of edges that span A and D. The permutation σ is defined in the following way.

(a) If $v \in D$, then let σ be the identity permutation.

(b) Otherwise, let $C_X := \left(\bigcup_{u \in \Gamma(v)} \{c_{X_t}(u)\} \right) \setminus \left(\bigcup_{u \in \Gamma(v)} \{c_{Y_t}(u)\} \right)$, similarly, let $C_Y := \left(\bigcup_{u \in \Gamma(v)} \{c_{Y_t}(u)\} \right) \setminus \left(\bigcup_{u \in \Gamma(v)} \{c_{X_t}(u)\} \right)$. Clearly, $C_X \cap C_Y = \emptyset$. Also observe that $|C_X|, |C_y| \leq d'(v)$. Without loss of generality we may assume that $|C_x| \leq |C_y|$. Then let C'_Y be an arbitrary subset of C_Y with cardinality $|C_X|$. Let $C_X = \{c_1, c_2, \ldots, c_r\}$ and $C'_Y = \{c'_1, c'_2, \ldots, c'_r\}$ be an arbitrary enumeration of the sets C_X and C'_Y. The permutation σ is defined via its cyclic decomposition as

$$\sigma := (c_1, c'_1)(c_2, c'_2) \ldots (c_r, c'_r),$$

which interchanges the colors in C_X and C'_Y, and fixes all other colors.

It is easy to see that the coupling event happens when $|D_t| = 0$. The cardinality of D_t can change at most one. First we consider the case when the size of D_t increases. This can only happen if the selected vertex, v, is in A. Then the permutation is selected in line (b), and c' must be in C_Y. Since $|C_Y| \leq d'(v)$, we get that

$$P(|D_{t+1}| = |D_t| + 1) \leq \frac{1}{n} \sum_{v \in A} \frac{d'(v)}{k} = \frac{m'}{nk}, \qquad (8.68)$$

where n is the number of vertices in G. When D_t decreases, then v is in D. In this case, the selected permutation, σ, is the identity permutation. The selected color c' is accepted if there is no $u \in \Gamma(v)$ such that $c_{X_t}(u) = c'$ or $c_{Y_t}(u) = c'$. Observe that there are at least

$$k - 2\Delta + d'(v)$$

such colors, since for $d'(v)$ number of neighbors, the color of the neighbors is the same in X_t and Y_t. Thus we get that

$$P(|D_{t+1}| = |D_t| - 1) \geq \frac{1}{n} \sum_{v \in D} \frac{k - 2\Delta + d'(d)}{k} = \frac{k - 2\Delta}{kn} \times |D_t| + \frac{m'}{kn}. \quad (8.69)$$

Since the probability that D_t increases is smaller than the probability that D_t deceases, the Markov chain couples rapidly. We can give an estimation of the expectation of $|D_t|$:

$$E(|D_{t+1}|) \leq \frac{m'}{nk}(|D_t| + 1) + \left(\frac{k - 2\Delta}{kn} \times |D_t| + \frac{m'}{kn} \right)(|D_t| - 1) +$$
$$\left(1 - \frac{k - 2\Delta}{kn} \times |D_t| - \frac{2m'}{kn} \right)|D_t| = \left(1 - \frac{k - 2\Delta}{kn} \right)|D_t|. \qquad (8.70)$$

That is,

$$E(|D_t|) \leq \left(1 - \frac{k - 2\Delta}{kn} \right)^t |D_0| \leq \left(1 - \frac{k - 2\Delta}{kn} \right)^t n. \qquad (8.71)$$

Since $|D_t|$ is a random variable on the non-negative integers, we have that

$$P(D_t \neq 0) \leq n \left(1 - \frac{k - 2\Delta}{kn} \right)^t \leq n e^{\frac{k - 2\Delta}{kn} t}. \qquad (8.72)$$

We get that $P(|D_t| \neq 0) \leq \varepsilon$ when $t \geq \frac{kn}{k - 2\Delta} \log \left(\frac{n}{\varepsilon} \right)$. Since the coupling time upper bounds the relaxation time, we get that the relaxation time is bounded by a polynomial of n and $-\log(\varepsilon)$, that is, the proposed Markov chain provides an FPAUS sampler. It is also easy to show that the number of k-colorings is in FPRAS if $k \geq 2\Delta + 1$ [102], see also Exercises 15 and 16.

8.3   Further results and open problems

8.3.1   Further results

- Mark Jerrum, Alastair Sinclair and Eric Vigoda showed that computing the permanent is in FPRAS for an arbitrary $n \times n$ matrix with non-negative weights [100].

- Catherine Greenhill proved that the swap Markov chain is rapidly mixing if the degree sequence satisfies the condition

$$3 \leq d_{\max} \leq \frac{1}{4}\sqrt{M},$$

  where $d_{\max}$ is the largest degree and M is the sum of the degrees [84]. Catherine Greenhill and Matteo Sfragara improved this result. They proved that the Markov chain is rapidly mixing if the degree sequence satisfies the condition

$$3 \leq d_{\max} \leq \frac{1}{3}\sqrt{M}.$$

  They also proved that the swap Markov chain is rapidly mixing on the realizations of directed degree sequences when all in-degrees and out-degrees are positive and bounded above by $\frac{1}{4}\sqrt{m}$, where m is the number of arcs, and not all in-degrees and out-degrees equal 1 [85].

- Regina Tyshkevich introduced the canonical decomposition of degree sequences [171]. Péter Erdős, István Miklós and Zoltán Toroczkai extended it to bipartite graphs, and proved that the swap Markov chain is rapidly mixing on the realizations of a bipartite degree sequence D, if the swap Markov chain is rapidly mixing on each degree sequence appearing in the decomposition of D [65]. Such degree sequences might be highly irregular and dense enough such that Greenhill's condition does not hold.

- Ivona Bezáková, Nayantara Bhatnagar and Eric Vigoda gave an FPAUS for the realizations of an arbitrary degree sequence [19]. Their method is based on Simulated Annealing [114].

- Colin Cooper, Martin Dyer, Catherine Greenhill and Andrew Handley proved that the swap Markov chain remains rapidly mixing when restricted to the connected realizations of regular graphs [46]. Actually, they proved a bit more: they proved that the swap Markov chain is rapidly mixing on connected regular graphs, when only those swap operations changing $(u_1, v_1), (u_2, v_2)$ to $(u_1, v_2), (u_2, v_1)$ are considered for which (u_1, u_2) is an edge.

- Milena Michail and Peter Winkler gave an FPAUS and FPRAS for approximately sampling and counting Eulerian orientations in any Eulerian unoriented graph [129].

- Eric Vigoda improved the bound of the Glauber dynamics Markov chain to $k > \frac{11}{6}\Delta$. Thomas P. Hayes and Eric Vigoda proved that the same Markov chain is rapidly mixing on the k-colorings of graphs for $k > (1 + \varepsilon)\Delta$ for all $\varepsilon > 0$, whenever $\Delta = \Omega(\log(n))$ and the graph does not contain any cycle shorter than 11 [93]. Martin Dyers and Catherine Greenhill introduced a slightly different Markov chain where an edge is sampled uniformly, and the colors of the vertices incident to the selected edge are changed. They obtained significantly better mixing time compared to the mixing time of the Glauber dynamics when $k = 2\Delta$ [58].

- The number of independent sets in a graph is a #P-complete counting problem, even if the maximum degree is 3 [59]. It is still possible to sample almost uniformly independent sets of a graph [123, 59].

8.3.2 Open problems

- Sukhamay Kundu proved that two tree degree sequences, $D = d_1, d_2, \ldots, d_n$ and $F = f_1, f_2, \ldots, f_n$, have edge-disjoint tree realizations if and only if their sum, $f_1 + d_1, f_2 + d_2, \ldots, f_n + d_n$, is a graphical degree sequence. It is unknown how to uniformly sample edge-disjoint realizations of D and F.

- Brooks' theorem says that a graph G can be colored with $k \geq \Delta$ colors if $\Delta \geq 3$ and the graph does not contain a complete graph K_{k+1} [28]. We do not know if FPRAS and FPAUS algorithms exist for approximate counting and sampling k-colorings of graphs when $k = \Delta$, the maximum degree of the graph.

- Although it is widely believed, it is unknown if the swap Markov chain is rapidly mixing on the realizations of arbitrary degree sequences.

- We know that counting perfect matchings in a bipartite graph is in FPRAS. However, it remains an open problem if counting the perfect matchings in simple graphs is in FPRAS. Daniel Štefankovič, Eric Vigoda and John Wimes showed that the Markov chain on perfect and almost perfect matchings of a simple graph might be torpidly mixing [182].

- Although there are FPRAS and FPAUS algorithms for approximately counting and sampling Eulerian orientations of an arbitrary unoriented graph, it is unknown if the same is true for Eulerian circuits of an arbitrary unoriented graph.

- There are polynomial running time algorithms to find a shortest reversal scenario between two signed permutations [89, 166]. The number of shortest reversal scenarios is conjectured to be in #P-complete. No efficient algorithms are known for approximately sampling or counting these scenarios. Sorting permutations by block-interchanges [43] is another example where the optimization problem can be solved in polynomial time, and the complexity of counting the shortest rearrangement scenarios is unknown.

- Sampath Kannan, Z. Sweedyk and Steve Mahaney gave a quasi-polynomial algorithm for approximately sampling and counting words with a given length from a regular grammar [106]. (A function in form $e^{\log^k(n)}$ is called quasi-polynomial.) Vivek Gore and his co-workers extended this result for context-free grammars [80] and they also presented an FPAUS for some restricted regular grammars. We do not know if FPAUS and FPRAS algorithms are available for arbitrary regular and context-free grammars.

8.4 Exercises

1. Using a rounding technique similar to that presented in Subsection 8.1.1, give a deterministic approximation algorithm for the following problem. The input is a set of weights, $A = \{w_1, w_2, \ldots, w_n\}$ and a weight W, the output is the subset

$$\arg\max_{S \subset A} \left\{ \sum_{w \in S} w \mid \sum_{w \in S} w \leq W \right\}.$$

The running time of the algorithm must be polynomial with n and $\frac{1}{\varepsilon}$, where $1 + \varepsilon$ is the approximation rate of the solution.

2. Prove Observation 2.

3. ∘ Prove that the number of trees realizing degree sequence $D = d_1, d_2, \ldots, d_n$ and in which v_i is adjacent to a prescribed leaf and v_j is also adjacent to a prescribed leaf is

$$\frac{(n-4)!}{(d_i - 2)!(d_j - 2)! \prod_{k \neq i,j} (d_k - 1)!}.$$

4. By computing the definite integral

$$\int_{x_1=0}^{1} \int_{x_2=x_1}^{1} \cdots \int_{x_n=x_{n-1}}^{1} 1 dx_n d_x n - 1 \ldots d_1,$$

give an alternative proof that the volume of the poset polytope of a total ordering of n elements is indeed $\frac{1}{n!}$.

5. * Find a graph $G = (V, E)$ and two matchings X and Y on it such that the canonical path between X and Y as described in Subsection 8.2.2 has length $\frac{3n}{5}$, where $n = |V|$.

6. Let $G = (V, E)$ be a simple graph, and let $w : E \to \mathbb{R}^+$ be a weight function, such that the maximum weight is a polynomial function of the number of vertices in G. Let $\mathcal{M}(G)$ denote the set of (not necessarily perfect) matchings of G. Develop an FPRAS estimating

$$\sum_{M \in \mathcal{M}(G)} \prod_{e \in M} w(e)$$

using the fact that there is a rapidly mixing Markov chain converging to the distribution

$$\pi(M) \propto \prod_{e \in M} w(e).$$

7. * Prove that for any bipartite degree sequence $D = \{d_{1,1}, d_{1,2}, \ldots, d_{1,n}\}, \{d_{2,1}, d_{2,2}, \ldots, d_{2,m}\}$ there exists a simple degree sequence $F = f_1, f_2, \ldots, f_{n,m}$ such that D and F have the same realizations, furthermore, a polynomial time computable bijection between the realizations exists.

8. Give an example for two directed graphs $\vec{G}_1$ and $\vec{G}_2$ having the same degree sequence D, however, $\vec{G}_1$ cannot be transformed into $\vec{G}_2$ using only swap operations. A swap operation in a directed graph deletes the presented edges (a, b) and (c, d) and adds the edges (a, d) and (c, b).

9. ∘ Construct two bipartite graphs, G_1 and G_2, with n-n vertices in their vertex classes which have the same degree sequence and for which $\Omega(n^2)$ swap operations are needed to transform G_1 into G_2.

10. Prove that the swap Markov chain is irreducible on the realizations of regular directed degree sequences.

11. Prove Theorem 92.

12. ∘ Prove that the Glauber dynamics Markov chain introduced in Subsection 8.2.6 is rapidly mixing when $k \geq 2\Delta$.

13. Prove that the Glauber dynamics Markov chain is irreducible when $k \geq \Delta + 2$.

14. * Give an example that the Glauber dynamics Markov chain might not be irreducible when $k = \Delta + 1$.

15. ∘ Let $G = (V, E)$ be a simple graph, and let $G' = G \setminus \{e\}$ for some $e \in E$. Let $\Omega_k(G)$ denote the set of k-colorings of G. Furthermore, let $k \geq 2\Delta + 1$, where Δ is the maximum degree in G. Show that

$$\frac{\Delta + 1}{\Delta + 2} \leq \frac{|\Omega_k(G)|}{|\Omega_k(G')|} \leq 1.$$

Design an algorithm that estimates this ratio via sampling k-colorings of G'.

16. It is easy to see that $|\Omega_k(G)|$ can be estimated as

$$|\Omega_k(G_0)| \prod_{i=0}^{m-1} \frac{|\Omega_k(G_{i+1})|}{|\Omega_k(G_i)|},$$

where each G_i contains one less edge than G_{i+1}, G_0 is the empty graph, and $G_m = G$ (that is, m is the number of edges in G). How well should each fraction be estimated to get an FPRAS for $|\Omega_k(G)|$?

8.5   Solutions

Exercise 3. Observe that the number of trees with the prescribed conditions is the number of trees realizing D', where D' is obtained from D by removing 1-1 from d_i and d_j and deleting two degree 1s.

Exercise 5. Let G be P_5, that is, the path on 5 vertices. Let X be the matching containing the first and third edges of P_5, and let Y be the matching containing the second and the fourth edges of P_5. In the canonical path from X to Y, the first edge is deleted in the first step. Then the second edge is added and the third is deleted in the second step. Finally, in the third step, the fourth edge of P_3 is added.

Exercise 7. Observe that

$$F = d_{1,1} + n, d_{1,2} + n, \ldots, d_{1,n} + n, d_{2,1}, d_{2,2}, \ldots, d_{2,m}$$

is an appropriate degree sequence. To see this, consider any realization $G = (U, V, E)$ of D, and add a complete graph k_n to the vertex set U. Clearly, the so obtained graph G' is a realization of F. Any realization of F can be obtained from G' by swaps. However, these swaps can swap edges between U and V, since G' is the complete graph on U and the empty graph on V. That is, any realization of F contains a complete graph on U, and has no edge inside V. Deleting K_n on U establishes the bijection between the realizations of F and D.

Exercise 9. Find appropriate $n/2$-regular graphs.

Exercise 12. When $k = 2\Delta$, the probability that $|D_t|$ is increasing is less than or equal to the probability that $|D_t|$ is decreasing. Model the change of $|D_t|$ as a random walk on the $[0, n]$ integer line, where n is the number of vertices in G, and estimate the time hitting 0.

Exercise 14. Let G be K_3. Then $\Delta = 2$, and there are $3! = 6$ possible 3-colorings. However, any of these colorings differ in at least 2 vertices, so there are no couple of colorings with a transition between them.

Exercise 15. First, observe that $\Omega_k(G) \subseteq \Omega_k(G')$, which proves the right inequality. Let e be (u, v). The next observation is that in any k-colorings in $\Omega_k(G') \setminus \Omega_k(G)$, the color of u is the color of v. There might be at least $k - \Delta \geq \Delta + 1$ colors to which the color of vertex u is changed to a color to get a coloring of G. On the other hand, any coloring in $\Omega_k(G)$ can be transformed into a coloring in $\Omega_k(G') \setminus \Omega_k(G)$ by changing the color of u to the color of v. Therefore, indeed

$$\frac{\Delta + 1}{\Delta + 2} \leq \frac{|\Omega_k(G')|}{|\Omega_k(G)|}.$$

Bibliography

[1] http://www.claymath.org/sites/default/files/pvsnp.pdf.

[2] van T. Aardenne-Ehrenfest and N. G. de Bruijn. *Wis- en Natuurkundig Tijdschrift*, volume 28, chapter Circuits and trees in oriented linear graphs, pages 203–217. 1951.

[3] M. Agrawal, N. Kayal, and N. Saxena. PRIMES is in P. *Annals of Mathematics*, 160(2):781–793, 2004.

[4] Y. Ajana, J.F. Lefebvre, E. Tillier, and N. El-Mabrouk. Exploring the set of all minimal sequences of reversals - An application to test the replication-directed reversal hypothesis. In R. Guigo and D. Gusfield, editors, *Proceedings of the 2nd International Workshop on Algorithms in Bioinformatics*, volume 2452 of *Lecture Notes in Computer Science*, pages 300–315. Springer, 2002.

[5] D.J. Aldous. Some inequalities for reversible Markov chains. *Journal of the London Mathematical Society (2)*, 25(3):564–576, 1982.

[6] D.J. Aldous. Random walks on finite groups and rapidly mixing Markov chains. In *Séminaire de Probabilites XVII*, volume 986 of *Lecture Notes in Mathematics*, pages 243–297. Springer, 1983.

[7] N. M. Amato, M. T. Goodrich, and E. A. Ramos. A randomized algorithm for triangulating a simple polygon in linear time. *Discrete and Computational Geometry*, 26(2):245–265, 2001.

[8] S. Arora and B. Barak. *Computational Complexity: A Modern Approach.* Cambridge University Press, 2009.

[9] L. Babai. Graph isomorphism in quasipolynomial time, 2015. `arXiv: 1512.03547`.

[10] L. Babai and E.M. Luks. Canonical labeling of graphs. In *Proceedings of the 15th Annual ACM Symposium on Theory of Computing*, pages 171–183, 1983.

[11] D.A. Bader, B.M.E. Moret, and M. Yan. A linear-time algorithm for computing inversion distance between signed permutations with an experimental study. *Journal of Computational Biology*, 8(5):483–491, 2001.

[12] I. Bárány and Z. Füredi. Computing the volume is difficult. In *Proceedings of the 18th Annual ACM Symposium on Theory of Computing*, pages 442–447, 1986.

[13] R. Barbanchon. On unique graph 3-colorability and parsimonious reductions in the plane. *Theoretical Computer Science*, 319:455–482, 2004.

[14] L.E. Baum and J.A. Egon. An inequality with applications to statistical estimation for probabilistic functions of a Markov process and to a model for ecology. *Bulletin of the American Mathematical Society*, 73:360–363, 1967.

[15] L.E. Baum and T. Petrie. Statistical inference for probabilistic functions of finite state Markov chains. *The Annals of Mathematical Statistics*, 37(6):1554–1563, 1966.

[16] L.E. Baum and G.R. Sell. Growth functions for transformations on manifolds. *Pacific Journal of Mathematics*, 27(2):211–227, 1968.

[17] A. Bergeron, J. Mixtacki, and J. Stoye. A unifying view of genome rearrangements. In *Proceedings of the 6th Workshop on Algorithms in Bioinformatics*, volume 4175 of *Lecture Notes in Computer Science*, pages 163–173. Springer, 2006.

[18] S.J. Berkowitz. On computing the determinant in small parallel time using a small number of processors. *Information Processing Letters*, 18:147–150, 1984.

[19] I. Bezáková, N. Bhatnagar, and E. Vigoda. Sampling binary contingency tables with a greedy start. *Random Structures and Algorithms*, 30(1–2):168–205, 2008.

[20] J.P.M. Binet. Mémoire sur un systeme de formules analytiques, et leur application á des considerations géométriques. *J. de l'Ecole Polytechnique IX, Cahier*, 16:280–287, 1815.

[21] P. Bose, J.F. Buss, and A. Lubiw. Pattern matching for permutations. *Information Processing Letters*, 65(5):277–283, 1998.

[22] G.E.P. Box and M.E. Muller. A note on the generation of random normal deviates. *The Annals of Mathematical Statistics*, 29(2):610–611, 1958.

[23] M.D.V. Braga, M. Sagot, C. Scornavacca, and E. Tannier. The solution space of sorting by reversals. In Mândoiu I. and A. Zelikovsky, editors, *Bioinformatics Research and Applications. ISBRA 2007*, volume 4463 of *Lecture Notes in Computer Science*, pages 293–304. Springer, Berlin, Heidelberg, 2007.

[24] M.D.V. Braga and J. Stoye. The solution space of sorting by DCJ. *Journal of Computational Biology*, 17(9):1145–1165, 2010.

[25] P. Brémaud. *Markov Chains: Gibbs Fields, Monte Carlo Simulation, and Queues*. Texts in Applied Mathematics. Springer, New York, 1999.

[26] G. Brightwell and P. Winkler. Counting linear extensions is #P-complete. In *Proceedings of the 23rd Annual ACM Symposium on Theory of Computing*, pages 175–181, 1991.

[27] G. Brightwell and P. Winkler. Note on counting Eulerian circuits, 2004. https://arxiv.org/pdf/cs/0405067.pdf.

[28] R. L. Brooks. On coloring the nodes of a network. *Proceedings of the Cambridge Philosophical Society*, 37:194–197, 1941.

[29] R. Bubley and M. Dyer. Path coupling: A technique for proving rapid mixing in Markov chains. In *Procccdings of the 38th Annual Symposium on Foundations of Computer Science*, pages 223–231, 1997.

[30] G. Buffon. Essai d'arithmétique morale. *Histoire naturelle, générale er particuliére*, Supplément 4:46–123, 1777.

[31] J.-Y. Cai. Holographic algorithms: Guest column. *SIGACT News*, 39(2):51–81, 2008.

[32] J-Y. Cai and X. Chen. *Complexity Dichotomies for Counting Problems: Volume 1, Boolean Domain*. Cambridge University Press, 2017.

[33] J-Y. Cai, H. Guo, and T. Williams. The complexity of counting edge colorings and a dichotomy for some higher domain Holant problems. *Research in the Mathematical Sciences*, 3:18, 2016.

[34] J-Y. Cai and P. Lu. Holographic algorithms: The power of dimensionality resolved. In L. Arge, C. Cachin, Jurdziński T., and A. Tarlecki, editors, *Proceedings of the 34th International Colloquium on Automata, Languages and Programming*, volume 4596 of *Lecture Notes in Computer Science*, pages 631–642, 2007.

[35] J-Y. Cai and P. Lu. On symmetric signatures in holographic algorithms. In W. Thomas and P. Weil, editors, *Proceedings of the 24th Annual Symposium on Theoretical Aspects of Computer Science*, volume 4393 of *Lecture Notes in Computer Science*, pages 429–440, 2007.

[36] J-Y. Cai and P. Lu. Basis collapse in holographic algorithms. *Computational Complexity*, 17(2):254–281, 2008.

[37] J.-Y. Cai, P. Lu, and M. Xia. Holographic algorithms by Fibonacci gates. *Linear Algebra and Its Applications*, 438(2):690–707, 2013.

[38] J.-Y. Cai, P. Lu, and M. Xia. Holographic algorithms with matchgates capture precisely tractable planar #csp. *SIAM Journal on Computing*, 46(3):853–889, 2017.

[39] A. Cauchy. Memoire sur le nombre de valeurs qu'une fonction peut obtenir. *J. de l'Ecole Polytechnique X*, pages 51–112, 1815.

[40] B. Chazelle. Triangulating a simple polygon in linear time. *Discrete and Computational Geometry*, 6(3):485–524, 1991.

[41] N. Chomsky. *Transformational Analysis*. PhD thesis, University of Pennsylvania, 1955.

[42] N. Chomsky. On certain formal properties of grammars. *Information and Control*, 2:137–167, 1959.

[43] D. A. Christie. Sorting permutations by block-interchanges. *Information Processing Letters*, 60:165–169, 1996.

[44] S. Cook. The complexity of theorem proving procedures. In *Proceedings of the 3rd Annual ACM Symposium on Theory of Computing*, pages 151–158, 1971.

[45] C. Cooper, M. Dyer, and C. Greenhill. Sampling regular graphs and a peer-to-peer network. *Combinatorics, Probability and Computing*, 16(4):557–593, 2007.

[46] C. Cooper, M. Dyer, C. Greenhill, and A. Handley. The flip Markov chain for connected regular graphs, 2017. arXiv:1701.03856.

[47] D. Coppersmith and S. Winograd. Matrix multiplication via arithmetic progressions. *Journal of Symbolic Computation*, 9(3):251–280, 1980.

[48] P. Creed. *Counting and Sampling Problems on Eulerian G]raphs, school = University of Edinburgh, year = 2010, OPTkey = , OPTtype = , OPTaddress = , OPTmonth = , OPTnote = , OPTannote = .* PhD thesis.

[49] P. Creed. Sampling Eulerian orientations of triangular lattice graphs. *Journal of Discrete Algorithms*, 7(2):168–180, 2009.

[50] É. Czabarka, A. Dutle, P.L. Erdős, and I. Miklós. On realizations of a joint degree matrix. *Discrete Applied Mathematics*, 181(30):283–288, 2014.

[51] A.M. Davie and A.J. Stothers. Improved bound for complexity of matrix multiplication. *Proceedings of the Royal Society of Edinburgh Section A*, 143(2):351–369, 2013.

[52] P. Diaconis and L. Saloff-Coste. Comparison theorems for reversible Markov chains. *The Annals of Applied Probability*, 3(2):696–730, 1993.

[53] P. Diaconis and L. Saloff-Coste. Logarithmic Sobolev inequalities for finite Markov chains. *The Annals of Applied Probability*, 6(3):695–750, 1996.

[54] P. Diaconis and D. Stroock. Geometric bounds for eigenvalues of Markov chains. *The Annals of Applied Probability*, 1(1):36–61, 1991.

[55] W. Doeblin. Esposé de la théorie des chaînes simple constantes de Markov á un nombre fini d'états. *Rev. Math. Union Interbalkan*, 2:77–105, 1938.

[56] M. Dyer. Approximate counting by dynamic programming. In *Proceedings of the 35th Annual ACM Symposium on Theory of Computing (STOC)*, pages 693–699, 2003.

[57] M. Dyer, L.A. Goldberg, C. Greenhill, and M. Jerrum. The relative complexity of approximate counting problems. *Algorithmica*, 38(3):471–500, 2004.

[58] M. Dyer and C. Greenhill. A more rapidly mixing Markov chain for graph colourings. *Random Structures and Algorithms*, 13:285–317, 1998.

[59] M. Dyer and C. Greenhill. On Markov chains for independent sets. *Journal of Algorithms*, 35(1):17–49, 2000.

[60] J. Edmonds. Paths, trees, and flowers. *Canadian Journal of Mathematics*, 17:449–467, 1965.

[61] G. Elekes. A geometric inequality and the complexity of computing volume. *Discrete and Computational Geometry*, 1:289–292, 1986.

[62] P.L. Erdős, Z.S. Kiss, I. Miklós, and L. Soukup. Approximate counting of graphical realizations. *PLoS ONE*, 10(7):e0131300, 2015.

[63] P. Erdős and T. Gallai. Gráfok előírt fokszámú pontokkal, (Graphs with prescribed degrees of vertices, in Hungarian). *Matematikai Lapok*, 11:264–274, 1960.

[64] P.L. Erdős, I. Miklós, and Z. Toroczkai. A decomposition based proof for fast mixing of a Markov chain over balanced realizations of a joint degree matrix. *SIAM Journal on Discrete Mathematics*, 29:481–499, 2015.

[65] P.L. Erdős, I. Miklós, and Z. Toroczkai. New classes of degree sequences with fast mixing swap Markov chain sampling. *Combinatorics, Probability and Computing*, 2017. https://doi.org/10.1017/S0963548317000499 Published online: 02 November 2017.

[66] D.K. Faddeev and V.N. Faddeeva. *Numerical Methods of Linear Algebra*. Freeman, San Francisco, 1963.

[67] R. Fagin. Generalized first-order spectra and polynomial time recognisable sets. In R. Karp, editor, *Complexity of Computation*, volume 7 of *SIAM-AMS Proceedings*, pages 43–73. American Mathematical Society, 1974.

[68] J. Felsenstein. Evolutionary trees from DNA sequences: A maximum likelihood approach. *Journal of Molecular Evolution*, 17(6):368–376, 1981.

[69] G.D. Forney. The Viterbi algorithm. *Proceedings of the IEEE*, 61:268–278, 1973.

[70] A. Frank. *Paths, flows, and VLSI-Layout*, chapter Packing paths, circuits and cuts: A survey. Springer, Berlin, 1990.

[71] A. Galanis, L.A. Goldberg, and M. Jerrum. A complexity trichotomy for approximately counting list H-colorings. *ACM Transactions on Computation Theory*, 9(2):1–22, 2017.

[72] M.R. Garey and D.S. Johnson. *A Guide to the Theory of NP–Completeness*. A Series of Books in the Mathematical Sciences. W. H. Freeman and Co., 1979.

[73] M.R. Garey, D.S. Johnson, and L. Stockmeyer. Some simplified np–complete graph problems. *Theoretical Computer Science*, 1:237–267, 1976.

[74] M.R. Garey, D.S. Johnson, and R.E. Tarjan. The planar Hamiltonian circuit problem is NP-complete. *SIAM Journal on Computing*, 5:704–714, 1976.

[75] Q. Ge and D. Štefankovič. The complexity of counting Eulerian tours in 4-regular graphs. *Algorithmica*, 63:588–601, 2012.

[76] R. Giegerich and C. Meyer. Algebraic dynamic programming. In H. Kirchner and C. Ringeissen, editors, *Algebraic Methodology and Software Technology. AMAST 2002*, volume 2422 of *Lecture Notes in Computer Science*, pages 349–364, 2002.

[77] G.H. Gloug and C.F. Van Loan. *Matrix Computations (3rd edition)*. Johns Hopkins, 1996.

[78] L.A. Goldberg and M. Jerrum. Counting unlabelled subtrees of a tree is #P-complete. *LMS Journal of Computation and Mathematics*, 3:117–124, 2000.

[79] L.A. Goldberg and M. Jerrum. The "Burnside process" converges slowly. *Combinatorics, Probability and Computing*, 11(1):21–34, 2002.

[80] V. Gore, M. Jerrum, S. Kannan, Z. Sweedyk, and S. Mahaney. A quasi-polynomial-time algorithm for sampling words from a context-free language. *Information and Computation*, 134:59–74, 1997.

[81] O. Gotoh. An improved algorithm for matching biological sequences. *Journal of Molecular Biology*, 162:705–708, 1982.

[82] C. Greenhill. The complexity of counting colourings and independent sets in sparse graphs and hypergraphs. *Computational Complexity*, 9(1):52–72, 2000.

[83] C. Greenhill. A polynomial bound on the mixing time of a Markov chain for sampling regular directed graphs. *Electronic Journal of Combinatorics*, 18(1):#P234, 2011.

[84] C. Greenhill. The switch Markov chain for sampling irregular graphs. In *Proceedings of the 26th ACM SIAM Symposium on Discrete Algorithms, New York-Philadelphia*, pages 1564–1572, 2015.

[85] C. Greenhill and M. Sfragara. The switch Markov chain for sampling irregular graphs and digraphs. *Theoretical Computer Science*, 719:1–20, 2018.

[86] L. Gross. Logarithmic Sobolev inequalities. *American Journal of Mathematics*, 97(4):1061–1083, 1975.

[87] D. Gusfield. *Algorithms on Strings, Trees, and Sequences: Computer Science and Computational Biology*, chapter Maximum parsimony, Steiner trees, and perfect phylogeny, page 470. Cambridge University Press, 1997.

[88] S.L. Hakimi. On realizability of a set of integers as degrees of the vertices of a linear graph. i. *Journal of the Society for Industrial and Applied Mathematics*, 10:496–506, 1962.

[89] S. Hannenhalli and P. Pevzner. Transforming cabbage into turnip (polynomial algorithm for sorting signed permutations by reversals). In *Proceedings of the 27th Annual Symposium on Theory of Computing*, pages 178–189, 1995.

[90] J.A. Hartigan. Minimum mutation fits to a given tree. *Biometrics*, 29:53–65, 1973.

[91] W.K. Hastings. Monte Carlo sampling methods using Markov chains and their applications. *Biometrika*, 57(1):97–109, 1970.

[92] V. Havel. A remark on the existence of finite graphs (in Czech). *Časopis pro pěstování matematiky*, 80:477–480, 1955.

[93] T.P. Hayes and E. Vigoda. A non-Markovian coupling for randomly sampling colorings. In *Proccedings of the 44th Annual IEEE Symposium on Foundations of Computer Science*, pages 618–627, 2003.

[94] J.E. Hopcroft and R.M. Karp. An $n^{5/2}$ algorithm for maximum matchings in bipartite graphs. *SIAM Journal on Computing*, 2(4):225–231, 1973.

[95] M. Huber. Fast perfect sampling from linear extensions. *Discrete Mathematics*, 306:420–428, 2006.

[96] H. B. Hunt, M. V. Marathe, V. Radhakrishnan, and R. E. Stearns. The complexity of planar counting problems. *SIAM Journal on Computing*, 27(4):1142–1167, 1998.

[97] M. Jerrum. Two-dimensional monomer-dimer systems are computationally intractable. *Journal of Statistical Physics*, 48(1/2):121–134, 1987.

[98] M. Jerrum. Counting trees in a graph is #P-complete. *Information Processing Letters*, 51:111–116, 1994.

[99] M. Jerrum and A. Sinclair. Approximating the permanent. *SIAM Journal on Computing*, 18(6):1149–1178, 1989.

[100] M. Jerrum, A. Sinclair, and E. Vigoda. A polynomial-time approximation algorithm for the permanent of a matrix with nonnegative entries. *Journal of the ACM*, 51(4):671–697, 2004.

[101] M. Jerrum and M. Snir. Some exact complexity results for straight-line computations over semirings. *Journal of the Association for Computing Machinery*, 29(3):874–897, 1982.

[102] M.R. Jerrum. A very simple algorithm for estimating the number of k-colourings of a low-degree graph. *Random Structures and Algorithms*, 7(2):157–165, 1995.

[103] M.R. Jerrum, L.G. Valiant, and V.V. Vazirani. Random generation of combinatorial structures from a uniform distribution. *Theoretical Computer Science*, 43(2–3):169–188, 1986.

[104] W. Just. Computational complexity of multiple sequence alignment with SP-score. *Journal of Computational Biology*, 8(6):615–623, 2001.

[105] R. Kannan, L. Lovász, and M. Simonovits. Random walks and an $O*(n^5)$ volume algorithm for convex bodies. *Random Structures and Algorithms*, 11(1):1–50, 1997.

[106] S. Kannan, Z. Sweedyk, and S. R. Mahaney. Counting and random generation of strings in regular languages. In *Proceedings of the 6th Annual ACM-SIAM Symposium on Discrete Algorithms*, pages 551–557, 1995.

[107] Shamir R. Kaplan, H. and R. Tarjan. A faster and simpler algorithm for sorting signed permutations by reversals. *SIAM Journal on Computing*, 29(3):880–892, 1999. First appeared in the Proceedings of the 8th Annual Symposium on Discrete Algorithms.

[108] R.M. Karp. *Complexity of Computer Computations*, chapter Reducibility among combinatorial problems, pages 85–103. Plenum, New York, 1972.

[109] A. Karzanov and L. Kachiyan. On the conductance of order Markov chains. *Order*, 8:7–15, 1991.

[110] P. W. Kasteleyn. Dimer statistics and phase transitions. *Journal of Mathematical Physics*, 4(2):287–293, 1963.

[111] P. W. Kasteleyn. *Graph Theory and Theoretical Physics*, chapter Graph theory and crystal physics, pages 43–110. Academic Press, New York, 1967.

[112] P.W. Kasteleyn. The statistics of dimers on a lattice. I. The number of dimer arrangements on a quadratic lattice. *Physica*, 27(12):1209–1225, 1961.

[113] G. Kirchhoff. Über die Auflösung der Gleichungen, auf welche man bei der untersuchung der linearen verteilung galvanischer Ströme geführt wird. *Annalen der Physic und Chemie*, 72:497–508, 1847.

[114] S. Kirkpatrick, C. D Gelatt Jr, and M. P. Vecchi. Optimization by simulated annealing. *Science*, 220(4598):671–680, 1983.

[115] B. Knudsen and J.J. Hein. Using stochastic context-free grammars and molecular evolution to predict RNA secondary structure. *Bioinformatics*, 15:446–454, 1999.

[116] J. B. Kruskal. On the shortest spanning subtree of a graph and the traveling salesman problem. *Proceedings of the American Mathematical Society*, 7:48–50, 1956.

[117] S. Kundu. Disjoint representation of tree realizable sequences. *SIAM Journal on Applied Mathematics*, 26(1):103–107, 1974.

[118] N. Linial. Hard enumeration problems in geometry and combinatorics. *SIAM Journal on Algebraic and Discrete Methods*, 7(2):331–335, 1986.

[119] M. Liskiewicz, M. Ogihara, and S. Toda. The complexity of counting self-avoiding walks in subgraphs of two-dimensional grids and hypercubes. *Theoretical Computer Science*, 304(1-3):129–156, 2003.

[120] C.H.C. Little. An extension of Kasteleyn's method of enumerating the 1-factors of planar graphs. In D.A. Holton, editor, *Combinatorial Mathematics*, volume 403 of *Lecture Notes in Mathematics*, pages 63–72. Springer, Berlin, Heidelberg, 1974.

[121] J. S. Liu. *Monte Carlo Strategies in Scientific Computing*. Springer Series in Statistics. Springer, New York, 1999.

[122] L. Lovász and S. Vempala. Simulated annealing in convex bodies and an $O^*(n^4)$ volume algorithm. *Journal of Computer and System Sciences*, 72(2):392–417, 2006.

[123] M. Luby and E. Vigoda. Fast convergence of the Glauber dynamics for sampling independent sets. *Random Structures and Algorithms*, 15:229–241.

[124] M. Mahajan and V. Vinay. Old algorithms, new insights. *SIAM Journal on Discrete Mathematics*, 12:474–490, 1999.

[125] R. Martin and D. Randall. Disjoint decomposition of Markov chains and sampling circuits in Cayley graphs. *Combinatorics, Probability and Computing*, 15:411–448, 2006.

[126] J.S. McCaskill. The equilibrium partition function and base pair binding probabilities for RNA secondary structure. *Biopolymers*, 29:1105–1119, 1990.

[127] Braga M.D.V. and Stoye J. Counting all DCJ sorting scenarios. In Ciccarelli F.D. and Miklós I., editors, *Proceedings of the 6th RECOMB Comparative Genomics Workshop*, volume 5817 of *Lecture Notes in Computer Science*, pages 36–47. Springer, Berlin, Heidelberg, 2009.

[128] N. Metropolis, A.W. Rosenbluth, M.N. Rosenbluth, A.H. Teller, and E. Teller. Equations of state calculations by fast computing machines. *Journal of Chemical Physics*, 21(6):1087–1092, 1953.

[129] M. Michail and P. Winkler. On the number of Eulerian orientations of a graph. *Algorithmica*, 16(4/5):402–414, 1996.

[130] I. Miklós, P.L. Erdős, and L. Soukup. Towards random uniform sampling of bipartite graphs with given degree sequence. *Electronic Journal of Combinatorics*, 20(1):P16, 2013.

[131] I. Miklós, B. Mélykúti, and K. Swenson. The Metropolized partial importance sampling MCMC mixes slowly on minimum reversal rearrangement paths. *ACM/IEEE Transactions on Computational Biology and Bioinformatics*, 4(7):763–767, 2010.

[132] I. Miklós, I.M. Meyer, and B. Nagy. Moments of the Boltzmann distribution for RNA secondary structures. *Bulletin of Mathematical Biology*, 67:1031–1047, 2005.

[133] I. Miklós and H. Smith. The computational complexity of calculating partition functions of optimal medians with hamming distance, 2017. https://arxiv.org/abs/1506.06107.

[134] I. Miklós and E. Tannier. Approximating the number of double cut-and-join scenarios. *Theoretical Computer Science*, 439:30–40, 2012.

[135] I. Miklós, E. Tannier, and Z.S. Kiss. On sampling SCJ rearrangement scenarios. *Theoretical Computer Science*, 552:83–98, 2014.

[136] B. Morris and A. Sinclair. Random walks on truncated cubes and sampling 0-1 knapsack solutions. *SIAM J. Comput.*, 34(1):195–226, 2004.

[137] S.B. Needleman and C.D. Wunch. A general method applicable to the search for similarities in the amino acid sequence of two proteins. *Journal of Molecular Biology*, 48(3):443–453, 1970.

[138] A. Ouangraoua and A. Bergeron. Parking functions, labeled trees and DCJ sorting scenarios. In Ciccarelli F.D. and Miklós I., editors, *Proceedings of the 6th RECOMB Comparative Genomics Workshop*, volume 5817 of *Lecture Notes in Computer Science*, pages 24–35. Springer, Berlin, Heidelberg, 2009.

[139] A. Ouangraoua and A. Bergeron. Combinatorial structure of genome rearrangements scenarios. *Journal of Computational Biology*, 17(9):1129–1144, 2010.

[140] C.H. Papadimitriou. *Computational Complexity*. Addison-Wesley, Reading, Mass., 1994.

[141] L. Pauling. The structure and entropy of ice and of other crystals with some randomness of atomic arrangement. *Journal of the American Chemical Society*, 57(12):2680–2684, 1935.

[142] G. Pólya. Aufgabe 424. *Arch. Math. Phys.*, 20:271, 1913.

[143] J.G. Propp and D.B. Wilson. Coupling from the past: A user's guide. In D. Aldous and J. Propp, editors, *Microsurveys in Discrete Probability*, volume 41 of *DIMACS Series in Discrete Mathematics and Theoretical Computer Science*, pages 181–192, 1998.

[144] J.S. Provan and M.O. Ball. The complexity of counting cuts and of computing the probability that a graph is connected. *SIAM Journal on Computing*, 12(4):777–788, 1983.

[145] R.L. Rabiner. A tutorial on hidden Markov models and selected applications in speech recognition. *Proceedings of the IEEE*, 77(2):257–286, 1989.

[146] N. Robertson, P.D. Seymour, and R. Thomas. Permanents, Pfaffian orientations, and even directed circuits. *The Annals of Mathematics*, 150:929–975, 1999.

[147] O.S. Rothaus. Diffusion on compact Riemannian manifolds and logarithmic Sobolev inequalities. *Journal of Functional Analysis*, 42(1):102–109, 1981.

[148] S. Saluja, K.V. Subrahmanyam, and M.N. Thakur. Descriptive complexity of #P functions. *Journal of Computer and Systems Sciences*, 50:493–505, 1995.

[149] P.A. Samuelson. A method of determining explicitly the coefficients of the characteristic equation. *The Annals of Mathematical Statistics*, 13:424–429, 1942.

[150] D. Sankoff and P. Rousseau. Locating the vertices of a Steiner tree in an arbitrary metric space. *Mathematical Programming*, 9:240–246, 1975.

[151] C.P. Schnorr. A lower bound on the number of additions in monotone computations. *Theoretical Computer Science*, 2(3):305–315, 1976.

[152] J. Schweinsberg. An $O(n^2)$ upper bound for the relaxation time of a Markov chain on cladograms. *Random Structures and Algorithms*, 20:59–70, 2001.

[153] P.H. Shellers. On the theory and computation of evolutionary distances. *SIAM Journal on Applied Mathematics*, 26(4):787–793, 1974.

[154] A. Siepel. An algorithm to enumerate sorting reversals for signed permutations. *Journal of Computational Biology*, 10(3–4):575–597, 2003.

[155] J. Simon. On the difference between one and many. In *Proceedings of the 4th International Colloquium on Automata, Languages and Programming*, volume 52 of *Lecture Notes in Computer Science*, pages 480–491. Springer-Verlag, 1977.

[156] A. Sinclair and M. Jerrum. Approximate counting, uniform generation and rapidly mixing Markov chains. *Information and Computation*, 82:133, 1989.

[157] A. J. Sinclair. *Algorithms for Random Generation and Counting: A Markov Chain Approach*, PhD Thesis, University of Edinburgh. Monograph in the series Progress in Theoretical Computer Science. Springer-Birkhäuser, Boston, 1993.

[158] A.J. Sinclair. Improved bounds for mixing rates of Markov chains and multicommodity flow. *Combinatorics, Probability and Computing*, 1(4):351–370, 1992.

[159] M. Sipser. *Introduction to the Theory of Computation*, page 99. PWS, 1996.

[160] M. Sipser. *Introduction to the Theory of Computation*. Cengage Learning; 3 edition, 2012.

[161] L.G. Stockmeyer. The complexity of approximate counting. In *Proceedings of the 15th Annual ACM Symposium on Theory of Computing*, pages 118–126.

[162] V. Strassen. Gaussian elimination is not optimal. *Numerische Mathematik*, 13(4):354–356, 1969.

[163] R.L. Stratonovich. Conditional Markov processes. *Theory of Probability and its Applications*, 5(2):156–178, 1960.

[164] S. Straub, T Thierauf, and F. Wagner. Counting the number of the perfect matchings in K_5-free graphs. In *Electronic Colloquium on Computational Complexity*. 2014.

[165] K.M. Swenson, G. Badr, and D. Sankoff. Listing all sorting reversals in quadratic time. *Algorithms for Molecular Biology*, 6:11, 2011.

[166] E. Tannier, A. Bergeron, and M.-F. Sagot. Advances on sorting by reversals. *Discrete Applied Mathematics*, 155(6–7):881–888, 2007.

[167] H. N. V. Temperley and Michael E. Fisher. Dimer problem in statistical mechanics: An exact result. *Philosophical Magazine*, 6(68):1061–1063, 1961.

[168] I. Tinoco, O.C. Uhlenbeck, and M.D. Levine. Estimation of secondary structure in ribonucleic acids. *Nature*, 230:362–367, 1971.

[169] I.J. Tinoco, P. Borer, B. Dengler, M. Levine, and O. Uhlenbeck. Improved estimation of secondary structure in ribonucleic acids. *Nature: New Biology*, 246:40–41, 1973.

[170] W. T. Tutte and C. A. B. Smith. On unicursal paths in a network of degree 4. *American Mathematical Monthly*, 48:233–237, 1941.

[171] R. I. Tyshkevich. [The canonical decomposition of a graph] (in Russian). *Doklady Akademii Nauk SSSR*, 24:677–679, 1980.

[172] A. Urbańska. Faster combinatorial algorithms for determinant and pfaffian. *Algorithmica*, 56:35–50, 2010.

[173] S.P. Vadhan. The complexity of counting in sparse, regular and planar graphs. *SIAM Journal on Computing*, 31(2):398–427, 2001.

[174] L. G. Valiant. Accidental algorithms. In *Proceedings of the 47th Annual IEEE Symposium on Foundations of Computer Science*, pages 509–517.

[175] L.G. Valiant. The complexity of computing the permanent. *Theoretical Computer Science*, 8(3):189–201, 1979.

[176] L.G. Valiant. The complexity of enumeration and reliability problems. *SIAM Journal on Computing*, 8(3):410–421, 1979.

[177] L.G. Valiant. Negation can be exponentially powerful. *Theoretical Computer Science*, 12:303–314, 1980.

[178] L.G. Valiant. Holographic algorithms. In *Proceedings of the 45th Annual IEEE Symposium on Foundations of Computer Science*, pages 306–315, 2004.

[179] D. L. Vertigan and D. J. A. Welsh. The computational complexity of the Tutte plane: The bipartite case. *Combinatorics, Probability and Computing*, 1(2):181–187, 1992.

[180] A.J. Viterbi. Error bounds for convolutional codes and an asymptotically optimum decoding algorithm. *IEEE Transactions on Information Theory*, 13(2):260–269, 1967.

[181] J. von Neumann. *Monte Carlo Method*, volume 12 of *National Bureau of Standards Applied Mathematics Series*, chapter 13. Various techniques used in connection with random digits, pages 36–38. Washington, D.C.: U.S. Government Printing Office, 1951.

[182] D. Štefankovič, E. Vigoda, and J. Wilmes. On counting perfect matchings in general graphs, 2017. https://arxiv.org/pdf/1712.07504.pdf.

[183] L. Wang and T. Jiang. On the complexity of multiple sequence alignment. *Journal of Computational Biology*, 1(4):337–348, 1994.

[184] M. Xia, P. Zhang, and W. Zhao. Computational complexity of counting problems on 3-regular planar graphs. *Theoretical Computer Science*, 384(1):111–125, 2007.

[185] S. Yancopoulos, O. Attie, and R. Friedberg. Efficient sorting of genomic permutations by translocation, inversion and block interchange. *Bioinformatics*, 21(6):3340–3346, 2005.

[186] C.H. zu Siederdissen, S.J. Prohaska, and P.F. Stadler. Algebraic dynamic programming over general data structures. *BMC Bioinformatics*, 16(Suppl 19):S2, 2015.

[187] M. Zucker and D. Sankoff. RNA secondary structures and their prediction. *Bulletin of Mathematical Biology*, 46:591–621, 1984.

Index